A Practical Approach to
LINEAR ALGEBRA

A Practical Approach to LINEAR ALGEBRA

Prabhat Choudhary

Oxford Book Company
Jaipur ▪ New Delhi

Oxford Book Company

Regd. Off. "Bony Residency", Gate No. 2
Opp. Tilak Public School, Vishweswariya Nagar,
Gopalpura Road, Jaipur - 302018, India
Ph: 0141 - 2761280, Telefax: 2761381
email - oxfordbookjpr@gmail.com
website - oxfordbookcompany.com

Delhi Office: 102, 1st Floor, Satyam House,
4327/3, Ansari Road, Darya Ganj,
New Delhi - 110002, India
Ph: 011-45652440

Edition, 2025

ISBN : 9788189473952

Price : ₹ 2495.00

Typeset by Shivangi Computers, Jaipur

Printed and bound in India by Thomson Press Ltd.,
New Delhi

Preface

Linear Algebra has occupied a very crucial place in Mathematics. Linear Algebra is a continuation of classical course in the light of the modern development in Science and Mathematics. We must emphasize that mathematics is not a spectator sport, and that in order to understand and appreciate mathematics it is necessary to do a great deal of personal cogitation and problem solving.

Scientific and engineering research is becoming increasingly dependent upon the development and implementation of efficient parallel algorithms. Linear algebra is an indispensable tool in such research and this paper attempts to collect and describe a selection of some of its more important parallel algorithms. The purpose is to review the current status and to provide an overall perspective of parallel algorithms for solving dense, banded, or block-structured problems arising in the major areas of direct solution of linear systems, least squares computations, eigenvalue and singular value computations, and rapid elliptic solvers. There is a widespread feeling that the non-linear world is very different, and it is usually studied as a sophisticated phenomenon of Interpolation between different approximately-Linear Regimes.

Prabhat Choudhary

Contents

Chapter 1

Basic Notions

VECTOR SPACES

A vector space V is a collection of objects, called vectors, along with two operations, addition of vectors and multiplication by *a* number (scalar), such that the following properties (the so–called axioms of *a* vector space) hold:

The first four properties deal with the addition of vector:

1. Commutativity: $v + w = w + v$ for all $v, w \in V$.
2. Associativity: $(u + v) + w = u + (v + w)$ for all $u, v, w \in V$.
3. Zero vector: there exists *a* special vector, denoted by 0 such that $v + 0 = v$ for all $v \in V$.
4. Additive inverse: For every vector $v \in V$ there exists *a* vector $w \in V$ such that $v + w = 0$. Such additive inverse is usually denoted as $-v$.

   The next two properties concern multiplication:

5. Multiplicative identity: $1v = v$ for all $v \in V$.
6. Multiplicative associativity: $(\alpha\beta)v = \alpha(\beta v)$ for all $v \in V$ and all $\in$ scalars α, β.

   And finally, two distributive properties, which connect multiplication and addition:

7. $\alpha(u + v) = \alpha u + \alpha v$ for all $u, v \in V$ and all scalars α.
8. $(\alpha + \beta)v = \alpha v + \beta v$ for all $v \in V$ and all scalars α, β.

Remark: The above properties seem hard to memorize, but it is not necessary. They are simply the familiar rules of algebraic manipulations with numbers.

The only new twist here is that you have to understand what operations you can apply to what objects. You can add vectors, and you can multiply *a* vector by *a* number (scalar). Of course, you can do with number all possible manipulations that you have learned before. But, you cannot multiply two vectors, or add *a* number to *a* vector.

Remark: It is not hard to show that zero vector 0 is unique. It is also easy to show that given $v \in V$ the inverse vector $-v$ is unique. In fact, properties can be deduced from the properties: they imply that $0 = 0v$ for any $v \in V$, and that $-v = (-1)v$.

If the scalars are the usual real numbers, we call the space V a real vector space. If the scalars are the complex numbers, *i.e.*, if we can multiply vectors by complex numbers, we call the space V a complex vector space.

Note, that any complex vector space is *a* real vector space as well (if we can multiply by complex numbers, we can multiply by real numbers), but not the other way around.

It is also possible to consider *a* situation when the scalars are elements of an arbitrary field $\mathbb{F}$.

In this case we say that V is *a* vector space over the field $\mathbb{F}$. Although many of the constructions in the book work for general fields, in this text we consider only real and complex vector spaces, *i.e.*, $\mathbb{F}$ is always either $\mathbb{R}$ or $\mathbb{C}$.

Example: The space $\mathbb{R}^n$ consists of all columns of size n,

$$v = \begin{pmatrix} v_1 \\ v_2 \\ \vdots \\ v_n \end{pmatrix}$$

whose entries are real numbers. Addition and multiplication are defined entrywise, *i.e.*,

$$\alpha \begin{pmatrix} v_1 \\ v_2 \\ \vdots \\ v_n \end{pmatrix} = \begin{pmatrix} \alpha v_1 \\ \alpha v_2 \\ \vdots \\ \alpha v_n \end{pmatrix}, \quad \begin{pmatrix} v_1 \\ v_2 \\ \vdots \\ v_n \end{pmatrix} + \begin{pmatrix} w_1 \\ w_2 \\ \vdots \\ w_n \end{pmatrix} = \begin{pmatrix} v_1 + w_1 \\ v_2 + w_2 \\ \vdots \\ v_n + w_n \end{pmatrix}$$

Example: The space $\mathbb{C}^n$ also consists of columns of size n, only the entries now are complex numbers. Addition and multiplication are defined exactly as in the case of $\mathbb{R}^n$, the only difference is that we can now multiply vectors by complex numbers, *i.e.*, $\mathbb{C}^n$ is *a* complex vector space.

Example: The space $M_{m \times n}$ (also denoted as $M_{m,n}$) of $m \times n$ matrices: the multiplication and addition are defined entrywise. If we allow only real entries (and so only multiplication only by reals), then we have *a* real vector space; if we allow complex entries and multiplication by complex numbers, we then have *a* complex vector space.

Example: The space $\mathbb{P}_n$ of polynomials of degree at most n, consists of all polynomials p of form

$$p(t) = a_0 + a_1 t + a_2 t^2 + ... + a_n t^n,$$

where t is the independent variable. Note, that some, or even all, coefficients ak can be 0.

In the case of real coefficients a_k we have *a* real vector space, complex coefficient give us *a* complex vector space.

Question: What are zero vectors in each of the above examples?

Matrix notation

An $m \times n$ matrix is *a* rectangular array with m rows and n columns. Elements of the array are called entries of the matrix.

It is often convenient to denote matrix entries by indexed letters is j, the first index denotes the number of the row, where the entry is a_{ij}, and the second one is the number of the column. For example

$$A = \left(a_{jk}\right)_{j=1,k=1}^{m,n} = \begin{pmatrix} a_{1,1} & a_{1,2} & \cdots & a_{1,n} \\ a_{2,1} & a_{2,2} & \cdots & a_{2,n} \\ \vdots & \vdots & & \vdots \\ a_{m,1} & a_{m,2} & \cdots & a_{m,n} \end{pmatrix}$$

is *a* general way to write an $m \times n$ matrix.

Very often for *a* matrix A the entry in row number j and column number k is denoted by $A_{j,k}$ or $(A)_{j,k}$, and sometimes as in example above the same letter but in lowercase is used for the matrix entries.

Given *a* matrix A, its transpose (or transposed matrix) A^T, is defined by transforming the rows of A into the columns. For example

$$\begin{pmatrix} 1 & 2 & 3 \\ 4 & 5 & 6 \end{pmatrix}^T = \begin{pmatrix} 1 & 4 \\ 2 & 5 \\ 3 & 6 \end{pmatrix}.$$

So, the columns of A^T are the rows of A and vise versa, the rows of A^T are the columns of A.

The formal definition is as follows: $(A^T)_{j,k} = (A)_{k,j}$ meaning that the entry of A^T in the row number j and column number k equals the entry of A in the row number k and row number j.

The transpose of *a* matrix has *a* very nice interpretation in terms of linear transformations, namely it gives the so called *adjoint* transformation. We will study this in detail later, but for now transposition will be just *a* useful formal operation.

One of the first uses of the transpose is that we can write *a* column vector $x \in \mathbb{R}^n$ as $x = (x_1, x_2, ..., x_n)^T$. If we put the column vertically, it will use significantly more space.

LINEAR COMBINATIONS, BASES.

Let V be *a* vector space, and let $v_1, v_2, ..., v_p \in V$ be *a* collection of vectors. *A* linear combination of vectors $v_1, v_2, ..., v_p$ is *a* sum of form

$$\alpha_1 v_1 + \alpha_2 v_2 + \ldots + \alpha_p v_p = \sum_{k=1}^{p} \alpha_k v_k.$$

Definition: *A* system of vectors $v_1, v_2, \ldots v_n \in V$ is called *a* basis (for the vector space V) if any vector $v \in V$ admits *a* unique representation as *a* linear combination

$$v = \alpha_1 v_1 + \alpha_2 v_2 + \ldots + \alpha_n v_n = \sum_{k=1}^{n} \alpha_k v_k.$$

The coefficients $\alpha_1, \alpha_2, \ldots, a_n$ are called coordinates of the vector v (in the basis, or with respect to the basis $v_1, v_2, \ldots, v_n$).

Another way to say that $v_1, v_2, \ldots, v_n$ is *a* basis is to say that the equation $x_1 v_1 + x_2 v_2 + \ldots + x_m v_n = v$ (with unknowns x_k) has *a* unique solution for arbitrary right side v.

Before discussing any properties of bases, let us give few examples, showing that such objects exist, and it makes sense to study them.

Example: The space V is $\mathbb{R}^n$. Consider vectors

$$e_1 = \begin{pmatrix} 1 \\ 0 \\ 0 \\ \vdots \\ 0 \end{pmatrix}, e_2 = \begin{pmatrix} 0 \\ 1 \\ 0 \\ \vdots \\ 0 \end{pmatrix}, e_3 = \begin{pmatrix} 0 \\ 0 \\ 1 \\ \vdots \\ 0 \end{pmatrix}, \ldots, e_n = \begin{pmatrix} 0 \\ 0 \\ 0 \\ \vdots \\ 1 \end{pmatrix},$$

(the vector e_k has all entries 0 except the entry number k, which is 1). The system of vectors $e_1, e_2, \ldots, e_n$ is *a* basis in R^n. Indeed, any vector

$$v = \begin{pmatrix} x_1 \\ x_2 \\ \vdots \\ x_n \end{pmatrix} \in \mathbb{R}^n$$

can be represented as the linear combination

$$v = x_1 e_1 + x_2 e_2 + \ldots x_n e_n = \sum_{k=1}^{n} x_k e_k$$

and this representation is unique. The system $e_1, e_2, \ldots, e_n \in \mathbb{R}^n$ is called the standard basis in $\mathbb{R}^n$.

Example: In this example the space is the space $\mathbb{P}^n$ of the polynomials of degree at most n. Consider vectors (polynomials) $e_0, e_1, e_2, \ldots, e_n \in \mathbb{P}^n$ defined by

$$e_0 = 1, e_1 = t, e_2 = t^2, e_3 = t^3, \ldots, e_n = t^n.$$

Clearly, any polynomial p, $p(t) = a_0 + a_1 t + a_2 t^2 + ... + a_n t^n$ admits *a* unique representation

$$p = a_0 e_0 + a_1 e_1 + ... + a_n e_n.$$

So the system $e_0, e_1, e_2, ..., e_n \in \mathbb{P}^n$ is *a* basis in $\mathbb{P}^n$. We will call it the standard basis in $\mathbb{P}^n$.

Remark: If *a* vector space V has *a* basis $v_1, v_2, ..., v_n$, then any vector v is uniquely defined by its co-effcients in the decomposition $v = \sum_{k=1}^{n} \alpha_k v_k$.

So, if we stack the coefficients α_k in *a* column, we can operate with them as if they were column vectors, *i.e.*, as with elements of $\mathbb{R}^n$.

Namely, if $v = \sum_{k=1}^{n} \alpha_k v_k$ and $w = \sum_{k=1}^{n} \beta_k v_k$, then

$$v + w = \sum_{k=1}^{n} \alpha_k v_k + \sum_{k=1}^{n} \beta_k v_k = \sum_{k=1}^{n} (\alpha_k + \beta_k) v_k,$$

i.e., to get the column of coordinates of the sum one just need to add the columns of coordinates of the summands.

Generating and Linearly Independent Systems. The definition of *a* basis says that any vector admits *a* unique representation as *a* linear combination. This statement is in fact two statements, namely that the representation exists and that it is unique. Let us analyse these two statements separately.

Definition: A system of vectors $v_1, v_2, ..., v_p \in V$ is called *a* generating system (also *a* spanning system, or *a* complete system) in V if any vector $v \in V$ admits representation as *a* linear combination

$$v = \alpha_1 v_1 + \alpha_2 v_2 + ... + \alpha_p v_p = \sum_{k=1}^{p} \alpha_k v_k$$

The only difference with the definition of *a* basis is that we do not assume that the representation above is unique. The words generating, spanning and complete here are synonyms. The term complete, because of my operator theory background.

Clearly, any basis is *a* generating (complete) system. Also, if we have *a* basis, say v_1, $v_2, ..., v_n$, and we add to it several vectors, say $v_{n+1}, ..., v_p$, then the new system will be *a* generating (complete) system. Indeed, we can represent any vector as *a* linear combination of the vectors $v_1, v_2, ..., v_n$, and just ignore the new ones (by putting corresponding coefficients $\alpha_k = 0$).

Now, let us turn our attention to the uniqueness. We do not want to worry about existence, so let us consider the zero vector 0, which always admits *a* representation as *a* linear combination.

Definition: *A* linear combination $\alpha_1 v_1 + \alpha_2 v_2 + ... +$ apvp is called trivial if $\alpha_k = 0\ \forall k$.

A trivial linear combination is always (for all choices of vectors $v_1, v_2, ..., v_p$) equal to 0, and that is probably the reason for the name.

Definition: *A* system of vectors $v_1, v_2, ..., v_p \in V$ is called linearly independent if only the trivial linear combination ($\sum_{k=1}^{p} \alpha_k v_k$ with $\alpha_k = 0\ \forall k$) of vectors $v_1, v_2, ..., v_p$ equals 0.

In other words, the system $v_1, v_2, ..., v_p$ is linearly independent i the equation $x_1 v_1 + x_2 v_2 + ... + x_p v_p = 0$ (with unknowns x_k) has only trivial solution $x_1 = x_2 = ... = x_p = 0$.

If *a* system is not linearly independent, it is called linearly dependent. By negating the definition of linear independence, we get the following

Definition: *A* system of vectors $v_1, v_2, ..., v_p$ is called linearly dependent if 0 can be represented as *a* nontrivial linear combination, $0 = \sum_{k=1}^{p} \alpha_k v_k$.

Non–trivial here means that at least one of the coefficient α_k is non–zero. This can be (and usually is) written as $\sum_{k=1}^{p} |\alpha_k| \neq 0$.

So, restating the definition we can say, that *a* system is linearly dependent if and only if there exist scalars $\alpha_1, \alpha_2, ..., \alpha_p$, $\sum_{k=1}^{p} |\alpha_k| \neq 0$ such that

$$\sum_{k=1}^{p} \alpha_k v_k = 0.$$

An alternative definition (in terms of equations) is that *a* system $v_1, v_2, ..., v_p$ is linearly dependent i the equation

$x_1 v_1 + x_2 v_2 + ... + x_p v_p = 0$

(with unknowns x_k) has *a* non–trivial solution. Non–trivial, once again again means that at least one of xk is different from 0, and it can be written as $\sum_{k=1}^{p} |x_k| \neq 0$.

The following proposition gives an alternative description of linearly dependent systems.

Proposition: A system of vectors $v_1, v_2, ..., v_p \in V$ is linearly dependent if and only if one of the vectors v_k can be represented as a linear combination of the other vectors,

$$v_k = \sum_{\substack{j=1 \\ j \neq k}}^{p} \beta_i v_j.$$

Proof. Suppose the system $v_1, v_2, ..., v_p$ is linearly dependent. Then there exist scalars α_k, $\sum_{k=1}^{p} |\alpha_k| \neq 0$ such that

$$\alpha_1 v_1 + \alpha_2 v_2 + \ldots + \alpha_p v_p = 0.$$

Let k be the index such that $\alpha_k \neq 0$. Then, moving all terms except $\alpha_k v_k$ to the right side we get

$$\alpha_k v_k \sum_{\substack{j=1 \\ j \neq k}}^{p} \alpha_j v_j.$$

Dividing both sides by ak we get with $\beta_j = -\alpha_j/\alpha_k$.

On the other hand, if holds, 0 can be represented as *a* non–trivial linear combination

$$v_k - \sum_{\substack{j=1 \\ j \neq k}}^{p} \beta_j v_j = 0$$

Obviously, any basis is *a* linearly independent system. Indeed, if *a* system $v_1, v_2, \ldots, v_n$ is *a* basis, 0 admits *a* unique representation

$$0 = \alpha_1 v_1 + \alpha_2 v_2 + \ldots + \alpha_n v_n = \sum_{k=1}^{n} \alpha_k v_k.$$

Since the trivial linear combination always gives 0, the trivial linear combination must be the only one giving 0.

So, as we already discussed, if *a* system is *a* basis it is *a* complete (generating) and linearly independent system. The following proposition shows that the converse implication is also true.

Proposition: A system of vectors $v_1, v_2, \ldots, v_n \in V$ is a basis if and only if it is linearly independent and complete (generating).

Proof: We already know that a basis is always linearly independent and complete, so in one direction the proposition is already proved.

Let us prove the other direction. Suppose a system $v_1, v_2, \ldots, v_n$ is linearly independent and complete. Take an arbitrary vector $v_2 V$. Since the system $v_1, v_2, \ldots, v_n$ is linearly complete (generating), v can be represented as

$$v = \alpha_1 v_1 + \alpha_2 v_2 + \ldots + \alpha_n v_n = \sum_{k=1}^{n} \alpha_k v_k.$$

We only need to show that this representation is unique.

Suppose v admits another representation

$$v = \sum_{k=1}^{n} \tilde{\alpha}_k v_k.$$

Then

$$\sum_{k=1}^{n}(\alpha_k-\tilde{\alpha}_k)v_k=\sum_{k=1}^{n}(\alpha_k v_k)-\sum_{k=1}^{n}\tilde{\alpha}_k v_k=v-v=0$$

Since the system is linearly independent, $\alpha_k - \tilde{\alpha}_k = 0\ \forall k$, and thus the representation $v = \alpha_1 v_1 + \alpha_2 v_2 + ... + \alpha_n v_n$ is unique.

Remark: In many textbooks *a* basis is defined as *a* complete and linearly independent system. Although this definition is more common than one presented in this text. It emphasizes the main property of *a* basis, namely that any vector admits *a* unique representation as *a* linear combination.

Proposition: Any (finite) generating system contains *a* basis.

Proof: Suppose $v_1, v_2, ..., v_p \in V$ is *a* generating (complete) set. If it is linearly independent, it is *a* basis, and we are done.

Suppose it is not linearly independent, *i.e.*, it is linearly dependent. Then there exists *a* vector v_k which can be represented as *a* linear combination of the vectors v_j, $j \neq k$.

Since v_k can be represented as *a* linear combination of vectors v_j, $j \neq k$, any linear combination of vectors $v_1, v_2, ..., v_p$ can be represented as *a* linear combination of the same vectors without v_k (*i.e.*, the vectors v_j, $1 \leq j \leq p, j = k$). So, if we delete the vector v_k, the new system will still be *a* complete one.

If the new system is linearly independent, we are done. If not, we repeat the procedure.

Repeating this procedure finitely many times we arrive to *a* linearly independent and complete system, because otherwise we delete all vectors and end up with an empty set.

So, any finite complete (generating) set contains *a* complete linearly independent subset, *i.e.*, *a* basis.

LINEAR TRANSFORMATIONS. MATRIX–VECTOR MULTIPLICATION

A transformation T from *a* set X to *a* set Y is *a* rule that for each argument (input) $x \in X$ assigns *a* value (output) $y = T(x) \in Y$. The set X is called the domain of T, and the set Y is called the target space or codomain of T. We write $T: X \to Y$ to say that T is *a* transformation with the domain X and the target space Y.

Definition: Let V, W be vector spaces. *A* transformation $T: V \to W$ is called linear if

1. $T(u + v) = T(u) + T(v)\ \forall u, v \in V$;
2. $T(\alpha v) = \alpha T(v)$ for all $v \in V$ and for all scalars a.

Properties 1 and 2 together are equivalent to the following one:

$T(\alpha u + \beta v) = \alpha T(u) + \beta T(v)$ for all $u, v \in V$ and for all scalars a, b.

Examples: You dealt with linear transformation before, may be without even suspecting it, as the examples below show.

Example: Differentiation: Let $V = \mathbb{P}_n$ (the set of polynomials of degree at most n), $W = \mathbb{P}_{n-1}$, and let $T: \mathbb{P}_n \to \mathbb{P}_{n-1}$ be the differentiation operator,

$$T(p) := p' \ \forall p \in \mathbb{P}_n.$$

Since $(f+g) = f + g$ and $(af)' = af'$, this is *a* linear transformation.

Example: Rotation: in this example $V = W = \mathbb{R}^2$ (the usual coordinate plane), and *a* transformation T_γ: $\mathbb{R}^2 \to \mathbb{R}^2$ takes *a* vector in $\mathbb{R}^2$ and rotates it counterclockwise by γ radians. Since T_γ rotates the plane as *a* whole, it rotates as *a* whole the parallelogram used to define *a* sum of two vectors (parallelogram law). Therefore the property 1 of linear transformation holds. It is also easy to see that the property 2 is also true.

Example: Reflection: in this example again $V = W = \mathbb{R}^2$, and the trans–formation T: $\mathbb{R}^2 \to \mathbb{R}^2$ is the reflection in the first coordinate axis. It can also be shown geometrically, that this transformation is linear, but we will use another way to show that.

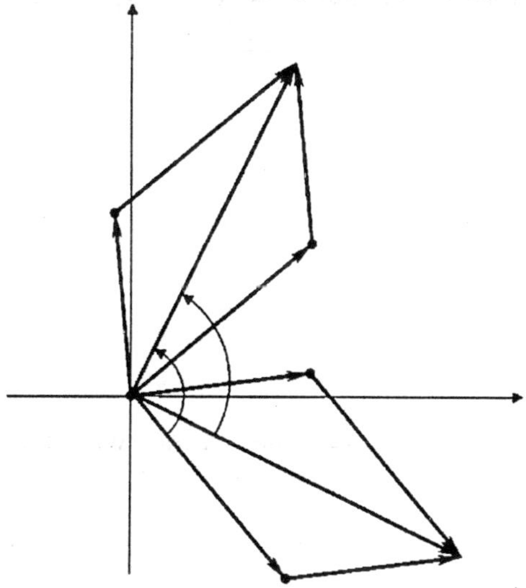

Fig. Rotation

Namely, it is easy to write *a* formula for T,

$$T\left(\begin{pmatrix} x_1 \\ x_2 \end{pmatrix}\right) = \begin{pmatrix} x_1 \\ -x_2 \end{pmatrix}$$

and from this formula it is easy to check that the transformation is linear.

Example: Let us investigate linear transformations T: $\mathbb{R} \to \mathbb{R}$. Any such transformation is given by the formula

$T(x) = ax$ where $a = T(1)$.

Indeed,

$$T(x) = T(x \times 1) = xT(1) = xa = ax.$$

So, any linear transformation of $\mathbb{R}$ is just *a* multiplication by *a* constant.

Linear transformations $\mathbb{R}^n \to \mathbb{R}^m$. Matrix–column multiplication: It turns out that *a* linear transformation T: $\mathbb{R}^n \to \mathbb{R}^m$ also can be represented as *a* multiplication, not by *a* number, but by *a* matrix.

Let us see how. Let $T: \mathbb{R}^n \to \mathbb{R}^m$ be *a* linear transformation. What information do we need to compute $T(x)$ for all vectors $x \in \mathbb{R}^n$? My claim is that it is sufficient how T acts on the standard basis $e_1, e_2, \ldots, e_n$ of R^n. Namely, it is sufficient to know n vectors in R^m (*i.e.*,, the vectors of size m),

$$a_1 = T(e_1), a_2 = T(e_2), \ldots, a_n = T(e_n).$$

Indeed, let

$$x = \begin{pmatrix} x_1 \\ x_2 \\ \vdots \\ x_n \end{pmatrix}.$$

Then $x = x_1e_1 + x_2e_2 + \ldots + x_ne_n = \sum_{k=1}^{n} x_ke_k$ and

$$\mathrm{T}(x) = T\left(\sum_{k=1}^{n} x_ke_k\right) = \sum_{k=1}^{n} T(x_ke_k) = \sum_{k=1}^{n} x_kT(e_k) = \sum_{k=1}^{n} x_ka_k.$$

So, if we join the vectors (columns) $a_1, a_2, \ldots, a_n$ together in *a* matrix

$$A = [a_1, a_2, \ldots, a_n]$$

(*ak* being the *k*th column of A, $k = 1, 2, \ldots, n$), this matrix contains all the information about T. Let us show how one should define the product of *a* matrix and *a* vector (column) to represent the transformation T as *a* product, $T(x) = Ax$. Let

$$A = \begin{pmatrix} a_{1,1} & a_{1,2} & \cdots & a_{1,n} \\ a_{2,1} & a_{2,2} & \cdots & a_{2,n} \\ \vdots & \vdots & & \vdots \\ a_{m,1} & a_{m,2} & \cdots & a_{m,n} \end{pmatrix}.$$

Recall, that the column number k of A is the vector a_k, *i.e.*,

$$a_k = \begin{pmatrix} a_{1,k} \\ a_{2,k} \\ \vdots \\ a_{m,k} \end{pmatrix}.$$

Then if we want $Ax = T(x)$ we get

$$Ax = \sum_{k=1}^{n} x_ka_k = x_1\begin{pmatrix} a_{1,1} \\ a_{2,1} \\ \vdots \\ a_{m,1} \end{pmatrix} + x_2\begin{pmatrix} a_{1,2} \\ a_{2,2} \\ \vdots \\ a_{m,2} \end{pmatrix} + \cdots + x_n\begin{pmatrix} a_{1,n} \\ a_{2,n} \\ \vdots \\ a_{m,n} \end{pmatrix}.$$

So, the matrix–vector multiplication should be performed by the following column by coordinate rule: Multiply each column of the matrix by the corresponding coordinate of the vector.

Example:

$$\begin{pmatrix} 1 & 2 & 3 \\ 3 & 2 & 1 \end{pmatrix}\begin{pmatrix} 1 \\ 2 \\ 3 \end{pmatrix} = 1\begin{pmatrix} 1 \\ 2 \end{pmatrix} + 2\begin{pmatrix} 2 \\ 2 \end{pmatrix} + 3\begin{pmatrix} 3 \\ 1 \end{pmatrix} = \begin{pmatrix} 14 \\ 10 \end{pmatrix}$$

The "column by coordinate " rule is very well adapted for parallel computing. It will be also very important in different theoretical constructions later.

However, when doing computations manually, it is more convenient to compute the result one entry at *a* time. This can be expressed as the following row by column rule:

To get the entry number k of the result, one need to multiply row number k of the matrix by the vector, that is, if $Ax = y$, then

$$yk = \sum_{j=1}^{n} a_{k,j}x_j, \quad k = 1, 2, \ldots m;$$

here x_j and y_k are coordinates of the vectors x and y respectively, and a_j,k are the entries of the matrix A.

Example:

$$\begin{pmatrix} 1 & 2 & 3 \\ 4 & 5 & 6 \end{pmatrix}\begin{pmatrix} 1 \\ 2 \\ 3 \end{pmatrix} = \begin{pmatrix} 1\cdot 1 + 2\cdot 2 + 3\cdot 3 \\ 4\cdot 1 + 5\cdot 2 + 6\cdot 3 \end{pmatrix} = \begin{pmatrix} 14 \\ 32 \end{pmatrix}$$

Linear transformations and generating sets: As we discussed above, linear transformation T (acting from $\mathbb{R}^n$ to $\mathbb{R}^m$) is completely defined by its values on the standard basis in $\mathbb{R}^n$. The fact that we consider the standard basis is not essential, one can consider any basis, even any generating (spanning) set. Namely, a linear transformation T: $V \rightarrow W$ is completely defined by its values on *a* generating set (in particular by its values on *a* basis). In particular, if $v_1, v_2, \ldots, v_n$ is *a* generating set (in particular, if it is *a* basis) in V, and T and T_1 are linear transformations T, T_1: $V \rightarrow W$ such that

$$Tv_k = T_1 v_k,\ k = 1, 2, \ldots, n$$

then $T = T_1$.

Conclusions

1. To get the matrix of *a* linear transformation T: $\mathrm{R}^n \rightarrow \mathrm{R}^m$ one needs to join the vectors $a_k = T\, e_k$ (where $e_1, e_2, \ldots, e_n$ is the standard basis in R^n) into *a* matrix: kth column of the matrix is a_k, $k = 1, 2, \ldots, n$.
2. If the matrix A of the linear transformation T is known, then $T(x)$ can be found by the matrix–vector multiplication, $T(x) = Ax$. To perform matrix–vector multiplication one can use either "column by coordinate" or "row by column" rule.

The latter seems more appropriate for manual computations. The former is well adapted for parallel computers, and will be used in different theoretical constructions.

For a linear transformation $T: \mathbb{R}^n \to \mathbb{R}^m$, its matrix is usually denoted as $[T]$. However, very often people do not distinguish between a linear transformation and its matrix, and use the same symbol for both. When it does not lead to confusion, we will also use the same symbol for a transformation and its matrix.

Since a linear transformation is essentially a multiplication, the notation T_v is often used instead of $T(v)$. We will also use this notation. Note that the usual order of algebraic operations apply, *i.e.*, $Tv + u$ means $T(v) + u$, not $T(v + u)$.

Remark: In the matrix–vector multiplication Ax the number of columns of the matrix A matrix must coincide with the size of the vector x, *i.e.*,, a vector in $\mathbb{R}^n$ can only be multiplied by an $m \times n$ matrix. It makes sense, since an $m \times n$ matrix defines a linear transformation $\mathbb{R}^n \to \mathbb{R}^m$, so vector x must belong to $\mathbb{R}^n$.

The easiest way to remember this is to remember that if performing multiplication you run out of some elements faster, then the multiplication is not defined. For example, if using the "row by column" rule you run out of row entries, but still have some unused entries in the vector, the multiplication is not defined. It is also not defined if you run out of vector's entries, but still have unused entries in the column.

COMPOSITION OF LINEAR TRANSFORMATIONS AND MATRIX MULTIPLICATION

Definition of the matrix multiplication: Knowing matrix–vector multiplication, one can easily guess what is the natural way to define the product AB of two matrices: Let us multiply by A each column of B (matrix–vector multiplication) and join the resulting column–vectors into a matrix. Formally, if $b_1, b_2, ..., b_r$ are the columns of B, then $Ab_1, Ab_2, ..., Ab_r$ are the columns of the matrix AB. Recalling the row by column rule for the matrix–vector multiplication we get the following row by column rule for the matrices the entry $(AB)_{j,k}$ (the entry in the row j and column k) of the product AB is defined by

$$(AB)_{j,k} = (\text{row } j \text{ of } A) \cdot (\text{column } k \text{ of } B)$$

Formally it can be rewritten as

$$(AB)_{j,k} = \sum_l a_{j,l} b_{l,k},$$

if $a_{j,k}$ and $b_{j,k}$ are entries of the matrices A and B respectively.

I intentionally did not speak about sizes of the matrices A and B, but if we recall the row by column rule for the matrix–vector multiplication, we can see that in order for the multiplication to be defined, the size of a row of A should be equal to the size of a column of B. In other words the product AB is defined if and only if A is an $m \times n$ and B is $n \times r$ matrix.

Motivation: Composition of linear transformations. Why are we using such *a* complicated rule of multiplication? Why don't we just multiply matrices entrywise? And the answer is, that the multiplication, as it is defined above, arises naturally from the composition of linear transformations. Suppose we have two linear transformations, T_1: $\mathbb{R}^n \to \mathbb{R}^m$ and T_2: $\mathbb{R}^r \to \mathbb{R}^n$. Define the composition $T = T_1\ T_2$ of the transformations T_1, T_2 as

$$T(x) = T_1(T_2(x))\ \forall x \in R^r.$$

Note that $T_1(x) \in R^n$. Since T_1: $\mathbb{R}^n \to \mathbb{R}^m$, the expression $T_1(T_2(x))$ is well defined and the result belongs to $\mathbb{R}^m$. So, T: $\mathbb{R}^r \to \mathbb{R}^m$.

It is easy to show that T is *a* linear transformation, so it is defined by an $m \times r$ matrix. How one can find this matrix, knowing the matrices of T_1 and T_2?

Let A be the matrix of T_1 and B be the matrix of T_2. As we discussed in the previous section, the columns of T are vectors $T(e_1)$, $T(e_2)$,..., $T(e_r)$, where e_1, e_2, ..., e_r is the standard basis in Rr. For $k = 1, 2, ..., r$ we have

$$T(e_k) = T_1(T_2(e_k)) = T_1(Be_k) = T_1(b_k) = Ab_k$$

(operators T_2 and T_1 are simply the multiplication by B and A respectively).

So, the columns of the matrix of T are Ab_1, Ab_2,..., Ab_r, and that is exactly how the matrix AB was defined!

Let us return to identifying again *a* linear transformation with its matrix. Since the matrix multiplication agrees with the composition, we can (and will) write $T_1\ T_2$ instead of T_1 T$_2$ and T_1T_{2x} instead of $T_1(T_2(x))$.

Note that in the composition T_1T_2 the transformation T_2 is applied first! The way to remember this is to see that in T_1T_2x the transformation T_2 meets x first.

Remark: There is another way of checking the dimensions of matrices in *a* product, different form the row by column rule: for *a* composition T_1T_2 to be defined it is necessary that T_{2x} belongs to the domain of T_1. If T_2 acts from some space, say $\mathbb{R}^r$ to $\mathbb{R}^n$, then T_1 must act from Rn to some space, say $\mathbb{R}^m$. So, in order for T_1T_2 to be defined the matrices of T_1 and T_2 should We will usually identify *a* linear transformation and its matrix, but in the next few paragraphs we will distinguish them be of sizes $m \times n$ and $n \times r$ respectively–the same condition as obtained from the row by column rule.

Example: Let T: $\mathbb{R}^2 \to \mathbb{R}^2$ be the reflection in the line $x_1 = 3x_2$. It is *a* linear transformation, so let us find its matrix. To find the matrix, we need to compute T_{e1} and T_{e2}. However, the direct computation of T_{e1} and T_{e2} involves significantly more trigonometry than *a* sane person is willing to remember.

An easier way to find the matrix of T is to represent it as *a* composition of simple linear transformation. Namely, let g be the angle between the x_1 axis and the line $x_1 = 3x_2$, and let T_0 be the reflection in the x_1–axis. Then to get the reflection T we can first rotate the plane by the angle $-g$, moving the line $x_1 = 3x_2$ to the x_1–axis, then reflect everything in the x_1-axis, and then rotate the plane by g, taking everything back. Formally it can be written as

$$T = R_g T_0 R_{-\gamma}$$

where R_g is the rotation by g. The matrix of T_0 is easy to compute,

$$T_0 = \begin{pmatrix} 1 & 0 \\ 0 & -1 \end{pmatrix},$$

the rotation matrices are known

$$R_\gamma = \begin{pmatrix} \cos\gamma & -\sin\gamma \\ \sin\gamma & \cos\gamma \end{pmatrix},$$

$$R_{-\gamma} = \begin{pmatrix} \cos(-\gamma) & -\sin(-\gamma) \\ \sin(-\gamma) & \cos(-\gamma) \end{pmatrix} = \begin{pmatrix} \cos\gamma & \sin\gamma \\ -\sin\gamma & \cos\gamma, \end{pmatrix}$$

To compute $\sin\gamma$ and $\cos\gamma$ take *a* vector in the line $x_1 = 3x_2$, say a vector $(3, 1)^T$. Then

$$\cos\gamma = \frac{\text{first coordinate}}{\text{length}} = \frac{3}{\sqrt{3^2+1^2}} = \frac{3}{\sqrt{10}}$$

and similarly

$$\sin\gamma = \frac{\text{second coordinate}}{\text{length}} = \frac{1}{\sqrt{3^2+1^2}} = \frac{1}{\sqrt{10}}$$

Gathering everything together we get

$$T = R_\gamma T_0 R_{-\gamma} = \frac{1}{\sqrt{10}}\begin{pmatrix} 3 & -1 \\ 1 & 3 \end{pmatrix}\begin{pmatrix} 1 & 0 \\ 0 & -1 \end{pmatrix}\frac{1}{\sqrt{10}}\begin{pmatrix} 3 & 1 \\ -1 & 3 \end{pmatrix}$$

$$= \frac{1}{10}\begin{pmatrix} 3 & -1 \\ 1 & 3 \end{pmatrix}\begin{pmatrix} 1 & 0 \\ 0 & -1 \end{pmatrix}\begin{pmatrix} 3 & 1 \\ -1 & 3 \end{pmatrix}$$

It remains only to perform matrix multiplication here to get the final result.

Properties of Matrix Multiplication.

Matrix multiplication enjoys *a* lot of properties, familiar to us from high school algebra:

1. *Associativity:* $A(BC) = (AB)C$, provided that either left or right side is well defined;
2. *Distributivity:* $A(B + C) = AB + AC$, $(A + B)C = AC + BC$, provided either left or right side of each equation is well defined;
3. One can take scalar multiplies out: $A(\alpha B) = \alpha AB$.

This properties are easy to prove. One should prove the corresponding properties for linear transformations, and they almost trivially follow from the definitions. The properties of linear transformations then imply the properties for the matrix multiplication.

The new twist here is that the commutativity fails: Matrix multiplication is non–commutative, *i.e.*, generally for matrices $AB = BA$.

One can see easily it would be unreasonable to expect the commutativity of matrix multiplication. Indeed, let A and B be matrices of sizes $m \times n$ and $n \times r$ respectively. Then the product AB is well defined, but if $m = r$, BA is not defined.

Even when both products are well defined, for example, when A and B are n×n (square) matrices, the multiplication is still non–commutative. If we just pick the matrices A and B at random, the chances are that $AB = BA$: we have to be very lucky to get $AB = BA$.

Transposed Matrices and Multiplication.

Given *a* matrix A, its transpose (or transposed matrix) A^T is defined by transforming

the rows of A into the columns. For example

$$\begin{pmatrix} 1 & 2 & 3 \\ 4 & 5 & 6 \end{pmatrix}^T = \begin{pmatrix} 1 & 4 \\ 2 & 5 \\ 3 & 6 \end{pmatrix}.$$

So, the columns of A^T are the rows of A and vise versa, the rows of A^T are the columns of A.

The formal definition is as follows: $(A^T)_{j,k} = (A)_{k,j}$ meaning that the entry of A^T in the row number j and column number k equals the entry of A in the row number k and row number j.

The transpose of *a* matrix has *a* very nice interpretation in terms of linear transformations, namely it gives the so–called adjoint transformation.

We will study this in detail later, but for now transposition will be just *a* useful formal operation.

One of the first uses of the transpose is that we can write *a* column vector $x \in R^n$ as $x = (x_1, x_2, ..., x_n)T$. If we put the column vertically, it will use significantly more space.

A simple analysis of the row by columns rule shows that

$$(AB)^T = B^T A^T,$$

i.e.,, when you take the transpose of the product, you change the order of the terms.

Trace and Matrix Multiplication.

For *a* square $(n \times n)$ matrix $A = (a_{j,k})$ its trace (denoted by trace A) is the sum of the diagonal entries

$$\text{trace } A = \sum_{k=1}^{n} a_{k,k}$$

Theorem: Let A and B be matrices of size $m \times n$ and $n \times m$ respectively (so the both products AB and BA are well defined). Then

$$\text{trace}(AB) = \text{trace}(BA)$$

There are essentially two ways of proving this theorem. One is to compute the diagonal entries of AB and of BA and compare their sums. This method requires some proficiency in manipulating sums in notation. If you are not comfortable with algebraic manipulations, there is another way. We can consider two linear transformations, T and T_1, acting from $M_{n\times m}$ to $\mathbb{R} = \mathbb{R}_1$ defined by

$$T(X) = \text{trace}(AX),\ T_1(X) = \text{trace}(XA)$$

To prove the theorem it is sufficient to show that $T = T_1$; the equality for $X = A$ gives the theorem. Since *a* linear transformation is completely defined by its values on *a* generating system, we need just to check the equality on some simple matrices, for example on matrices $X_{j,k}$, which has all entries 0 except the entry 1 in the intersection of *j*th column and *k*th row.

INVERTIBLE TRANSFORMATIONS AND MATRICES. ISOMORPHISMS

IDENTITY TRANSFORMATION AND IDENTITY MATRIX

Among all linear transformations, there is *a* special one, the identity transformation (operator) I, $Ix = x$, $\forall x$. To be precise, there are infinitely many identity transformations: for any vector space V, there is the identity transformation $I = I_V : V \to V$, $I_V x = x$, $\forall x \in V$. However, when it is does not lead to the confusion we will use the same symbol I for all identity operators (transformations). We will use the notation IV only we want to emphasize in what space the transformation is acting. Clearly, if I: $\mathbb{R}^n \to \mathbb{R}^n$ is the identity transformation in Rn, its matrix is an $n \times n$ matrix

$$I = I_n = \begin{pmatrix} 1 & 0 & \cdots & 0 \\ 0 & 1 & \cdots & 0 \\ \vdots & \vdots & \ddots & \vdots \\ 0 & 0 & \cdots & 1 \end{pmatrix}$$

(1 on the main diagonal and 0 everywhere else). When we want to emphasize the size of the matrix, we use the notation I_n; otherwise we just use I. Clearly, for an arbitrary linear transformation A, the equalities

$$AI = A,\ IA = A$$

hold (whenever the product is defined).

INVERTIBLE TRANSFORMATIONS

Definition: Let A: $V \to W$ be *a* linear transformation. We say that the transformation A is left invertible if there exist *a* transformation B: $W \to V$ such that $BA = I$ ($I = I_V$ here). The transformation A is called right invertible if there exists *a* linear transformation C: $W \to V$ such that

$AC = I$ (here $I = I_W$).

The transformations B and C are called left and right inverses of A. Note, that we did not assume the uniqueness of B or C here, and generally left and right inverses are not unique.

Definition: A linear transformation A: $V \to W$ is called invertible if it is both right and left invertible.

Theorem. If *a* linear transformation A: $V \to W$ is invertible, then its left and right inverses B and C are unique and coincide.

Corollary: A transformation A: $V \to W$ is invertible if and only if there erty is used as the exists *a* unique linear transformation (denoted A^{-1}), A^{-1}: $W \to V$ such definition of an

$$A^{-1}A = I_V,\ AA^{-1} = I_W.$$

The transformation A^{-1} is called the inverse of A.

Proof: Let $BA = I$ and $AC = I$. Then

$$BAC = B(AC) = BI = B.$$

On the other hand

$$BAC = (BA)C = IC = C,$$

and therefore $B = C$.

Suppose for some transformation B_1 we have $B_1A = I$. Repeating the above reasoning with B_1 instead of B we get $B_1 = C$. Therefore the left inverse B is unique. The uniqueness of C is proved similarly.

Definition: A matrix is called invertible (resp. left invertible, right invertible) if the corresponding linear transformation is invertible (resp. left invertible, right invertible).

Theorem: asserts that *a* matrix A is invertible if there exists *a* unique matrix A^{-1} such that $A^{-1}A = I$, $AA^{-1} = I$. The matrix A^{-1} is called (surprise) the inverse of A.

Examples:

1. The identity transformation (matrix) is invertible, $I^{-1} = I$;
2. The rotation R_g

$$R_\gamma = \begin{pmatrix} \cos\gamma & -\sin\gamma \\ \sin\gamma & \cos\gamma \end{pmatrix}$$

is invertible, and the inverse is given by $(R_\gamma)^{-1} = R_{-\gamma}$. This equality is clear from the geometric description of Rg, and it also can be checked by the matrix multiplication;

3. The column (1, 1)*T* is left invertible but not right invertible. One of the possible left inverses in the row (1/2, 1/2).

   To show that this matrix is not right invertible, we just notice that there are more than one left inverse. Exercise: describe all left inverses of this matrix.

4. The row (1, 1) is right invertible, but not left invertible. The column $(1/2, 1/2)^T$ is *a* possible right inverse.

Remark: An invertible matrix must be square ($n \times n$). Moreover, if a square matrix A has either left of right inverse, it is invertible. So, it is sufficient to check only one of the identities $AA^{-1} = I$, $A^{-1}A = I$.

This fact will be proved later. Until we prove this fact, we will not use it. I presented it here only to stop trying wrong directions.

Properties of the Inverse Transformation

Theorem: (Inverse of the product). If linear transformations A and B are invertible (and such that the product AB is defined), then the product AB is invertible and

$$(AB)^{-1} = B^{-1}A^{-1}$$

(note the change of the order!)

Proof: Direct computation shows:

$$(AB)(B^{-1}A^{-1}) = A(BB^{-1})A^{-1} = AIA^{-1} = AA^{-1} = I$$

and similarly

$$(B^{-1}A^{-1})(AB) = B^{-1}(A^{-1}A)B = B^{-1}IB = B^{-1}B = I$$

Remark: The invertibility of the product AB does not imply the in-vertibility of the factors A and B (can you think of an example?). However, if one of the factors (either A or B) and the product AB are invertible, then the second factor is also invertible.

Theorem: (Inverse of A^T). If *a* matrix A is invertible, then A^T is also invertible and

$$(A^T)^{-1} = (A^{-1})T$$

Proof: Using $(AB)^T = B^T A^T$ we get

$(A^{-1})^T A^T = (AA^{-1})^T = I^T = I$, and similarly

$A^T (A^{-1})^T = (A^{-1}A)^T = I^T = I$.

And finally, if A is invertible, then A^{-1} is also invertible, $(A^{-1})^{-1} = A$. So, let us summarize the main properties of the inverse:

1. If A is invertible, then A^{-1} is also invertible, $(A^{-1})^{-1} = A$;
2. If A and B are invertible and the product AB is defined, then AB is invertible and $(AB)^{-1} = B^{-1}A^{-1}$.
3. If A is invertible, then A^T is also invertible and $(A^T)^{-1} = (A^{-1})T$.

ISOMORPHISM ISOMORPHIC SPACES.

An invertible linear transformation $A: V \to W$ is called an isomorphism. We did not introduce anything new here, it is just another name for the object we already studied.

Two vector spaces V and W are called isomorphic (denoted $V \cong W$) if there is an isomorphism $A: V \to W$.

Isomorphic spaces can be considered as di erent representation of the same space, meaning that all properties and constructions involving vector space operations are preserved under isomorphism.

The theorem below illustrates this statement.

Theorem: Let $A: V \to W$ be an isomorphism, and let $v_1, v_2, ..., v_n$ be *a* basis in V. Then the system $Av_1, Av_2, ..., Av_n$ is *a* basis in W.

Remark: In the above theorem one can replace "basis" by "linearly independent", or "generating", or "linearly dependent"–all these properties are preserved under isomorphisms.

Remark: If A is an isomorphism, then so is A^{-1}. Therefore in the above theorem we can state that $v_1, v_2, ..., v_n$ is *a* basis if and only if $Av_1, Av_2, ..., Av_n$ is *a* basis.

The inverse to the Theorem is also true

Theorem: Let $A: V \to W$ be *a* linear map, and let $v_1, v_2, ..., v_n$ and $w_1, w_2, ..., w_n$ are bases in V and W respectively. *If* $Av_k = w_k$, $k = 1,2,..., n$, then A is an isomorphism.

Proof: Define the inverse transformation A^{-1} by $A^{-1}w_k = v_k$, $k = 1, 2,..., n$ (as we know, *a* linear transformation is defined by its values on *a* basis).

Examples:

1. Let A: $\mathbb{R}^{n+1} \to \mathbb{P}_n$ ($\mathbb{P}_n$ is the set of polynomials $\sum_{k=0}^{n} a_k t^k$ of degree at most n) is defined by

   $Ae_1 = 1$, $Ae_2 = t, ..., Ae_n = t^{n-1}$, $Ae_{n+1} = t^n$ By Theorem A is an isomorphism, so $\mathbb{P}^n \sim = \mathbb{R}^{n+1}$.

2. Let V be *a* (real) vector space with *a* basis $v_1, v_2, ..., v_n$. Define transformation A: $\mathbb{R}^n \to V$ by

   $$Ae_k = v_k, k = 1, 2,..., n,$$

   where $e_1, e_2, ..., e_n$ is the standard basis in $\mathbb{R}^n$. Again by Theorem A is an isomorphism, so $V \cong$ Rn.

3. $M_{2\times 3} \cong \mathbb{R}^6$;

4. More generally, $M_{m\times n} \cong R^{m \cdot n}$

Invertibility and equations

Theorem: Let $A: V \to W$ be *a* linear transformation. Then A is invertible if and only if for any right side $b \in W$ the equation

$$Ax = b$$

has *a* unique solution $x \in V$.

Proof: Suppose A is invertible. Then $x = A^{-1}b$ solves the equation $Ax = b$. To show that the solution is unique, suppose that for some other vector $x_1 \in V$

$$Ax_1 = b$$

Multiplying this identity by A^{-1} from the left we get

$$A^{-1}Ax = A^{-1}b,$$

and therefore $x_1 = A^{-1}b = x$. Note that both identities, $AA^{-1} = I$ and $A^{-1}A = I$ were used here. Let us now suppose that the equation $Ax = b$ has a unique solution x for any $b \in W$. Let us use symbol y instead of b. We know that given $y \in W$ the equation

$$Ax = y$$

has a unique solution $x \in V$. Let us call this solution $B(y)$.

Let us check that B is a linear transformation. We need to show that

$$B(\alpha y_1 + \beta y_2) = \alpha\beta(y_1) + \beta B(y_2).$$

Let $x_k := B(y_k)$, $k = 1, 2$, *i.e.*, $Ax_k = y_k$, $k = 1, 2$.

Then

$$A(\alpha x_1 + \beta x_2) = \alpha Ax_1 + \beta Ax_2 = \alpha_y 2 + \beta y_2,$$

which means

$$B(\alpha y_1 + \beta y_2) = \alpha B(y_1) + \beta B(y_2).$$

Corollary: An $m \times n$ matrix is invertible if and only if its columns form a basis in $\mathbb{R}^m$.

SUBSPACES

A subspace of a vector space V is a subset $V_0 \subset V$ of V which is closed under the vector addition and multiplication by scalars, *i.e.*,

1. If $v \in V_0$ then $av \in V_0$ for all scalars a.
2. For any $u, v \in V_0$ the sum $u + v \in V_0$.

Again, the conditions 1 and 2 can be replaced by the following one:

$$\alpha u + bv \in V_0 \text{ for all } u, v \in V_0, \text{ and for all scalars } \alpha, \beta.$$

Note, that a subspace $V_0 \subset V$ with the operations (vector addition and multiplication by scalars) inherited from V is a vector space. Indeed, because all operations are inherited from the vector space V they must satisfy all eight axioms of the vector space. The only thing that could possibly go wrong, is that the result of some operation does not belong to V_0. But the definition of a subspace prohibits this!

Now let us consider some examples:

1. Trivial subspaces of a space V, namely V itself and $\{0\}$ (the subspace consisting only of zero vector). Note, that the empty set $\varnothing$ is not a vector space, since it does not contain a zero vector, so it is not a subspace. With each linear transformation $A : V \rightarrow W$ we can associate the following two subspaces:
2. The null space, or kernel of A, which is denoted as Null A or Ker A and consists of all vectors $v \in V$ such that $A_v = 0$.
3. The range Ran A is defined as the set of all vectors $w \in W$ which can be represented as $w = A_v$ for some $v \in V$.

If A is a matrix, *i.e.*, A: $\mathbb{R}^m \rightarrow \mathbb{R}^n$, then recalling column by coordinate rule of the matrix–vector multiplication, we can see that any vector $w \in$ Ran A can be represented as *a* linear combination of columns of the matrix A. That explains why the term column space (and notation Col A) is often used for the range of the matrix. So, for *a* matrix A, the notation Col A is often used instead of Ran A.

And now the last Example.

4. Given *a* system of vectors $v_1, v_2,..., v_r \in V$ its linear span (some-times called simply span) $\mathcal{L}\{v_1, v_2,..., v_r\}$ is the collection of all vectors $v \in V$ that can be represented as *a* linear combination $v = \alpha_1 v_1 + \alpha_2 v_2 +... + \alpha_r v_r$ of vectors $v_1, v_2,..., v_r$. The notation span$\{v_1, v_2,..., v_r\}$ is also used instead of $\mathcal{L}\{v_1, v_2,..., v_r\}$

It is easy to check that in all of these examples we indeed have subspaces.

APPLICATION TO COMPUTER GRAPHICS

In this section we give some ideas of how linear algebra is used in computer graphics. We will not go into the details, but just explain some ideas. In particular we explain why manipulation with 3 dimensional images are reduced to multiplications of 4 × 4 matrices.

2-Dimensional Manipulation

The $x - y$ plane (more precisely, *a* rectangle there) is *a* good model of *a* computer monitor. Any object on *a* monitor is represented as *a* collection of pixels, each pixel is assigned *a* specific colour.

Position of each pixel is determined by the column and row, which play role of x and y coordinates on the plane. So *a* rectangle on *a* plane with $x - y$ coordinates is *a* good model for *a* computer screen: and *a* graphical object is just *a* collection of points.

Remark: There are two types of graphical objects: bitmap objects, where every pixel of an object is described, and vector object, where we describe only *critical points*, and graphic engine connects them to reconstruct the object. *A* (digital) photo is *a* good example of *a* bitmap object: every pixel of it is described.

Bitmap object can contain *a* lot of points, so manipulations with bitmaps require *a* lot of computing power. Anybody who has edited digital photos in *a* bitmap manipulation programme, like Adobe Photoshop, knows that one needs quite *a* powerful computer, and even with modern and powerful computers manipulations can take some time.

That is the reason that most of the objects, appearing on *a* computer screen are vector ones: the computer only needs to memorize critical points.

For example, to describe *a* polygon, one needs only to give the coordinates of its vertices, and which vertex is connected with which. Of course, not all objects on *a* computer screen can be represented as polygons, some, like letters, have curved smooth boundaries.

But there are standard methods allowing one to draw smooth curves through *a* collection of points. For us *a* graphical object will be *a* collection of points (either wireframe model, or bitmap) and we would like to show how one can perform some manipulations with such objects. The simplest transformation is *a* translation (shift), where each point (vector) v is translated by a, *i.e.*, the vector v is replaced by $v + a$ (notation $v \mapsto v + a$ is used for this). *A* vector addition is very well adapted to the computers, so the translation is easy to implement.

Note, that the translation is not *a* linear transformation (if $a \neq 0$): while it preserves the straight lines, it does not preserve 0. All other transformation used in computer graphics are linear. The first one that comes to mind is rotation. The rotation by γ around the origin 0 is given by the multiplication by the rotation matrix R_γ we discussed above,

$$R_\gamma = \begin{pmatrix} \cos\gamma & -\sin\gamma \\ \sin\gamma & \cos\gamma \end{pmatrix}$$

If we want to rotate around a point *a*, we first need to translate the picture by-*a*, moving the point *a* to 0, then rotate around 0 (multiply by R_γ) and then translate everything back by *a*. Another very useful transformation is scaling, given by *a* matrix

$$\begin{pmatrix} a & 0 \\ 0 & b \end{pmatrix},$$

$a, b \geq 0$. If $a = b$ it is uniform scaling which enlarges (reduces) an object, preserving its shape. If $a \neq b$ then x and y coordinates scale di erently; the object becomes "taller" or "wider". Another often used transformation is reflection: for example the matrix

$$\begin{pmatrix} 1 & 0 \\ 0 & -1 \end{pmatrix}$$

defines the reflection through *x*-axis. We will show later in the book, that any linear transformation in $\mathbb{R}^2$ can be represented either as *a* composition of scaling rotations and reflections. However it is sometimes convenient to consider some di erent transformations, like the shear transformation, given by the matrix

$$\begin{pmatrix} 1 & \cos\varphi \\ 0 & 1 \end{pmatrix}$$

This transformation makes all objects slanted, the horizontal lines remain horizontal, but vertical lines go to the slanted lines at the angle *j* to the horizontal ones.

3-Dimensional Graphics

Three–dimensional graphics is more complicated. First we need to be able to manipulate 3–dimensional objects, and then we need to represent it on 2-dimensional plane (monitor). The manipulations with 3-dimensional objects is pretty straightforward, we have the same basic transformations:

Translation, reflection through *a* plane, scaling, rotation. Matrices of these transformations are very similar to the matrices of their 2 × 2 counterparts. For example the matrices

$$\begin{pmatrix} 1 & 0 & 0 \\ 0 & 1 & 0 \\ 0 & 0 & -1 \end{pmatrix}, \begin{pmatrix} a & 0 & 0 \\ 0 & b & 0 \\ 0 & 0 & c \end{pmatrix}, \begin{pmatrix} \cos\gamma & -\sin\gamma & 0 \\ \sin\gamma & \cos\gamma & 0 \\ 0 & 0 & 1 \end{pmatrix},$$

represent respectively reflection through $x - y$ plane, scaling, and rotation around z-axis.

Note, that the above rotation is essentially 2-dimensional transformation, it does not change z coordinate.

Similarly, one can write matrices for the other 2 elementary rotations around x and around y axes. It will be shown later that *a* rotation around an arbitrary axis can be represented as *a* composition of elementary rotations.

So, we know how to manipulate 3-dimensional objects. Let us now discuss how to represent such objects on *a* 2-dimensional plane.

The simplest way is to project it to *a* plane, say to the $x - y$ plane. To perform such projection one just needs to replace z coordinate by 0, the matrix of this projection is

$$\begin{pmatrix} 1 & 0 & 0 \\ 0 & 1 & 0 \\ 0 & 0 & 0 \end{pmatrix},$$

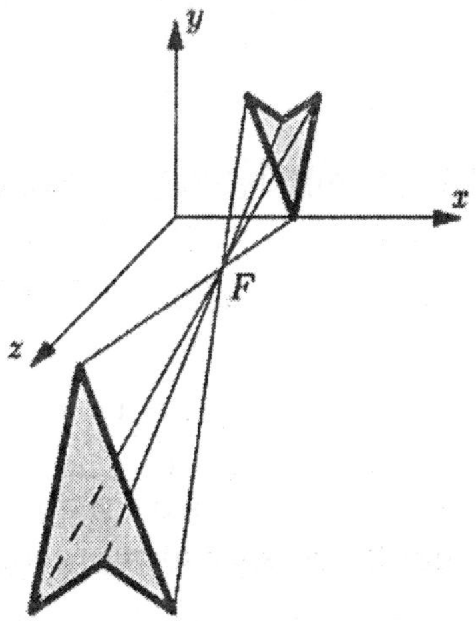

Fig. Perspective Projection onto $x - y$ plane: F is the centre (focal point) of the projection

Such method is often used in technical illustrations. Rotating an object and projecting it is equivalent to looking at it from di erent points. However, this method does not give *a* very realistic picture, because it does not take into account the perspective, the fact that the objects that are further away look smaller.

To get *a* more realistic picture one needs to use the so-called perspective projection. To define *a* perspective projection one needs to pick *a* point the centre of projection or the focal point) and a plane to project onto. Then each point in $\mathbb{R}^3$ is projected into a point on the plane such that the point, its image and the centre of the projection lie on the same line. This is exactly how a camera works, and it is a reasonable first approximation of how our eyes work.

Let us get a formula for the projection. Assume that the focal point is $(0, 0, d)^T$ and that we are projecting onto x–y plane. Consider a point $v = (x, y, z)^T$, and let

$$v^* = (x^*, y^*, 0)^T$$

be its projection, we get that

$$\frac{x^*}{d} = \frac{x}{d-z},$$

so

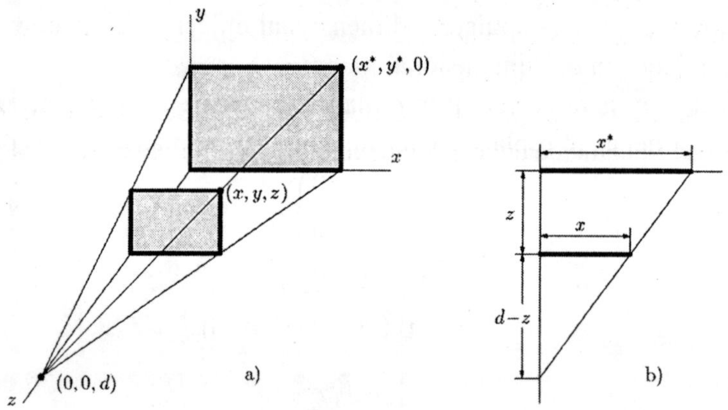

Fig. Finding Coordinates x^*, y^* of the Perspective Projection of the Point $(x, y, z)^T$

$$x^* = \frac{xd}{d-z} = \frac{x}{1-z/d}$$

and similarly

$$y^* = \frac{y}{1-z/d}.$$

Note, that this formula also works if $z > d$ and if $z < 0$: you can draw the corresponding similar triangles to check it. Thus the perspective projection maps a point (x, y, z) to the point $\left(\frac{x}{1-z/d}, \frac{y}{1-z/d}, 0\right)^T$.

This transformation is definitely not linear (because of z in the denominator). However it is still possible to represent it as a linear transformation. To do this let us introduce the so–called homogeneous coordinates.

In the homogeneous coordinates, every point in $\mathbb{R}^3$ is represented by 4 coordinates, the last, 4th coordinate playing role of the scaling coe cient. Thus, to get usual 3-dimensional coordinates of the vector $v = (x, y, z)^T$ from its homogeneous coordinates $(x_1, x_2, x_3, x_4)^T$

one needs to divide all entries by the last coordinate x_4 and take the first 3 coordinates 3 (if $x_4 = 0$ this recipe does not work, so we assume that the case $x_4 = 0$ corresponds to the point at infinity).

Thus in homogeneous coordinates the vector v^* can be represented as $(x, y, 0, 1 - z/d)^T$, so in homogeneous coordinates the perspective projection.

If we multiply homogeneous coordinates of *a* point in $\mathbb{R}^2$ by *a* non–zero scalar, we do not change the point. In other words, in homogeneous coordinates *a* point in $\mathbb{R}^3$ is represented by *a* line through 0 in $\mathbb{R}^4$.
is *a* linear transformation:

$$\begin{pmatrix} x \\ y \\ 0 \\ 1 - z/d \end{pmatrix} = \begin{pmatrix} 1 & 0 & 0 & 0 \\ 0 & 1 & 0 & 0 \\ 0 & 0 & 0 & 0 \\ 0 & 0 & -1/d & 1 \end{pmatrix} \begin{pmatrix} x \\ y \\ z \\ 1 \end{pmatrix}.$$

Note that in the homogeneous coordinates the translation is also *a* linear transformation:

$$\begin{pmatrix} x + a_1 \\ y + a_2 \\ z + a_3 \\ 1 \end{pmatrix} = \begin{pmatrix} 1 & 0 & 0 & a_1 \\ 0 & 1 & 0 & a_2 \\ 0 & 0 & 0 & a_3 \\ 0 & 0 & 0 & 1 \end{pmatrix} \begin{pmatrix} x \\ y \\ z \\ 1 \end{pmatrix}.$$

But what happen if the centre of projection is not *a* point $(0, 0, d)^T$ but some arbitrary point $(d_1, d_2, d_3)^T$. Then we first need to apply the translation by $-(d_1, d_2, 0)T$ to move the centre to $(0, 0, d_3)T$ while preserving the *x-y* plane, apply the projection, and then move everything back translating it by $(d_1, d_2, 0)^T$.

Similarly, if the plane we project to is not *x–y* plane, we move it to the *x–y* plane by using rotations and translations, and so on.

All these operations are just multiplications by 4 × 4 matrices. That explains why modern graphic cards have 4 × 4 matrix operations embedded in the processor.

Of course, here we only touched the mathematics behind 3–dimensional graphics, there is much more.

For example, how to determine which parts of the object are visible and which are hidden, how to make realistic lighting, shades, etc.

Chapter 2

Systems of Linear Equations

Different Faces of Linear Systems

There exist several points of view on what a system of linear equations, or in short a linear system is. The first one is, that it is simply a collection of m linear equations with n unknowns $x_1, x_2, ..., x_n$,

$$\begin{cases} a_{11}x_1 + a_{12}x_2 + \dots + a_{1n}x_n = b_1 \\ a_{21}x_1 + a_{22}x_2 + \dots + a_{2n}x_n = b_2 \\ \dots \\ a_{m1}x_1 + a_{m2}x_2 + \dots + a_{mn}x_n = b_m \end{cases}$$

To solve the system is to find all n–tuples of numbers $x_1, x_2, ..., x_n$ which satisfy all m equations simultaneously.

If we denote $x := (x_1, x_2, ..., x_n)^T \in \mathbb{R}^n$, $b = (b_1, b_2, ..., b_m)^T \in \mathbb{R}^m$, and

$$A = \begin{pmatrix} a_{1,1} & a_{1,2} & \cdots & a_{1,n} \\ a_{2,1} & a_{2,2} & \cdots & a_{2,n} \\ \vdots & \vdots & & \vdots \\ a_{m,1} & a_{m,2} & \cdots & a_{m,n} \end{pmatrix},$$

then the above linear system can be written in the matrix form (as a *matrix vector equation*)

$$Ax = b.$$

To solve the above equation is to find all vectors $x \in R^n$ satisfying $Ax = b$, and finally, recalling the "column by coordinate" rule of the matrixvector multiplication, we can write the system as a *vector equation*

$$x_1a_1 + x_2a_2 + ... + x_na_n = b,$$

where a_k is the kth column of the matrix A, $a_k = (a_{1k}, a_{2,k}, ..., a_{m,k})^T$, $k = 1, 2, ..., n$.

Note, these three examples are essentially just different representations of the same mathematical object.

Before explaining how to solve a linear system, let us notice that it does not matter what we call the unknowns, x_k, y_k or something else. So, all the information necessary to solve the system is contained in the matrix A, which is called the *coefficient matrix* of the system and in the vector (right side) b. Hence, all the information we need is contained in the following matrix

$$\left(\begin{array}{cccc|c} a_{11} & a_{12} & \dots & a_1n & b_1 \\ a_{21} & a_{22} & \dots & a_2n & b_2 \\ \vdots & \vdots & & \vdots & \vdots \\ a_{m1} & a_{m2} & \dots & a_{mn} & b_m \end{array}\right)$$

which is obtained by attaching the column b to the matrix A. This matrix is called the augmented matrix of the system. We will usually put the vertical line separating A and b to distinguish between the augmented matrix and the coefficient matrix.

Solution of a Linear System. Echelon and Reduced Echelon Forms

Linear system are solved by the *Gauss–Jordan elimination* (which is sometimes called row reductio*n*). By performing operations on rows of the augmented matrix of the system (*i.e.*, on the equations), we reduce it to a simple form, the so–called *echelon form*. When the system is in the echelon form, one can easily write the solution.

Row operations. There are three types of row operations we use:

1. Row exchange: interchange two rows of the matrix;
2. Scaling: multiply a row by a non–zero scalar a;
3. Row replacement: replace a row # k by its sum with a constant multiple of a row # j; all other rows remain intact;

It is clear that the operations 1 and 2 do not change the set of solutions of the system; they essentially do not change the system. As for the operation 3, one can easily see that it does not lose solutions.

Namely, let a "new" system be obtained from an "old" one by a row operation of type 3. Then any solution of the "old" system is a solution of the "new" one.

To see that we do not gain anything extra, *i.e.*, that any solution of the "new" system is also a solution of the "old" one, we just notice that row operation of type 3 are *reversible*, *i.e.*, the "old' system also can be obtained from the "new" one by applying a row operation of type 3.

Row operations and multiplication by elementary matrices. There is another, more "advanced" explanation why the above row operations are legal.

Namely, every row operation is equivalent to the multiplication of the matrix from the left by one of the special elementary matrices. Namely, the multiplication by the matrix

$$
\begin{array}{c}
\\ \\ \\ \\ k \\ \\ \\ \\
\end{array}
\begin{pmatrix}
1 & & & \vdots & & & \vdots & & \\
& \ddots & & \vdots & & & \vdots & 0 & \\
& & 1 & \vdots & & & \vdots & & \\
& & & 0 & \cdots & \cdots & 1 & & \\
& & & \vdots & & & \vdots & & \\
\cdots & \cdots & \cdots & 1 & \cdots & \cdots & 0 & & \\
& & 0 & & & & & 1 & \\
& & & & & & & & \ddots \\
& & & & & & & & & 1
\end{pmatrix}
$$

with columns j and k marked above,

just interchanges the rows number j and number k.

$$
\begin{array}{c}
\\ \\ \\ \\ k \\ \\ \\
\end{array}
\begin{pmatrix}
1 & & & \vdots & & & \\
& \ddots & & \vdots & & 0 & \\
& & 1 & 0 & & & \\
\cdots & \cdots & 0 & a & 0 & & \\
& & & 0 & 1 & & \\
0 & & & & & \ddots & 0 \\
& & & & & 0 & 1
\end{pmatrix}
$$

Multiplication by the matrix multiplies the row number k by a. Finally, multiplication by the matrix

$$
\begin{array}{c}
\\ \\ j \\ \\ k \\ \\ \\
\end{array}
\begin{pmatrix}
1 & & \vdots & & \vdots & & \\
& \ddots & \vdots & & \vdots & 0 & \\
\cdots & \cdots & 1 & \cdots & 0 & & \\
& & \vdots & & \vdots & & \\
\cdots & \cdots & a & \cdots & 1 & & \\
0 & & & & & \ddots & \\
& & & & & & 1
\end{pmatrix}
$$

A way to describe (or to remember) these elementary matrices: they are obtained from I by applying the corresponding row operation to it adds to the row # k row # j multiplied by a, and leaves all other rows intact. To see, that the multiplication by these matrices works as advertised, one can just see how the multiplications act on vectors (columns).

Note that all these matrices are invertible (compare with reversibility of row operations). The inverse of the first matrix is the matrix itself. To get the inverse of the second one, one just replaces a by 1/a. And finally, the inverse of the third matrix is obtained by replacing a by –a. To see that the inverses are indeed obtained this way, one again can simply check how they act on columns.

So, performing a row operation on the augmented matrix of the system $Ax = b$ is equivalent to the multiplication of the system (from the left) by a special invertible matrix E. Left multiplying the equality $Ax = b$ by E we get that any solution of the equation

$$Ax = b$$

is also a solution of

$$EAx = Eb.$$

Multiplying this equation (from the left) by E^{-1} we get that any of its solutions is a solution of the equation

$$E^{-1}EAx = E^{-1}Eb,$$

which is the original equation $Ax = b$. So, a row operation does not change the solution set of a system.

Row reduction. The main step of row reduction consists of three sub–steps:

1. Find the leftmost non–zero column of the matrix;
2. Make sure, by applying row operations of type 2, if necessary, that the first (the upper) entry of this column is non–zero. This entry will be called the pivot entry or simply the pivot;
3. "Kill" (*i.e.,* make them 0) all non–zero entries below the pivot by adding (subtracting) an appropriate multiple of the first row from the rows number 2, 3, ..., m.

We apply the main step to a matrix, then we leave the first row alone and apply the main step to rows 2, ..., m, then to rows 3, ..., m, etc.

The point to remember is that after we subtract a multiple of a row from all rows below it (step 3), we leave it alone and do not change it in any way, not even interchange it with another row.

After applying the main step finitely many times (at most m), we get what is called the *echelon form* of the matrix.

An example of row reduction. Let us consider the following linear system:

$$\begin{cases} x_1 + 2x_2 + 3x_3 = 1 \\ 3x_1 + 2x_2 + x_3 = 7 \\ 2x_1 + x_2 + 2x_3 = 1 \end{cases}$$

The augmented matrix of the system is

$$\left(\begin{array}{ccc|c} 1 & 2 & 3 & 1 \\ 3 & 2 & 1 & 7 \\ 2 & 1 & 2 & 1 \end{array}\right)$$

Operate R_2, 2(–3), R_3, (–2), we get:

$$\left(\begin{array}{ccc|c} 1 & 2 & 3 & 1 \\ 3 & 2 & 1 & 7 \\ 2 & 1 & 2 & 1 \end{array}\right)\begin{matrix} \\ -3R_1 \\ -2R_1 \end{matrix} \sim \left(\begin{array}{ccc|c} 1 & 2 & 3 & 1 \\ 0 & -4 & -8 & 4 \\ 0 & -3 & -4 & -1 \end{array}\right)$$

Operate $R_2\left(-\frac{1}{4}\right)$, we get

$$\left(\begin{array}{ccc|c} 1 & 2 & 3 & 1 \\ 0 & 1 & 2 & -1 \\ 0 & -3 & -4 & -1 \end{array}\right)$$

Operate, $R_{3,2}$ (3), we obtain

$$\left(\begin{array}{ccc|c} 1 & 2 & 3 & 1 \\ 0 & 1 & 2 & -1 \\ 0 & -3 & -4 & -1 \end{array}\right) -3R_2 \quad \sim \quad \left(\begin{array}{ccc|c} 1 & 2 & 3 & 1 \\ 0 & 1 & 2 & -1 \\ 0 & 0 & 2 & -4 \end{array}\right)$$

Now we can use the so called back *substitution* to solve the system. Namely, from the last row (equatio*n*) we get $x_3 = -2$. Then from the second equation we get

$$x_2 = -1 - 2x_3 = -1 - 2(-2) = 3,$$

and finally, from the first row (equatio*n*)

$$x_1 = 1 - 2x_2 - 3x_3 = 1 - 6 + 6 = 1.$$

So, the solution is

$$\begin{cases} x_1 = 1 \\ x_2 = 3 \\ x_3 = -2, \end{cases}$$

or in vector form

$$x = \begin{pmatrix} 1 \\ 3 \\ -2 \end{pmatrix}$$

or $x = (1, 3, -2)^T$. We can check the solution by multiplying Ax, where A is the coefficient matrix.

Instead of using back substitution, we can do row reduction from down to top, killing all the entries above the main diagonal of the coefficient matrix: we start by multiplying the last row by 1/2, and the rest is pretty self–explanatory:

$$\left(\begin{array}{ccc|c} 1 & 2 & 3 & 1 \\ 0 & 1 & 2 & -1 \\ 0 & 0 & 1 & -2 \end{array}\right)^{\substack{-3R_3 \\ -2R_3}} \sim \left(\begin{array}{ccc|c} 1 & 2 & 0 & 7 \\ 0 & 1 & 0 & 3 \\ 0 & 0 & 1 & -2 \end{array}\right)^{-2R_2} \sim \left(\begin{array}{ccc|c} 1 & 0 & 0 & 1 \\ 0 & 1 & 0 & 3 \\ 0 & 0 & 1 & -2 \end{array}\right).$$

and we just read the solution $x = (1, 3, -2)^T$ o the augmented matrix.

Echelon form. *A* matrix is in echelon form if it satisfies the following two conditions:

1. All zero rows (*i.e.,*, the rows with all entries equal 0), if any, are below all non–zero entries.

For a non–zero row, let us call the leftmost non–zero entry the leading entry. Then the second property of the echelon form can be formulated as follows:

2. For any non–zero row its leading entry is strictly to the right of the leading entry in the previous row.

The leading entry in each row in echelon form is also called pivot entry, Pivots: leading (rightmost non–zero entries) in a row. or simply *pivot*, because these entries are exactly the pivots we used in the row reduction.

A particular case of the echelon form is the so–called *triangular* form. We got this form in our example above. In this form the coefficient matrix is square ($n \times n$), all its entries on the main diagonal are non–zero, and all the entries below the main diagonal are zero. The right side, *i.e.,* the rightmost column of the augmented matrix can be arbitrary.

After the backward phase of the row reduction, we get what the socalled *reduced echelon form* of the matrix: coefficient matrix equal I, as in the above example, is a particular case of the reduced echelon form.

The general definition is as follows: we say that a matrix is in the reduced echelon form, if it is in the echelon form and

3. All pivot entries are equal 1;
4. All entries above the pivots are 0. Note, that all entries below the pivots are also 0 because of the echelon form.

To get reduced echelon form from echelon form, we work from the bottom to the top and from the right to the left, using row replacement to kill all entries above the pivots.

An example of the reduced echelon form is the system with the coefficient matrix equal *I*. In this case, one just reads the solution from the reduced echelon form. In general case, one can also easily read the solution from the reduced echelon form. For example, let the reduced echelon form of the system (augmented matrix) be

$$\left(\begin{array}{ccccc|c} \boxed{1} & 2 & 0 & 0 & 0 & 1 \\ 0 & 0 & \boxed{1} & 5 & 0 & 2 \\ 0 & 0 & 0 & 0 & \boxed{1} & 3 \end{array}\right);$$

here we boxed the pivots. The idea is to move the variables, corresponding to the columns without pivot (the so–called free variables) to the right side.

Then we can just write the solution.

$$\begin{cases} x_1 = 1 - 2x_2 \\ x_2 \text{ is free} \\ x_3 = 2 - 5x_4 \\ x_4 \text{ is free} \\ x_5 = 3 \end{cases}$$

or in the vector form

$$x = \begin{pmatrix} 1-2x_2 \\ x_2 \\ 1-5x_4 \\ x_4 \\ 3 \end{pmatrix} = \begin{pmatrix} 1 \\ 0 \\ 1 \\ 0 \\ 3 \end{pmatrix} + x_2 \begin{pmatrix} -2 \\ 1 \\ 0 \\ 0 \\ 0 \end{pmatrix} + x_4 \begin{pmatrix} 0 \\ 0 \\ -5 \\ 1 \\ 0 \end{pmatrix}$$

One can also find the solution from the echelon form by using back substitution: the idea is to work from bottom to top, moving all free variables to the right side.

Analyzing the Pivots

All questions about existence of a solution and it uniqueness can be answered by analyzing pivots in the echelon (reduced echelo*n*) form of the augmented matrix of the system. First of all, let us investigate the question of when when is the equation $Ax = b$ inconsistent, *i.e.,* when it does not have a solution. The answer follows immediately, if one just thinks about it:

a system is inconsistent (does not have a solution) if and only if there is a pivot in the last row of an echelon form of the augmented matrix, *i.e.,* I an echelon form of the augmented matrix has a row 00 ... 0 b, $b \neq 0$ in it.

Indeed, such a row correspond to the equation $0x_1 + 0x_2 + ... + x_n = b \neq 0$ that does not have a solution. If we don't have such a row, we just make the reduced echelon form and then read the solution off.

Now, three more statements. Note, they all deal with the coefficient matrix, and not with the augmented matrix of the system.

1. A solution (if it exists) is unique i there are no free variables, that is if and only if the echelon form of the coefficient matrix has a pivot in every column;
2. Equation $Ax = b$ is consistent for all right sides b if and only if the echelon form of the coefficient matrix has a pivot in every row.
3. Equation $Ax = b$ has a unique solution for any right side b if and only if echelon form of the coefficient matrix A has a pivot in every column and every row.

The first statement is trivial, because free variables are responsible for all non–uniqueness. I should only emphasize that this statement does not *say anything* about the existence.

The second statement is a tiny bit more complicated. If we have a pivot in every row of the coefficient matrix, we cannot have the pivot in the last column of the *augmented matrix*, so the system is always consistent, no matter what the right side b is.

Let us show that if we have a zero row in the echelon form of the coefficient matrix A, then we can pick a right side b such that the system Ax = b is not consistent. Let A_e echelon form of the coefficient matrix A. Then

$$A_e = EA,$$

where E is the product of elementary matrices, corresponding to the row operations, $E = EN, ..., E_2, E_1$. If A_e has a zero row, then the last row is also zero. Therefore, if we put be $= (0, ..., 0, 1)^T$ (all entries are 0, except the last one), then the equation

$$A_e x = b_e$$

does not have a solution. Multiplying this equation by E^{-1} from the left, an recalling that $E^{-1}A_e = A$, we get that the equation

$$Ax = E^{-1}b_e$$

does not have a solution.

Finally, statement 3 immediately follows from statements 1 and 2.

From the above analysis of pivots we get several very important corollaries. The main observation. In echelon form, any row and any column have no more than 1 pivot in it (it can have 0 pivots)

Corollaries about Linear Independence and Bases

Questions as to when a system of vectors in $\mathbb{R}^n$ is a basis, a linearly independent or a spanning system, can be easily answered by the row reduction.

Proposition. Let us have a system of vectors $v_1, v_2, ..., v_m \in \mathbb{R}^n$, *and let* $A = [v_1, v_2, ..., v_m]$ *be an* $n \times m$ *matrix with columns* $v_1, v_2, ..., v_m$. *Then*

1. *The system* $v_1, v_2, ..., v_m$ *is linearly independent i echelon form of A has a pivot in every column;*
2. *The system* $v_1, v_2, ..., v_m$ *is complete in* $\mathbb{R}^n$ *(spanning, generating) i echelon form of A has a pivot in every row;*
3. *The system* $v_1, v_2, ..., v_m$ *is a basis in* $\mathbb{R}^n$ *i echelon form of A has a pivot in every column and in every row.*

Proof. The system $v_1, v_2, ..., v_m \in \mathbb{R}^m$ is linearly independent if and only if the equation

$$x_1v_1 + x_2v_2 + ... + x_mv_m = 0$$

has the unique (trivial) solution $x_1 = x_2 = ... = x_m = 0$, or equivalently, the equation $Ax = 0$ has unique solution $x = 0$. By statement 1 above, it happens if and only if there is a pivot in every column of the matrix.

Similarly, the system $v_1, v_2, ..., v_m \in \mathbb{R}^m$ is complete in $\mathbb{R}^n$ if and only if the equation

$$x_1v_1 + x_2v_2 + ... + x_mv_m = b$$

has a solution for any right side $b \in \mathbb{R}^n$. By statement 2 above, it happens if and only if there is a pivot in every column in echelon form of the matrix.

And finally, the system $v_1, v_2, ..., v_m \in \mathbb{R}^m$ is a basis in $\mathbb{R}^n$ if and only if the equation

$$x_1v_1 + x_2v_2 + ... + x_mv_m = b$$

has unique solution for any right side $b \in \mathbb{R}^n$. By statement 3 this happens if and only if there is a pivot in every column and in every row of echelon form of A.

Proposition. Any linearly independent system of vectors in $\mathbb{R}^n$ *cannot have more than n vectors in it.*

Proof. Let a system $v_1, v_2, ..., v_m \in \mathbb{R}^n$ be linearly independent, and let $A = [v_1, v_2, ..., v_m]$ be the $n \times m$ matrix with columns $v_1, v_2, ..., v_m$. By Proposition echelon form of A must have a pivot in every column, which is impossible if $m > n$ (number of pivots cannot be more than number of rows).

Proposition. Any two bases in a vector space V have the same number of vectors in them.

Proof. Let $v_1, v_2, ..., v_n$ and $w_1, w_2, ..., w_m$ be two different bases in V. Without loss of generality we can assume that $n \leq m$. Consider an isomorphism $A : \mathbb{R}^n \to V$ defined by

$$Ae_k = v_k,\ k = 1, 2, ... n,$$

where $e_1, e_2, ..., e_n$ is the standard basis in $\mathbb{R}^n$.

Since A^{-1} is also an isomorphism, the system $A^{-1}w_1, A^{-1}w_2, ..., A^{-1}w_m$ is a basis. So it is linearly independent, $m \leq n$. Together with the assumption $n \leq m$ this implies that $m = n$.

The statement below is a particular case of the above proposition.

Proposition. Any basis in $\mathbb{R}^n$ must have exactly n vectors in it.

Proof. This fact follows immediately from the previous proposition, but there is also a direct proof. Let $v_1, v_2, ..., v_m$ be a basis in $\mathbb{R}^n$ and let A be the $n \times m$ matrix with columns $v_1, v_2, ..., v_m$. The fact that the system is a basis, means that the equation

$$Ax = b$$

has a unique solution for any (all possible) right side b. The existence means that there is a pivot in every row (of a reduced echelon form of the matrix), hence the number of pivots is exactly n. The uniqueness mean that there is pivot in every column of the coefficient matrix (its echelon form), so

$$m = \text{number of columns} = \text{number of pivots} = n$$

Proposition. Any spanning (generating) set in $\mathbb{R}^n$ must have at least n vectors.

Proof. Let $v_1, v_2, ..., v_m$ be a complete system in $\mathbb{R}^n$, and let A be $n \times m$ matrix with columns $v_1, v_2, ..., v_m$. Statement 2 of Proposition implies that echelon form of A has a pivot in every row. Since number of pivots cannot exceed the number of rows, $n \leq m$.

Corollaries About Invertible Matrices

Proposition. A matrix A is invertible if and only if its echelon form has pivot in every column and every row.

Proof. As it was discussed in the beginning of the section, the equation $Ax = b$ has a unique solution for any right side b if and only if the echelon form of A has pivot in every row and every column. But, we know, that the matrix (linear transformation) A is invertible if and only if the equation $Ax = b$ has a unique solution for any possible right side b.

There is also an alternative proof. We know that a matrix is invertible if and only if its columns form a basis in. Proposition above states that it happens if and only if there is a pivot in every row and every column.

The above proposition immediately implies the following

Corollary. An invertible matrix must be square ($n \times n$).

Proposition. If a square ($n \times n$) matrix is left invertible, or if it is right right invertible, then it is invertible. In other words, to check the invertibility of a square matrix A it is sucient to check only one of the conditions $AA^{-1} = 1$, $A^{-1}A = I$.

Note, that this proposition applies only to square matrices!

Proof. We know that matrix A is invertible if and only if the equation $Ax = b$ has a unique solution for any right side b. This happens if and only if the echelon form of the matrix A has pivots in every row and in every column.

If a matrix A is left invertible, the equation $Ax = 0$ has unique solution $x = 0$. Indeed, if b is a left inverse of A (*i.e.*, $BA = I$), and x satisfies

$$Ax = 0,$$

then multiplying this identity by B from the left we get $x = 0$, so the solution is unique. Therefore, the echelon form of A has pivots in every row. If the matrix A is square ($n \times n$), the echelon form also has pivots in every column, so the matrix is invertible.

If a matrix A is right invertible, and C is its right inverse ($AC = I$), then for $x = Cb$, b

$\in \mathbb{R}^n$

$$Ax = ACb = Ib = b.$$

Therefore, for any right side b the equation $Ax = b$ has a solution $x = Cb$. Thus, echelon form of A has pivots in every row. If A is square, it also has a pivot in every column, so A is invertible.

FINDING A^{-1} BY ROW REDUCTION

As it was discussed above, an invertible matrix must be square, and its echelon form must have pivots in every row and every column. Therefore reduced echelon form of an invertible matrix is the identity matrix I. Therefore,

Any invertible matrix is row equivalent (i.e. can be reduced by row operations) to the identity matrix.

Now let us state a simple algorithm of finding the inverse of an $n \times n$ matrix:

1. Form an *augmented* $n \times 2n$ matrix $(A|\ I)$ by writing the $n \times n$ identity matrix right of A;
2. Performing row operations on the augmented matrix transform A to the identity matrix I;
3. The matrix I that we added will be automatically transformed to A^{-1};
4. If it is impossible to transform A to the identity by row operations, A is not invertible

There are several possible explanations of the above algorithm. The first, a na¨ÿve one, is as follows: we know that (for an invertible A) the vector $A^{-1}b$ is the solution of the equation $Ax = b$. So to find the column number k of A^{-1} we need to find the solution of $Ax = ek$, where $e_1, e_2, ..., e_n$ is the standard basis in Rn. The above algorithm just solves the equations

$$Ax = e_k,\ k = 1, 2, ..., n$$

simultaneously!

Let us also present another, more "advanced" explanation. As we discussed above, every row operation can be realized as a left multiplication by an elementary matrix. Let $E_1, E_2, ..., E_N$ be the elementary matrices corresponding to the row operation we performed, and let $E = E_N \cdots E_2E_1$ be their product.1 We know that the row operations transform A to identity, *i.e.*, $EA = I$, so $E = A^{-1}$. But the same row operations transform the augmented matrix $(A| I)$ to $(EA \mid E) = (I \mid A^{-1})$.

This "advanced" explanation using elementary matrices implies an important proposition that will be often used later.

Theorem. Any invertible matrix can be represented as a product of elementary matrices.

Proof. As we discussed in the previous paragraph,

$$A^{-1} = EN \cdots E_2E_1,$$

so

$$A = (A^{-1})^{-1} = E_1^{-1}E_2^{-1}\ldots E_N^{-1}.$$

Suppose we want to find the inverse of the matrix

$$\begin{pmatrix} 1 & 4 & -2 \\ -2 & -7 & 7 \\ 3 & 11 & -6 \end{pmatrix}.$$

Augmenting the identity matrix to it and performing row reduction we get

$$\left(\begin{array}{ccc|ccc} 1 & 4 & -2 & 1 & 0 & 0 \\ -2 & -7 & 7 & 0 & 1 & 0 \\ 3 & 11 & -6 & 0 & 0 & 1 \end{array}\right)\begin{matrix} \\ +2R_1 \\ -3R_1 \end{matrix} \sim \left(\begin{array}{ccc|ccc} 1 & 4 & -2 & 1 & 0 & 0 \\ 0 & 1 & 3 & 2 & 1 & 0 \\ 0 & -1 & 0 & -3 & 0 & 1 \end{array}\right)_{+R_2} \sim$$

$$\left(\begin{array}{ccc|ccc} 1 & 4 & -2 & 1 & 0 & 0 \\ 0 & 1 & 3 & 2 & 1 & 0 \\ 0 & 0 & 3 & -1 & 1 & 1 \end{array}\right)^{\times 3} \sim \left(\begin{array}{ccc|ccc} 3 & 12 & -6 & 3 & 0 & 0 \\ 0 & 1 & 3 & 2 & 1 & 0 \\ 0 & 0 & 3 & -1 & 1 & 1 \end{array}\right)^{\begin{matrix}+2R_2\\ -R_3\end{matrix}} \sim$$

Here in the last row operation we multiplied the first row by 3 to avoid fractions in the backward phase of row reduction. Continuing with the row reduction we get

$$\left(\begin{array}{ccc|ccc} 3 & 12 & 0 & 1 & 2 & 2 \\ 0 & 1 & 0 & 3 & 0 & -1 \\ 0 & 0 & 3 & -1 & 1 & 1 \end{array}\right)^{-12R_2} \sim \left(\begin{array}{ccc|ccc} 3 & 0 & 0 & -35 & 2 & 14 \\ 0 & 1 & 0 & 3 & 0 & -1 \\ 0 & 0 & 3 & -1 & 1 & 1 \end{array}\right)$$

Dividing the first and the last row by 3 we get the inverse matrix

$$\begin{pmatrix} -35/3 & 2/3 & 14/3 \\ 3 & 0 & -1 \\ -1/3 & 1/3 & 1/3 \end{pmatrix}$$

DIMENSION, FINITE–DIMENSIONAL SPACES

Definition. The dimension dim V of a vector space V is the number of vectors in a basis.

For a vector space consisting only of zero vector 0 we put dim $V = 0$. If V does not have a (finite) basis, we put dim $V = \infty$. If dim V is finite, we call the space V *finite–dimensional*; otherwise we call it *infinite–dimensional.*

Proposition asserts that the dimension is well defined, *i.e.,* that it does not depend on the choice of a basis.

This immediately implies the following

Proposition. A vector space V is finite–dimensional if and only if it has a finite spanning system.

Suppose, that we have a system of vectors in a finite–dimensional vector space, and we want to check if it is a basis (or if it is linearly independent, or if it is complete)? Probably the simplest way is to use an isomorphism A : $V \to \mathbb{R}^n$, $n = \dim E$ to move the problem to $\mathbb{R}^n$, where all such questions can be answered by row reduction (studying pivots).

Note, that if dim $V = n$, then there always exists an isomorphism A : V $\to \mathbb{R}^n$. Indeed, if dim $V = n$ then there exists a basis

$$v_1, v_2, \ldots, v_n \in V,$$

and one can define an isomorphism

$$A : V \to \mathbb{R}^n$$

by

$$Av_k = e_k, k = 1, 2, \ldots, n.$$

Proposition. Any linearly independent system in a finite–dimensional vector space V cannot have more than dimV vectors in it.

Proof. Let $v_1, v_2, \ldots, v_m \in V$ be a linearly independent system, and let $A : V \to \mathbb{R}^n$ be an isomorphism. Then $Av_1, Av_2, \ldots, Av_m$ is a linearly independent system in $\mathbb{R}^n$, and by Proposition $m \leq n$.

Proposition. Any generating system in a finite–dimensional vector space V must have at least dimV vectors in it.

Proof. Let $v_1, v_2, \ldots, v_m \in V$ be a complete system in V, and let

$$A : V \to \mathbb{R}^n$$

be an isomorphism. Then $Av_1, Av_2, \ldots, Av_m$ is a complete system in $\mathbb{R}^n$, and by Proposition

$$m \geq n.$$

Proposition. Any linearly independent system of vectors in a finitedimensional space can be extended to a basis, i.e., if $v_1, v_2, \ldots, v_r$ *are linearly independent vectors in a finite–dimensional vector space V then one can find vectors* $v_r+1, v_r+ 2 \ldots, v_n$ *such that the system of vectors* $v_1, v_2, \ldots, v_n$ *is a basis in V.*

Proof. Let $n = \dim V$ and let $r < n$ (if $r = n$ then the system $v_1, v_2, ..., v_r$ is already a basis, and the case $r > n$ is impossible). Take any vector not belonging to span$\{v_1, v_2, ..., v_r\}$ and call it $v_r + 1$ (one can always do that because the system $v_1, v_2, ..., v_r$ is not generating). The system $v_1, v_2, ..., v_r, v_r + 1$ is linearly independent. Repeat the procedure with the new system to get vector $v_r + 2$, and so on.

We will stop the process when we get a generating system. Note, that the process cannot continue infinitely, because a linearly independent system of vectors in V cannot have more than $n = \dim V$ vectors.

General Solution of a Linear System

In this short section we discuss the structure of the general solution (*i.e.*, of the solution set) of a linear system. We call a system $Ax = b$ homogeneous, if the right side, $b = 0$, *i.e.*, a homogeneous system is a system of form $Ax = 0$. With each system

$$Ax = b$$

we can associate a homogeneous system just by putting $b = 0$.

Theorem. (General solution of a linear equation). *Let a vector* x_1 *satisfy the equation* $Ax = b$, *and let* H *be the set of all solutions of the associated homogeneous system*

$$Ax = 0.$$

Then the set

$$\{x = x_1 + x_h : xh \in H\}$$

is the set of all solutions of the equation $Ax = b$.

In other words, this theorem can be stated as

$$\boxed{\begin{array}{l}\text{General solution}\\ \text{of } Ax = b\end{array}} = \boxed{\begin{array}{l}\text{A particular solution}\\ \text{of } Ax = b\end{array}} + \boxed{\begin{array}{l}\text{General solution}\\ \text{of } Ax = 0\end{array}}.$$

Proof. Fix a vector x_1 satisfying

$$Ax_1 = b.$$

Let a vector xh satisfy

$$Axh = 0.$$

Then for

$$x = x_1 + x_h,$$

we have

$$Ax = A(x_1 + x_h) = Ax_1 + Ax_h = b + 0 = b,$$

so any x of form

$$x = x_1 + x_h, x_h \in H$$

is a solution of

$$Ax = b.$$

Now let x be satisfy $Ax = b$. Then for

$$x_h := x - x_1$$

we get

$$Ax_h = A(x - x_1) = Ax - Ax_1 = b - b = 0,$$

so

$$x_h \in H.$$

Therefore any solution x of $Ax = b$ can be represented as $x = x_1 + x_h$ with some $x_h \in H$.

The power of this theorem is in its generality. It applies to all linear equations, we do not have to assume here that vector spaces are finitedimensional. You will meet this theorem in differential equations, integral equations, partial differential equations, etc. Besides showing the structure of the solution set, this theorem allows one to separate investigation of uniqueness from the study of existence. Namely, to study uniqueness, we only need to analyse uniqueness of the homogeneous equation $Ax = 0$, which always has a solution.

There is an immediate application in this course: this theorem allows us to check a solution of a system $Ax = b$. For example, consider a system

$$\begin{pmatrix} 2 & 3 & 1 & 4 & -9 \\ 1 & 1 & 1 & 1 & -3 \\ 1 & 1 & 1 & 2 & -5 \\ 2 & 2 & 2 & 2 & -8 \end{pmatrix} x = \begin{pmatrix} 17 \\ 6 \\ 8 \\ 14 \end{pmatrix}.$$

Performing row reduction one can find the solution of this system

$$x = \begin{pmatrix} 3 \\ 1 \\ 0 \\ 2 \\ 0 \end{pmatrix} + x_3 \begin{pmatrix} -2 \\ 1 \\ 1 \\ 0 \\ 0 \end{pmatrix} + x_5 \begin{pmatrix} 2 \\ -1 \\ 0 \\ 2 \\ 1 \end{pmatrix}, x_3, x_5 \in \mathbb{R}$$

The parameters x_3, x_5 can be denoted here by any other letters, t and s, for example; we keeping notation x_3 and x_5 here only to remind us that they came from the corresponding free variables.

Now, let us suppose, that we are just given this solution, and we want to check whether or not it is correct. Of course, we can repeat the row operations, but this is too time consuming. Moreover, if the solution was obtained by some non–standard method, it can look differently from what we get from the row reduction. For example the formula

$$x = \begin{pmatrix} 3 \\ 1 \\ 0 \\ 2 \\ 0 \end{pmatrix} + s \begin{pmatrix} -2 \\ 1 \\ 1 \\ 0 \\ 0 \end{pmatrix} + t \begin{pmatrix} 0 \\ 0 \\ 1 \\ 2 \\ 1 \end{pmatrix}, s, t \in \mathbb{R}$$

gives the same set as (can you say why?); here we just replaced the last vector by its sum with the second one. So, this formula is different from the solution we got from the row reduction, but it is nevertheless correct.

The simplest way to check that give us correct solutions, is to check that the first vector $(3, 1, 0, 2, 0)^T$ satisfies the equation $Ax = b$, and that the other two (the ones with the parameters x_3 and x_5 or s and t in front of them) should satisfy the associated homogeneous equation $Ax = 0$.

If this checks out, we will be assured that any vector x defined is indeed a solution. Note, that this method of checking the solution does not guarantee that gives us all the solutions. For example, if we just somehow miss the term with x_2, the above method of checking will still work fine. What comes to mind, is to count the pivots again. In this example, if one does row operations, the number of pivots is 3. So indeed, there should be 2 free variables, and it looks like we did not miss anything.

To be able to *prove* this, we will need new notions of fundamental subspaces and of rank of a matrix. Systems of linear equations example, one does not have to perform all row operations to check that there are only 2 free variables, and that formulas both give correct general solution.

FUNDAMENTAL SUBSPACES OF A MATRIX

A : $V \to W$ we can associate two subspaces, namely, its kernel, or null space

$$\text{Ker}A = \text{Null } A := \{v \in V : Av = 0\} \subset \text{V, and its range}$$

$$\text{Ran } A = \{w \in W : w = A_v \text{ for some } v \in V\} \subset W.$$

In other words, the kernel Ker A is the solution set of the homogeneous equation $Ax = 0$, and the range Ran A is exactly the set of all right sides $b \in W$ for which the equation $Ax = b$ has a solution.

If A is an $m \times n$ matrix, *i.e.*, a mapping from $\mathbb{R}^n$ to $\mathbb{R}^m$, then it follows from the "column by coordinate" rule of the matrix multiplication that any vector $w \in$ Ran A can be represented as a linear combination of columns of A. This explains the name column space (notation Col A), which is often used instead of Ran A.

If A is a matrix, then in addition to Ran A and Ker A one can also consider the range and kernel for the transposed matrix A^T. Often the term row space is used for Ran A^T and the term left null space is used for KerA^T (but usually no special notation is introduced).

The four subspaces Ran A, Ker A, Ran A^T, Ker A^T are called the *fundamental subspaces* of the matrix A. In this section we will study important relations between the dimensions of the four fundamental subspaces. We will need the following definition, which is one of the fundamental notions of Linear Algebra.

Definition. Given a linear transformation (matrix) a its rank, rank$A$, is the dimension of the range of A

$$\text{rank}A := \dim \text{Ran } A.$$

Computing Fundamental Subspaces and Rank

To compute the fundamental subspaces and rank of a matrix, one needs to do echelon reduction. Namely, let a be the matrix, and A_e be its echelon form

1. The pivot columns of the original matrix a (*i.e.,* the columns where after row operations we will have pivots in the echelon form) give us a basis (one of many possible) in Ran A.
2. The pivot rows of the echelon from A_e give us a basis in the row space. Of course, it is possible just to transpose the matrix, and then do row operations. But if we already have the echelon form of A, say by computing Ran A, then we get Ran AT for free.
3. To find a basis in the null space Ker A one needs to solve the homogeneous equation $Ax = 0$: the details will be seen from the example below.

Example. Consider a matrix

$$\begin{pmatrix} 1 & 1 & 2 & 2 & 1 \\ 2 & 2 & 1 & 1 & 1 \\ 3 & 3 & 3 & 3 & 2 \\ 1 & 1 & -1 & -1 & 0 \end{pmatrix}.$$

Performing row operations we get the echelon form

$$\begin{pmatrix} \boxed{1} & 1 & 2 & 2 & 1 \\ 0 & 0 & \boxed{-3} & -3 & -1 \\ 0 & 0 & 0 & 0 & 0 \\ 0 & 0 & 0 & 0 & 0 \end{pmatrix}$$

(the pivots are boxed here). So, the columns 1 and 3 of the original matrix, *i.e.,* the columns

$$\begin{pmatrix} 1 \\ 2 \\ 3 \\ 1 \end{pmatrix}, \begin{pmatrix} 2 \\ 2 \\ 3 \\ -1 \end{pmatrix}$$

give us a basis in Ran A. We also get a basis for the row space RanAT for free: the first and second row of the *echelon form* of A, *i.e.,* the vectors

$$\begin{pmatrix} 1 \\ 1 \\ 2 \\ 2 \\ 1 \end{pmatrix}, \begin{pmatrix} 0 \\ 0 \\ -3 \\ -3 \\ -1 \end{pmatrix}$$

(we put the vectors vertically here. The question of whether to put vectors here vertically as columns, or horizontally as rows is is really a matter of convention. Our reason for putting them vertically is that although we call RanA^T the row space we define it as a column space of A^T)

To compute the basis in the null space *Ker A* we need to solve the equation $Ax = 0$. Compute the *reduced* echelon form of A, which in this example is

$$\begin{pmatrix} \boxed{1} & 1 & 0 & 0 & 1/3 \\ 0 & 0 & \boxed{1} & 1 & 1/3 \\ 0 & 0 & 0 & 0 & 0 \\ 0 & 0 & 0 & 0 & 0 \end{pmatrix}.$$

Note, that when solving the homogeneous equation $Ax = 0$, it is not necessary to write the whole augmented matrix, it is sucient to work with the coefficient matrix. Indeed, in this case the last column of the augmented matrix is the column of zeroes, which does not change under row operations. So, we can just keep this column in mind, without actually writing it. Keeping this last zero column in mind, we can read the solution 0 the reduced echelon form above:

$$\begin{cases} x_1 = -x_2 - \frac{1}{3}x_5, \\ x_2 \text{is free.} \\ x_3 = -x_4 - \frac{1}{3}x_5 \\ x_4 \text{is free,} \\ x_5 \text{ is free} \end{cases}$$

or, in the vector form

$$x = \begin{pmatrix} -x_2 - \frac{1}{3}x_5 \\ x_2 \\ -x_4 - \frac{1}{3}x_5 \\ x_4 \\ x_5 \end{pmatrix} = x_2 \begin{pmatrix} -1 \\ 1 \\ 0 \\ 0 \\ 0 \end{pmatrix} + x_4 \begin{pmatrix} 0 \\ 0 \\ -1 \\ 1 \\ 0 \end{pmatrix} + x_5 \begin{pmatrix} -1/3 \\ 0 \\ -1/3 \\ 0 \\ 1 \end{pmatrix}$$

The vectors at each free variable, *i.e.*, in our case the vectors

$$\begin{pmatrix} -1 \\ 1 \\ 0 \\ 0 \\ 0 \end{pmatrix}, \begin{pmatrix} 0 \\ 0 \\ -1 \\ 1 \\ 0 \end{pmatrix}, \begin{pmatrix} -1/3 \\ 0 \\ -1/3 \\ 0 \\ 1 \end{pmatrix}$$

form a basis in Ker$A$.

Unfortunately, there is no shortcut for finding a basis in KerA^T, one must solve the equation $A^T x = 0$. Unfortunately, the knowledge of the echelon form of a does not help here.

Explanation of the Computing Bases in the Fundamental Subspaces

So, why do the above methods indeed give us bases in the fundamental subspaces?

The *null space* KerA. The case of the null space Ker$A$ is probably the simplest one: since we solved the equation $Ax = 0$, *i.e.*, found all the solutions, then any vector in Ker A is a linear combination of the vectors we obtained. Thus, the vectors we obtained form a spanning system in Ker A. To see that the system is linearly independent, let us multiply each vector by the corresponding free variable and add everything. Then for each free variable x_k, the entry number k of the resulting vector is exactly x_k, so the only way this vector (the linear combination) can be 0 is when all free variables are 0.

***The column space* Ran A.** Let us now explain why the method for finding a basis in the column space Ran A works. First of all, notice that the pivot columns of the reduced echelon form are of a form a basis in Ran A_{re}. Since row operations are just left multiplications by invertible atrices, they do not change linear independence. Therefore, the pivot columns of the *original* matrix A are linearly independent.

Let us now show that the pivot columns of a span the column space of A. Let v_1, v_2, ..., v_r be the pivot columns of A, and let v be an arbitrary column of A. We want to show that v can be represented as a linear combination of the pivot columns $v_1, v_2, ..., v_r$,

$$v = a_1v_1 + a_2v_2 + ... + a_rv_r.$$

the reduced echelon form A_{re} is obtained from A by the left multiplication

$$A_{re} = EA,$$

where E is a product of elementary matrices, so E is an invertible matrix. The vectors $Ev_1, Ev_2, ..., Ev_r$ are the pivot columns of A_{re}, and the column v of a is transformed to the column Ev of A_{re}. Since the pivot columns of Ar_e form a basis in RanA_{re}, vector Ev can be represented as a linear combination

$$Ev = \alpha_1 Ev_1 + \alpha_2 Ev_2 + ...+ \alpha_r Ev_r.$$

Multiplying this equality by E^{-1} from the left we get the representation

$$v = \alpha_1 v_1 + \alpha_2 v_2 + ...+ \alpha_r v_r,$$

so indeed the pivot columns of A span Ran A.

***The row space* Ran A^T.** It is easy to see that the pivot rows of the echelon form Ae of a are linearly independent. Indeed, let $w_1, w_2, ..., w_r$ be the transposed (since we agreed always to put vectors vertically) pivot rows of Ae. Suppose

$$\alpha_1 w_1 + \alpha_2 w_2 + ... + \alpha_r w_r = 0.$$

Consider the first non zero entry of w_1. Since for all other vectors w_2, w_3, ...,wr the corresponding entries equal 0 (by the definition of echelon form), we can conclude that $\alpha_1 = 0$. So we can just ignore the first term in the sum.

Consider now the first non–zero entry of w_2. The corresponding entries of the vectors $w_3, ..., w_r$ are 0, so $\alpha_2 = 0$. Repeating this procedure, we get that $\alpha_k = 0 \ \forall k = 1, 2, ..., r$.

To see that vectors $w_1, w_2, ..., w_r$ span the row space, one can notice that *row operations do not change the row space*. This can be obtained directly from analyzing row operations, but we present here a more formal way to demonstrate this fact.

For a transformation A and a set X let us denote by $A(X)$ the set of all elements y which can represented as $y = A(x), x \in X$,

$$A(X) := \{y = A(x) : x \in X\} .$$

If a is an $m \times n$ matrix, and A_e is its echelon form, A_e is obtained from A be left multiplication

$$A_e = EA,$$

where E is an $m \times m$ invertible matrix (the product of the corresponding elementary matrices). Then

$$\text{Ran } A_e^T = \text{Ran}(A^T E^T) = A^T (\text{Ran } E^T) = A^T (\mathbb{R}^m) = \text{Ran } A^T ,$$

so indeed $\text{Ran}A^T = \text{Ran } A_e^T$

THE RANK THEOREM, DIMENSIONS OF FUNDAMENTAL SUBSPACES

There are many applications in which one needs to find a basis in column space or in the null space of a matrix. For example, as it was shown above, solving a homogeneous equation $Ax = 0$ amounts to finding a basis in the null space KerA. Finding a basis in the column space means simply extracting a basis from a spanning set, by removing unnecessary vectors (columns).

However, the most important application of the above methods of computing bases of fundamental subspaces is the relations between their dimensions. Theorem

$$\text{rank}A = \text{rank}A^T .$$

This theorem is often stated as follows:

The *column rank* of a matrix coincides with its *row rank.*

The proof of this theorem is trivial, since dimensions of both Ran A and RanA^T are equal to the number of pivots in the echelon form of A. The following theorem is gives us important relations between dimensions of the fundamental spaces. It is often also called the Rank Theorem

Theorem. Let A be an $m \times n$ matrix, i.e., a linear transformation from $\mathbb{R}^n$ to $\mathbb{R}^m$. Then

1. dim Ker A + dim Ran A = dim Ker A + rank $A = n$ *(dimension of the domain of A).*
2. dim Ker A^T + dimRan A^T = dim Ker A^T + rank A^T = dimKer A^T + rank$A^T = m$ *(dimension of the target space of A).*

Proof. The proof, modulo the above algorithms of finding bases in the fundamental subspaces, is almost trivial. The first statement is simply the fact that the number of free variables (dimKer A) plus the number of basic variables (*i.e.,* the number of pivots, *i.e.,* rank A) adds up to the number of columns (*i.e.,* to n). The second statement, if one takes into account that rank A = rank A^T is simply the first statement applied to A^T.

As an application of the above theorem, there we considered a system

$$\begin{pmatrix} 2 & 3 & 1 & 4 & -9 \\ 1 & 1 & 1 & 1 & -3 \\ 1 & 1 & 1 & 2 & -5 \\ 2 & 2 & 2 & 3 & -8 \end{pmatrix} x = \begin{pmatrix} 17 \\ 6 \\ 8 \\ 14 \end{pmatrix}.$$

and we claimed that its general solution given by

$$x = \begin{pmatrix} 3 \\ 1 \\ 0 \\ 2 \\ 0 \end{pmatrix} + x_3 \begin{pmatrix} -2 \\ 1 \\ 1 \\ 0 \\ 0 \end{pmatrix} + x_5 \begin{pmatrix} 2 \\ -1 \\ 0 \\ 2 \\ 1 \end{pmatrix}, \; x_3, x_5 \in \mathbb{R},$$

or by

$$x = \begin{pmatrix} 3 \\ 1 \\ 0 \\ 2 \\ 0 \end{pmatrix} + s \begin{pmatrix} -2 \\ 1 \\ 1 \\ 0 \\ 0 \end{pmatrix} + t \begin{pmatrix} 0 \\ 0 \\ 1 \\ 2 \\ 1 \end{pmatrix} \quad , s, t \in \mathbb{R}$$

A vector x given by either formula is indeed a solution of the equation. But, how can we guarantee that any of the formulas describe *all* solutions?

First of all, we know that in either formula, the last 2 vectors (the ones multiplied by the parameters) belong to Ker A. It is easy to see that in either case both vectors are linearly independent (two vectors are linearly dependent if and only if one is a multiple of the other).

Now, let us count dimensions: interchanging the first and the second rows and performing first round of row operations

$$\begin{matrix} \\ -2R_1 \\ -R_1 \\ -2R_1 \end{matrix} \begin{pmatrix} 1 & 1 & 1 & 1 & -3 \\ 2 & 3 & 1 & 4 & -9 \\ 1 & 1 & 1 & 2 & -5 \\ 2 & 2 & 2 & 3 & -8 \end{pmatrix} \sim \begin{pmatrix} 1 & 1 & 1 & 1 & -3 \\ 0 & 1 & -1 & 2 & -3 \\ 0 & 0 & 0 & 1 & -2 \\ 0 & 0 & 0 & 1 & -2 \end{pmatrix}$$

we see that there are three pivots already, sc rank A ≥ 3. (Actually, we already can see that the rank is 3, but it is enough just to have the estimate here). By Theorem, rank$A$ + dim Ker $A = 5$, hence dim Ker $A \leq 2$, and therefore there cannot be more than 2 linearly independent vectors in Ker$A$. Therefore, last 2 vectors in either formula form a basis in Ker$A$, so either formula give all solutions of the equation.

An important corollary of the rank theorem, is the following theorem connecting existence and uniqueness for linear equations.

Theorem. Let A be an an $m \times n$ matrix. Then the equation

$$Ax = b$$

has a solution for every $b \in \mathbb{R}^m$ if and only if the dual equation

$$A^T x = 0$$

has a unique (only the trivial) solution. (Note, that in the second equation we have A^T, not A).

Proof. The proof follows immediately from Theorem by counting the dimensions.

There is a very nice geometric interpretation of the second rank theorem. Namely, statement 1 of the theorem says, that if a transformation a : $\mathbb{R}^n \to \mathbb{R}^m$ has trivial kernel ($KerA = \{0\}$), then the dimensions of the domain R^n and of the range Ran A coincide. If the kernel is non–trivial, then the transformation "kills" dimKer$A$ dimensions, so dimRan$A$ = n – dim Ker A.

Representation of a Linear Transformation in Arbitrary Bases, Change of Coordinates Formula

The material we have learned about linear transformations and their matrices can be easily extended to transformations in abstract vector spaces with finite bases. In this section we will distinguish between a linear transformation T and its matrix, the reason being that we consider different bases, so a linear transformation can have different matrix representation.

Coordinate vector. Let V be a vector space with a basis

$$B := \{b_1, b_2, ..., b_n\}.$$

Any vector $v \in V$ admits a unique representation as a linear combination

$$v = x_1b_1 + x_2b_2 + ... + x_nb_n = \sum_{k=1}^{n} x_k b_k.$$

The numbers $x_1, x_2, ..., x_n$ are called the coordinates of the vector v in the basis B. It is convenient to join these coordinates into the so–called coordinate vector of v relative to the basis B, which is the column vector

$$[v]_B := \begin{pmatrix} x_1 \\ x_2 \\ \vdots \\ x_4 \end{pmatrix} \in \mathbb{R}^n.$$

Note that the mapping

$$v \to [v]_B$$

is an isomorphism between V and $\mathbb{R}^n$. It transforms the basis $v_1, v_2, ..., v_n$ to the standard basis $e_1, e_2, ..., e_n$ in $\mathbb{R}^n$.

Matrix of a linear transformation. Let $T : V \to W$ be a linear transformation, and let

$$a = \{a_1, a_2, ..., a_n\}, B := \{b_1, b_2, ..., b_m\}$$

be bases in V and W respectively.

A matrix of the transformation T in (or with respect to) the bases a and b is an $m \times n$ matrix, denoted by $[T]_{BA}$, which relates the coordinate vectors $[Tv]_B$ and $[v]_A$,

$$[Tv]_B = [T]_{BA} [v]_A;$$

notice the balance of symbols A and B here: this is the reason we put the first basis A into the second position.

The matrix $[T]_{BA}$ is easy to find: its kth column is just the coordinate vector $[Tak]_B$ (compare this with finding the matrix of a linear transformation from $\mathbb{R}^n$ to $\mathbb{R}^m$).

As in the case of standard bases, composition of linear transformations is equivalent to multiplication of their matrices: one only has to be a bit more careful about bases. Namely, let $T_1 : x \to Y$ and $T_2 : Y \to Z$ be linear transformation, and let A, B and C be bases in X, Y and Z respectively.

The for the composition $T = T_2T_1$,

$$T : x \to Z,\ Tx := T_2(T_1(x))$$

we have

$$[T]_{CA} = [T_2T_1]_{CA} = [T_2]_{CB}\,[T_1]_{BA}$$

(notice again the balance of indices here).

The proof here goes exactly as in the case of $\mathbb{R}^n$ spaces with standard bases, so we do not repeat it here. Another possibility is to transfer everything to the spaces $\mathbb{R}^n$ via the coordinate isomorphisms $v \to [v]_B$. Then one does not need any proof, everything follows from the results about matrix multiplication.

Change of Coordinate Matrix. Let us have two bases

$$A = \{a_1, a_2, ..., a_n\}$$

and

$$b = \{b_1, b_2, ..., b_n\}$$

in a vector space V. Consider the identity transformation $I = I_V$ and its matrix $[I]_{BA}$ in these bases. By the definition

$$[v]_B = [I]_{BA}\,[v]\,A\ ,\ \forall v \in V.$$

i.e., for any vector $v \in V$ the matrix $[I]_{BA}$ transforms its coordinates in the basis a into coordinates in the basis B. The matrix $[I]_{BA}$ is often called the *change of coordinates* (from the basis A to the basis B) matrix.

The matrix $[I]_{BA}$ is easy to compute: according to the general rule of finding the matrix of a linear transformation, its kth column is the coordinate representation $[ak]_B$ of kth element of the basis A Note that

$$[I]_{AB} = ([I]\ _{BA})^{-1},$$

(follows immediately from the multiplication of matrices rule), so any change of coordinate matrix is always invertible.

An example: change of coordinates from the standard basis. Let our space V be $\mathbb{R}^n$, and let us have a basis $B = \{b_1, b_2, ..., b_n\}$ there. We also have the standard basis

$$S = \{e_1, e_2, ..., e_n\}$$

there. The change of coordinates matrix $[I]_{SB}$ is easy to compute:

$$[I]_{SB} = [b_1, b_2, ..., b_n] =: B,$$

i.e., it is just the matrix B whose kth column is the vector (colum*n*) v_k. And in the other direction

$$[I]_{BS} = ([I]\ SB\ )^{-1} = B^{-1}.$$

For example, consider a basis

$$B = \left\{ \begin{pmatrix} 1 \\ 2 \end{pmatrix}, \begin{pmatrix} 2 \\ 1 \end{pmatrix} \right\}$$

in $\mathbb{R}^2$, and let S denote the standard basis there. Then

$$[I]_{SB} = \begin{pmatrix} 1 & 2 \\ 2 & 1 \end{pmatrix} =: B$$

and

$$S[I]_{BS} = [I]_{SB}^{-1} = B^{-1} = \frac{1}{3}\begin{pmatrix} -1 & 2 \\ 2 & -1 \end{pmatrix}$$

(we know how to compute inverses, and it is also easy to check that the above matrix is indeed the inverse of B)

An example: going through the standard basis. In the space of polynomials of degree at most 1 we have bases

$$A = \{1,\ 1 + x\}, \text{ and } B = \{1 + 2x,\ 1 - 2x\},$$

and we want to find the change of coordinate matrix $[I]_{BA}$.

Of course, we can always take vectors from the basis A and try to decompose them in the basis B; it involves solving linear systems, and we know how to do that.

However, *I* think the following way is simpler. In P_1 we also have the standard basis $S = \{1, x\}$, and for this basis

$$[I]_{SA} = \begin{pmatrix} 1 & 1 \\ 0 & 1 \end{pmatrix} =: A,\quad [I]_{SA} = \begin{pmatrix} 1 & 1 \\ 2 & -2 \end{pmatrix} =: B,$$

and taking the inverses

$$[I]_{AS} = A^{-1} = \begin{pmatrix} 1 & -1 \\ 0 & 1 \end{pmatrix},\ [I]_{BS} = B^{-1} = \frac{1}{4}\begin{pmatrix} 2 & 1 \\ 2 & -1 \end{pmatrix}.$$

Then

$$[I]_{BA} = [I]_{BS}[I]_{SA} = B^{-1}A = \frac{1}{4}\begin{pmatrix} 2 & 1 \\ 2 & -1 \end{pmatrix}\begin{pmatrix} 1 & 1 \\ 0 & 1 \end{pmatrix}$$

and

Notice the balance of indices here. [*I*]

Matrix of a transformation and change of coordinates. Let $T : V \to W$ be a linear transformation, and let A, $\widetilde{A}$ be two bases in V and let B, $\widetilde{B}$ be two bases in W. Suppose we know the matrix $[T]_{BA}$, and we would like to find the matrix representation with respect to new bases $\widetilde{A}$, $\widetilde{B}$, *i.e.*, the matrix $[T]_{\widetilde{B}\widetilde{A}}$. The rule is very simple:

to get the matrix in the "new" bases one has to surround the matrix in the "old" bases by change of coordinates matrices.

We did not mention here what change of coordinate matrix should go where, because we don't have any choice if we follow the balance of indices rule. Namely, matrix representation of a linear transformation changes according to the formula Notice the balance of indices.

$$[T]_{\tilde{B}\tilde{A}} = [I]_{\tilde{B}B}[T]_{BA}[I]_{A\tilde{A}}$$

The proof can be done just by analyzing what each of the matrices does.

Case of one basis: similar matrices. Let V be a vector space and let $A = \{a_1, a_2, ..., a_n\}$ be a basis in V. Consider a linear transformation $T: V \to V$ and let $[T]_{AA}$ be its matrix in this basis (we use the same basis for "inputs" and "outputs")

The case when we use the same basis for "inputs" and "outputs" is very important (because in this case we can multiply a matrix by itself), so let us study this case a bit more carefully. Notice, that very often in this $[T]_A$ is often used instead of $[T]_{AA}$. It is shorter, but two index notation is better adapted to the balance of indices rule. case the shorter notation $[T]_A$ is used instead of $[T]_{AA}$. However, the two index notation $[T]$ is better adapted to the balance of indices rule, so I recommend using it (or at least always keep it in mind) when doing change of coordinates.

Let $B = \{b_1, b_2, ..., bn\}$ be another basis in V. By the change of coordinate rule above

$$[T]_{BB} = [I]_{BA}[T]_{AA}[I]_{AB}$$

Recalling that

$$[I]_{BA} = [I]_{AB}^{-1}$$

and denoting $Q := [I]_{AB}$, we can rewrite the above formula as

$$[T]_{BB} = Q^{-1}[T]_{AA}\, Q.$$

This gives a motivation for the following definition

Definition. We say that a matrix A is similar to a matrix b if there exists an invertible matrix Q such that $A = Q^{-1}BQ$.

Since an invertible matrix must be square, it follows from counting dimensions, that similar matrices A and B have to be square and of the same size. If a is similar to B, *i.e.*, if $A = Q\ ^1BQ$, then

$$B = QAQ^{-1} = (Q^{-1})^{-1}A(Q^{-1})$$

(since Q^{-1} is invertible), therefore B is similar to A. So, we can just say that A and B are *similar*.

The above reasoning shows, that it does not matter where to put Q and where Q^{-1}: one can use the formula $A = QBQ^{-1}$ in the definition of similarity.

The above discussion shows, that one can treat similar matrices as different matrix representation of the same linear operator (transformation).

Chapter 3

Matrics

Introduction

A rectangular array of numbers of the form

$$\begin{pmatrix} a_{11} & \cdots & a_{1n} \\ \vdots & & \vdots \\ a_{m1} & \cdots & a_{mn} \end{pmatrix}$$

is called an $m \times n$ matrix, with m rows and n columns. We count rows from the top and columns from the left. Hence

$$(ai_1 \ldots a_{in}) \text{ and } \begin{pmatrix} a_{ij} \\ \vdots \\ a_{mj} \end{pmatrix}$$

represent respectively the i-th row and the j-th column of the matrix (1), and a_{ij} represents the entry in the matrix (1) on the i-th row and j-th column.

Example. Consider the 3×4 matrix

$$\begin{pmatrix} 2 & 4 & 3 & -1 \\ 3 & 1 & 5 & 2 \\ -1 & 0 & 7 & 6 \end{pmatrix}.$$

Here

$$\text{and } \begin{pmatrix} 3 \\ 5 \\ 7 \end{pmatrix}$$

represent respectively the 2-nd row and the 3-rd column of the matrix, and 5 represents the entry in the matrix on the 2-nd row and 3-rd column.

We now consider the question of arithmetic involving matrices. First of all, let us study the problem of addition. A reasonable theory can be derived from the following definition.

Definition. Suppose that the two matrices

$$A = \begin{pmatrix} a_{11} & \cdots & a_{1n} \\ \vdots & & \vdots \\ a_{m1} & \cdots & a_{mn} \end{pmatrix} \text{ and } B = \begin{pmatrix} b_{11} & \cdots & b_{1n} \\ \vdots & & \vdots \\ b_{m1} & \cdots & b_{mn} \end{pmatrix}$$

both have m rows and n columns. Then we write

$$A+B = \begin{pmatrix} a_{11}+b_{11} & \cdots & a_{1n}+b_{1n} \\ \vdots & & \vdots \\ a_{m1}+b_{m1} & \cdots & a_{mn}+b_{mn} \end{pmatrix}$$

and call this the sum of the two matrices A and B.

Example. Suppose that

$$A = \begin{pmatrix} 2 & 4 & 3 & -1 \\ 3 & 1 & 5 & 2 \\ -1 & 0 & 7 & 6 \end{pmatrix} \text{ and } B = \begin{pmatrix} 1 & 2 & -2 & 7 \\ 0 & 2 & 4 & -1 \\ -2 & 1 & 3 & 3 \end{pmatrix}.$$

Then

$$A+B = \begin{pmatrix} 2+1 & 4+2 & 3-2 & -1+7 \\ 3+0 & 1+2 & 5+4 & 2-1 \\ -1-1 & 0+1 & 7+3 & 6+3 \end{pmatrix} \text{ and } \begin{pmatrix} 3 & 6 & 1 & 6 \\ 3 & 3 & 9 & 1 \\ -1 & 1 & 10 & 9 \end{pmatrix}.$$

Example. We do not have a definition for adding" the matrices

$$\begin{pmatrix} 2 & 4 & 3 & -1 \\ -1 & 0 & 7 & 6 \end{pmatrix} \text{ and } \begin{pmatrix} 2 & 4 & 3 \\ 3 & 1 & 5 \\ -1 & 0 & 7 \end{pmatrix}$$

Proposition. (Matrix Addition) Suppose that A, B, C are $m \times n$ matrices. Suppose further that O represents the $m \times n$ matrix with all entries zero. Then

(a) $A + B = B + A$;

(b) $A + (B + C) = (A + B) + C$;

(c) $A + O = A$; and

(d) there is an $m \times n$ matrix A' such that $A + A' = O$.

Proof. Parts (a) – (c) are easy consequences of ordinary addition, as matrix addition is simply entry-wise addition. For part (d), we can consider the matrix A' obtained from A by multiplying each entry of A by -1.

The theory of multiplication is rather more complicated, and includes multiplication of a matrix by a scalar as well as multiplication of two matrices.

We first study the simpler case of multiplication by scalars.

Definition. Suppose that the matrix

$$A = \begin{pmatrix} a_{11} & \cdots & a_{1n} \\ \vdots & & \vdots \\ a_{m1} & \cdots & a_{mn} \end{pmatrix}$$

has m rows and n columns, and that $c \in \mathbb{R}$. Then we write

$$cA = \begin{pmatrix} ca_{11} & \cdots & ca_{1n} \\ \vdots & & \vdots \\ ca_{m1} & \cdots & ca_{mn} \end{pmatrix}$$

and call this the product of the matrix A by the scalar c.

Example. Suppose that

$$A = \begin{pmatrix} 2 & 4 & 3 & -1 \\ 3 & 1 & 5 & 2 \\ -1 & 0 & 7 & 6 \end{pmatrix}.$$

Then

$$2A = \begin{pmatrix} 4 & 8 & 6 & -2 \\ 6 & 2 & 10 & 4 \\ -2 & 0 & 14 & 12 \end{pmatrix}.$$

Proposition. (Multiplication By Scalar) Suppose that A, B are $m \times n$ matrices, and that $c, d \in \mathbb{R}$. Suppose further that O represents the $m \times n$ matrix with all entries zero. Then

(a) $c(A + B) = cA + cB$;

(b) $(c + d)A = cA + dA$;

(c) $0A = 0$; and

(d) $c(dA) = (cd)A$.

Proof. These are all easy consequences of ordinary multiplication, as multiplication by scalar c is simply entry-wise multiplication by the number c.

The question of multiplication of two matrices is rather more complicated. To motivate this, let us consider the representation of a system of linear equations

$$a_{11}x_1 + \cdots + a_{1n}x_n = b_1,$$
$$\vdots$$
$$a_{m1}x_1 + \cdots + a_{mn}x_n = b_1,$$

in the form $Ax = b$, where

$$A = \begin{pmatrix} a_{11} & \cdots & a_{1n} \\ \vdots & & \vdots \\ a_{m1} & \cdots & a_{mn} \end{pmatrix} \text{ and } b = \begin{pmatrix} b_1 \\ \vdots \\ b_m \end{pmatrix}$$

represent the coeffcients and

$$x = \begin{pmatrix} x_1 \\ \vdots \\ x_n \end{pmatrix}$$

represents the variables. This can be written in full matrix notation by

$$\begin{pmatrix} a_{11} & \cdots & a_{1n} \\ \vdots & & \vdots \\ a_{m1} & \cdots & a_{mn} \end{pmatrix} \begin{pmatrix} x_1 \\ \vdots \\ x_n \end{pmatrix} = \begin{pmatrix} b_1 \\ \vdots \\ b_n \end{pmatrix}$$

Can you work out the meaning of this representation?
Now let us define matrix multiplication more formally.
Definition. Suppose that

$$A = \begin{pmatrix} a_{11} & \cdots & a_{1n} \\ \vdots & & \vdots \\ a_{m1} & \cdots & a_{mn} \end{pmatrix} \text{ and } B = \begin{pmatrix} b_{11} & \cdots & b_{1p} \\ \vdots & & \vdots \\ b_{n1} & \cdots & b_{np} \end{pmatrix}$$

are respectively an $m \times n$ matrix and an $n \times p$ matrix. Then the matrix product AB is given by the $m \times p$ matrix

$$AB = \begin{pmatrix} q_{11} & \cdots & q_{1p} \\ \vdots & & \vdots \\ q_{m1} & \cdots & q_{mp} \end{pmatrix}$$

where for every $i = 1, ..., m$ and $j = 1, ..., p$, we have

$$q_{ij} = \sum_{k=1}^{n} a_{ik} b_{kj} = a_{i1} b_{1j} + ... + a_{inbnj}.$$

Remark. Note first of all that the number of columns of the first matrix must be equal to the numberof rows of the second matrix. On the other hand, for a simple way to work out q_{ij} , the entry in the i-throw and j-th column of AB, we observe that the i-th row of A and the j-th column of B are respectively

$$(a_{i1} \cdots a_{in}) \text{ and } \begin{pmatrix} b_{1\mathrm{j}} \\ \vdots \\ b_{\mathrm{nj}} \end{pmatrix}.$$

We now multiply the corresponding entries – from a_{i1} with b_{1j}, and so on, until a_{in} with b_{nj} – and then add these products to obtain q_{ij}.
Example. Consider the matrices

$$A=\begin{pmatrix} 2 & 4 & 3 & -1 \\ 3 & 1 & 5 & 2 \\ -1 & 0 & 7 & 6 \end{pmatrix} \text{ and } B=\begin{pmatrix} 1 & 4 \\ 2 & 3 \\ 0 & -2 \\ 3 & 1 \end{pmatrix}.$$

Note that A is a 3×4 matrix and B is a 4×2 matrix, so that the product AB is a 3×2 matrix. Let us calculate the product

$$AB=\begin{pmatrix} q_{11} & q_{12} \\ q_{21} & q_{22} \\ q_{31} & q_{32} \end{pmatrix}.$$

Consider first of all q_{11}. To calculate this, we need the 1-st row of A and the 1-st column of B, so let us cover up all unnecessary information, so that

$$\begin{pmatrix} 2 & 4 & 3 & -1 \\ \times & \times & \times & \times \\ \times & \times & \times & \times \end{pmatrix}\begin{pmatrix} 1 & \times \\ 2 & \times \\ 0 & \times \\ 3 & \times \end{pmatrix}=\begin{pmatrix} q_{11} & q_{12} \\ \times & \times \\ \times & \times \end{pmatrix}.$$

From the definition, we have

$$q_{11} = 2.1 + 4.2 + 3.0 + (-1) .3 = 2 + 8 + 0 -3 = 7.$$

Consider next q_{12}. To calculate this, we need the 1-st row of A and the 2-nd column of B, so let us cover up all unnecessary information, so that

$$\begin{pmatrix} 2 & 4 & 3 & -1 \\ \times & \times & \times & \times \\ \times & \times & \times & \times \end{pmatrix}\begin{pmatrix} \times & 4 \\ \times & 3 \\ \times & -2 \\ \times & 1 \end{pmatrix}=\begin{pmatrix} \times & q_{12} \\ \times & \times \\ \times & \times \end{pmatrix}.$$

From the definition, we have

$$q_{12} = 2.4 + 4.3 + 3\ (-2) + (-1)\ .1 = 8 + 12 - 6 -1 = 13.$$

Consider next q_{21}. To calculate this, we need the 2-nd row of A and the 1-st column of B, so let us cover up all unnecessary information, so that

$$\begin{pmatrix} \times & \times & \times & \times \\ 3 & 1 & 5 & 2 \\ \times & \times & \times & \times \end{pmatrix}\begin{pmatrix} 1 & \times \\ 2 & \times \\ 0 & \times \\ 3 & \times \end{pmatrix}=\begin{pmatrix} \times & \times \\ q_{21} & \times \\ \times & \times \end{pmatrix}.$$

From the definition, we have

$$q_{21} = 3.1 + 1.2 + 5.\ 0 + 2\ .3 = 3 + 2 + 0 + 6 = 11$$

Consider next q_{22}. To calculate this, we need the 2-nd row of A and the 2-nd column of B, so let us cover up all unnecessary information, so that

$$\begin{pmatrix} \times & \times & \times & \times \\ 3 & 1 & 5 & 2 \\ \times & \times & \times & \times \end{pmatrix} \begin{pmatrix} \times & 4 \\ \times & 3 \\ \times & -2 \\ 3 & 1 \end{pmatrix} = \begin{pmatrix} \times & \times \\ \times & q_{22} \\ \times & \times \end{pmatrix}.$$

From the definition, we have

$$q_{22} = 3.4 + 1.3 + 5.(-2) + 2.1 = 12 + 3 - 10 + 2 = 7.$$

Consider next q_{31}. To calculate this, we need the 3-rd row of A and the 1-st column of B, so let us cover up all unnecessary information, so that

$$\begin{pmatrix} \times & \times & \times & v \\ \times & \times & \times & \times \\ -1 & 0 & 7 & 6 \end{pmatrix} \begin{pmatrix} 1 & \times \\ 2 & \times \\ 0 & \times \\ 3 & \times \end{pmatrix} = \begin{pmatrix} \times & \times \\ \times & \times \\ q_{31} & \times \end{pmatrix}.$$

From the definition, we have

$$q_{31} = (-1).1 + 0.2 + 7.0 + 6.3 = -1 + 0 + 0 + 18 = 17.$$

Consider finally q_{32}. To calculate this, we need the 3-rd row of A and the 2-nd column of B, so let us cover up all unnecessary information, so that

$$\begin{pmatrix} \times & \times & \times & \times \\ \times & \times & \times & \times \\ -1 & 0 & 7 & 6 \end{pmatrix} \begin{pmatrix} \times & 4 \\ \times & 3 \\ \times & -2 \\ \times & 1 \end{pmatrix} = \begin{pmatrix} \times & \times \\ \times & \times \\ \times & q_{32} \end{pmatrix}.$$

From the definition, we have

$$q_{32} = (-1).4 + 0.3 + 7\ (-2) + 6.1 = -4 + 0 + (-14) + 6 = -12.$$

We therefore conclude that

$$AB = \begin{pmatrix} 2 & 4 & 3 & -1 \\ 3 & 1 & 5 & 2 \\ -1 & 0 & 7 & 6 \end{pmatrix} \begin{pmatrix} 1 & 4 \\ 2 & 3 \\ 0 & -2 \\ 3 & 1 \end{pmatrix} = \begin{pmatrix} 7 & 13 \\ 11 & 7 \\ 17 & -12 \end{pmatrix}.$$

Example. Consider again the matrices

$$A = \begin{pmatrix} 2 & 4 & 3 & -1 \\ 3 & 1 & 5 & 2 \\ -1 & 0 & 7 & 6 \end{pmatrix} \text{ and } B = \begin{pmatrix} 1 & 4 \\ 2 & 3 \\ 0 & -2 \\ 3 & 1 \end{pmatrix}.$$

Note that B is a 4×2 matrix and A is a 3×4 matrix, so that we do not have a definition for the "product" BA. We leave the proofs of the following results as exercises for the interested reader.

Proposition. (Associative Law) Suppose that A is an mn matrix, B is an np matrix and C is an $p \times r$ matrix. Then $A(BC) = (AB)C$.

Proposition (Distributive Laws)

(*a*) Suppose that A is an $m \times n$ matrix and B and C are $n \times p$ matrices. Then $A(B + C) = AB + AC$.

(*b*) Suppose that A and B are $m \times n$ matrices and C is an $n \times p$ matrix. Then $(A + B)C = AC + BC$.

Proposition. Suppose that A is an $m \times n$ matrix, B is an $n \times p$ matrix, and that $c \in \mathbb{R}$. Then $c(AB) = (cA)B = A(cB)$.

Systems of Linear Equations

Note that the system of linear equations can be written in matrix form as

$$Ax = b,$$

where the matrices A, x and b are given. We shall establish the following important result.

Proposition. Every system of linear equations of the form has either no solution, one solution or infinitely many solutions.

Proof. Clearly the system (2) has either no solution, exactly one solution, or more than one solution. It remains to show that if the system (2) has two distinct solutions, then it must have infinitely many solutions. Suppose that $x = u$ and $x = v$ represent two distinct solutions. Then

$$Au = b \text{ and } Av = b,$$

so that

$$A(u - v) = Au - Av = b - b = 0,$$

where 0 is the zero $m \times 1$ matrix. It now follows that for every $c \in \mathbb{R}$, we have

$$A(u + c(u - v)) = Au + A(c(u - v)) = Au + c(A(u - v)) = b + c) = b;$$

so that $x = u + c(u - v)$ is a solution for every $c \in \mathbb{R}$. Clearly we have infinitely many solutions.

Inversion of Matrices

We shall deal with square matrices, those where the number of rows equals the number of columns.

Definition. The $n \times n$ matrix

$$I_n = \begin{pmatrix} a_{11} & \cdots & a_{1n} \\ \vdots & & \vdots \\ a_{n1} & \cdots & a_{nn} \end{pmatrix},$$

where

$$a_{ij=}\begin{cases}1 \text{ if } i=j,\\ 0 \text{ if } i\neq j,\end{cases}$$

is called the identity matrix of order n.

Remark. Note that

$$I_1 = (1) \text{ and } i_4 = \begin{pmatrix}1 & 0 & 0 & 0\\ 0 & 1 & 0 & 0\\ 0 & 0 & 1 & 0\\ 0 & 0 & 0 & 1\end{pmatrix}$$

The following result is relatively easy to check. It shows that the identity matrix I_n acts as the identity for multiplication of $n \times n$ matrices.

Proposition. For every $n \times n$ matrix A, we have $AI_n = I_nA = A$.

This raises the following question: Given an $n \times n$ matrix A, is it possible to find another $n \times n$ matrix B such that $AB = BA - I_n$?

However, we shall be content with nding such a matrix B if it exists. We shall relate the existence of such a matrix B to some properties of the matrix A.

Definition. An $n \times n$ matrix A is said to be invertible if there exists an $n \times n$ matrix B such that $AB = BA = I_n$. In this case, we say that B is the inverse of A and write $B = A^{-1}$.

Proposition. Suppose that A is an invertible $n \times n$ matrix. Then its inverse A^{-1} is unique.

Proof. Suppose that B satises the requirements for being the inverse of A. Then $AB = BA = I_n$. It follows that

$$A^{-1} = A^{-1}I_n = A^{-1}(AB) = (A^{-1}A)B = I_nB = B.$$

Hence the inverse A^{-1} is unique.

Proposition. Suppose that A and B are invertible $n \times n$ matrices. Then $(AB)\ B^{-1} = A^{-1}$.

Proof. In view of the uniqueness of inverse, it is sucient to show that $B^{-1}A^{-1}$ satises the requirements for being the inverse of AB. Note that

$$(AB)(B^{-1}A^{-1}) = A(B(B^{-1}A^{-1})) = A((BB^{-1})A^{-1})$$
$$= A(I_nA^{-1}) = AA^{-1} = I_n$$

and

$(B^{-1}A^{-1})(AB) = B^{-1}(A^{-1}(AB)) = B^{-1}((A^{-1}A)B) = B^{-1}(I_nB)\ B^{-1}B = I_n$ as required.

Proposition. Suppose that A is an invertible $n \times n$ matrix. Then

$$(A^{-1})^{-1} = A.$$

Proof. Note that both $(A^{-1})^{-1}$ and A satisfy the requirements for being the inverse of A^{-1}. Equality follows from the uniqueness of inverse.

Application to Matrix Multiplication

In this section, we shall discuss an application of invertible matrices. Detailed discussion of the technique involved will be covered.

Definition. An $n \times n$ matrix

$$A = \begin{pmatrix} a_{11} & \cdots & a_{1n} \\ \vdots & & \vdots \\ a_{n1} & \cdots & a_{nn} \end{pmatrix},$$

where $a_{ij} = 0$ whenever $i \neq j$, is called a diagonal matrix of order n.

Example. The 3×3 matrices

$$\begin{pmatrix} 1 & 0 & 0 \\ 0 & 2 & 0 \\ 0 & 0 & 0 \end{pmatrix} \text{ and } \begin{pmatrix} 0 & 0 & 0 \\ 0 & 0 & 0 \\ 0 & 0 & 0 \end{pmatrix}$$

are both diagonal.

Given an $n \times n$ matrix A, it is usually rather complicated to calculate

$$A^k = \underbrace{A \ldots A}_{k}$$

However, the calculation is rather simple when A is a diagonal matrix, as we shall see in the following example.

Example. Consider the 3×3 matrix

$$A = \begin{pmatrix} 17 & -10 & -5 \\ 45 & -28 & -15 \\ -30 & 20 & 12 \end{pmatrix}.$$

Suppose that we wish to calculate A^{98}. It can be checked that if we take

$$P = \begin{pmatrix} 1 & 1 & 2 \\ 3 & 0 & 3 \\ -2 & 3 & 0 \end{pmatrix},$$

then

$$P^{-1} = \begin{pmatrix} -3 & 2 & 1 \\ -2 & 4/3 & 1 \\ 3 & -5/3 & -1 \end{pmatrix}.$$

Furthermore, if we write

$$D = \begin{pmatrix} -3 & 0 & 0 \\ 0 & 2 & 0 \\ 0 & 0 & 2 \end{pmatrix}.$$

then it can be checked that $A = PDP^{-1}$, so that

$$A^{98} = \underbrace{(PD^{-1})\cdots(PDP^{-1})}_{98} = PD^{98}P^{-1} = P\begin{pmatrix} 3^{98} & 0 & 0 \\ 0 & 2^{98} & 0 \\ 0 & 0 & 2^{98} \end{pmatrix}P^{-1}.$$

This is much simpler than calculating A^{98} directly. Note that this example is only an illustration. We have not discussed here how the matrices P and D are found.

Finding Inverses by Elementary Row Operations

In this section, we shall discuss a technique by which we can nd the inverse of a square matrix, if the inverse exists. Before we discuss this technique, let us recall the three elementary row operations we discussed in the previous chapter. These are: (1) interchanging two rows; (2) adding a multiple of one row to another row; and (3) multiplying one row by a non-zero constant.

Example. **Consider the matrices**

$$A = \begin{pmatrix} a_{11} & a_{12} & a_{13} \\ a_{21} & a_{22} & a_{23} \\ a_{31} & a_{32} & a_{33} \end{pmatrix} \text{ and } I_3 = \begin{pmatrix} 1 & 0 & 0 \\ 0 & 0 & 0 \\ 0 & 0 & 1 \end{pmatrix}.$$

- Let us interchange rows 1 and 2 of A and do likewise for I_3. We obtain respectively

$$\begin{pmatrix} a_{21} & a_{22} & a_{23} \\ a_{11} & a_{22} & a_{13} \\ a_{31} & a_{32} & a_{33} \end{pmatrix} \text{ and } \begin{pmatrix} 0 & 1 & 0 \\ 1 & 0 & 0 \\ 0 & 0 & 1 \end{pmatrix}.$$

Note that

$$\begin{pmatrix} a_{21} & a_{22} & a_{23} \\ a_{11} & a_{12} & a_{13} \\ a_{31} & a_{32} & a_{33} \end{pmatrix} = \begin{pmatrix} 0 & 1 & 0 \\ 1 & 0 & 0 \\ 0 & 0 & 1 \end{pmatrix}\begin{pmatrix} a_{11} & a_{12} & a_{13} \\ a_{21} & a_{22} & a_{23} \\ a_{31} & a_{32} & a_{33} \end{pmatrix}.$$

- Let us interchange rows 2 and 3 of A and do likewise for I_3. We obtain respectively

$$\begin{pmatrix} a_{11} & a_{12} & a_{13} \\ a_{31} & a_{32} & a_{33} \\ a_{21} & a_{22} & a_{23} \end{pmatrix} \text{and} \begin{pmatrix} 1 & 0 & 0 \\ 0 & 0 & 1 \\ 0 & 1 & 0 \end{pmatrix}.$$

Note that

$$\begin{pmatrix} a_{11} & a_{12} & a_{13} \\ a_{31} & a_{32} & a_{33} \\ a_{21} & a_{22} & a_{23} \end{pmatrix} \text{and} \begin{pmatrix} 1 & 0 & 0 \\ 0 & 0 & 1 \\ 0 & 1 & 0 \end{pmatrix}\begin{pmatrix} a_{11} & a_{12} & a_{13} \\ a_{21} & a_{22} & a_{23} \\ a_{31} & a_{32} & a_{33} \end{pmatrix}.$$

- Let us add 3 times row 1 to row 2 of A and do likewise for I_3. We obtain respectively

$$\begin{pmatrix} a_{11} & a_{12} & a_{13} \\ 3a_{11}+a_{21} & 3a_{12}+a_{22} & 3a_{13}+a_{23} \\ a_{31} & a_{32} & a_{33} \end{pmatrix} \text{ and } \begin{pmatrix} 1 & 0 & 0 \\ 3 & 1 & 0 \\ 0 & 0 & 1 \end{pmatrix}$$

Note that

$$\begin{pmatrix} a_{11} & a_{12} & a_{13} \\ 3a_{11}+a_{21} & 3a_{12}+a_{22} & 3a_{13}+a_{23} \\ a_{31} & a_{32} & a_{33} \end{pmatrix} = \begin{pmatrix} 1 & 0 & 0 \\ 3 & 1 & 0 \\ 0 & 0 & 1 \end{pmatrix}\begin{pmatrix} a_{11} & a_{12} & a_{13} \\ a_{21} & a_{22} & a_{23} \\ a_{31} & a_{32} & a_{33} \end{pmatrix}.$$

- Let us add –2 times row 3 to row 1 of A and do likewise for I_3. We obtain respectively

$$\begin{pmatrix} -2a_{31}+a_{11} & -2a_{32}+a_{12} & -2a_{33}+a_{13} \\ a_{21} & a_{22} & a_{23} \\ a_{31} & a_{32} & a_{33} \end{pmatrix} \text{ and } \begin{pmatrix} 1 & 0 & -2 \\ 0 & 1 & 0 \\ 0 & 0 & 1 \end{pmatrix}.$$

Note that

$$\begin{pmatrix} -2a_{31}+a_{11} & -2a_{32}+a_{12} & -2a_{33}+a_{13} \\ a_{21} & a_{22} & a_{23} \\ a_{31} & a_{32} & a_{33} \end{pmatrix} = \begin{pmatrix} 1 & 0 & -2 \\ 0 & 1 & 0 \\ 0 & 0 & 1 \end{pmatrix}\begin{pmatrix} a_{11} & a_{12} & a_{13} \\ a_{21} & a_{22} & a_{23} \\ a_{31} & a_{32} & a_{33} \end{pmatrix}$$

- Let us multiply row 2 of A by 5 and do likewise for I_3. We obtain respectively

$$\begin{pmatrix} a_{11} & a_{12} & a_{13} \\ 5a_{21} & 5a_{22} & 5a_{23} \\ a_{31} & a_{32} & a_{33} \end{pmatrix} \text{ and } \begin{pmatrix} 1 & 0 & 0 \\ 0 & 5 & 0 \\ 0 & 0 & 1 \end{pmatrix}.$$

Note that

$$\begin{pmatrix} a_{11} & a_{12} & a_{13} \\ 5a_{21} & 5a_{22} & 5a_{23} \\ a_{31} & a_{32} & a_{33} \end{pmatrix} = \begin{pmatrix} 1 & 0 & 0 \\ 0 & 5 & 0 \\ 0 & 0 & 1 \end{pmatrix}\begin{pmatrix} a_{11} & a_{12} & a_{13} \\ a_{21} & a_{22} & a_{23} \\ a_{31} & a_{32} & a_{33} \end{pmatrix}.$$

- Let us multiply row 3 of A by – 1 and do likewise for I_3. We obtain respectively

$$\begin{pmatrix} a_{11} & a_{12} & a_{13} \\ a_{21} & a_{22} & a_{23} \\ -a_{31} & -a_{32} & -a_{33} \end{pmatrix} \text{ and } \begin{pmatrix} 1 & 0 & 0 \\ 0 & 1 & 0 \\ 0 & 0 & -1 \end{pmatrix}.$$

Note that

$$\begin{pmatrix} a_{11} & a_{12} & a_{13} \\ a_{21} & a_{22} & a_{23} \\ -a_{31} & -a_{32} & -a_{33} \end{pmatrix} = \begin{pmatrix} 1 & 0 & 0 \\ 0 & 1 & 0 \\ 0 & 0 & -1 \end{pmatrix}\begin{pmatrix} a_{11} & a_{12} & a_{13} \\ a_{21} & a_{22} & a_{23} \\ a_{31} & a_{32} & a_{33} \end{pmatrix}.$$

Let us now consider the problem in general.

Definition. By an elementary $n \times n$ matrix, we mean an $n \times n$ matrix obtained from I_n by an elementary row operation.

We state without proof the following important result. The interested reader may wish to construct a proof, taking into account the different types of elementary row operations.

Proposition. Suppose that A is an $n \times n$ matrix, and suppose that B is obtained from A by an elementary row operation. Suppose further that E is an elementary matrix obtained from I_n by the same elementary row operation. Then $B = EA$.

We now adopt the following strategy. Consider an $n \times n$ matrix A. Suppose that it is possible to reduce the matrix A by a sequence $\alpha_1, \alpha_2, ..., \alpha_k$ of elementary row operations to the identity matrix I_n. If $E_1, E_2, ..., E_k$ are respectively the elementary $n \times n$ matrices obtained from I_n by the same elementary row operations $\alpha_1, \alpha_2, ..., \alpha_k$ then

$$I_n = E_k \ldots E_2E_1A$$

We therefore must have

$$A^{-1} = E_k \ldots E_2E_1 = E_k \ldots E_2E_1I_n.$$

It follows that the inverse A^{-1} can be obtained from I_n by performing the same elementary row operations $\alpha_1, \alpha_2 .... \alpha_k$. Since we are performing the same elementary row operations on A and I_n, it makes sense to put them side by side. The process can then be described pictorially by

$$(A \mid I_n) \xrightarrow{\alpha_1} (E_1A \mid E_1I_n)$$

$$\xrightarrow{\alpha_2} (E_2E_1A \mid E_2E_1I_n)$$

$$\xrightarrow{\alpha_3} \cdots$$

$$\xrightarrow{\alpha_k} (E_k \cdots E_2E_1A \mid E_k \cdots E_2E_1I_n) = (I_n \mid A^{-1}).$$

In other words, we consider an array with the matrix A on the left and the matrix In on the right. We now perform elementary row operations on the array and try to reduce the left hand half to the matrix I_n. If we succeed in doing so, then the right hand half of the array gives the inverse A^{-1}.

Example. Consider the matrix

$$A = \begin{pmatrix} 1 & 1 & 2 \\ 3 & 0 & 3 \\ -2 & 3 & 0 \end{pmatrix}.$$

To find A^{-1}, we consider the array

$$(A \mid I_3) = \begin{pmatrix} 1 & 1 & 2 & 1 & 0 & 0 \\ 3 & 0 & 3 & 0 & 1 & 0 \\ -2 & 3 & 0 & 0 & 0 & 1 \end{pmatrix}$$

We now perform elementary row operations on this array and try to reduce the left hand half to the matrix I_3. Note that if we succeed, then the final array is clearly in reduced row echelon form. We therefore follow the same procedure as reducing an array to reduced row echelon form. Adding – 3 times row 1 to row 2, we obtain

$$\begin{pmatrix} 1 & 1 & 2 & 1 & 0 & 0 \\ 0 & -3 & -3 & -3 & 1 & 0 \\ -2 & 3 & 0 & 0 & 0 & 1 \end{pmatrix}.$$

Adding 2 times row 1 to row 3, we obtain

$$\begin{pmatrix} 1 & 1 & 2 & 1 & 0 & 0 \\ 0 & -3 & -3 & -3 & 1 & 0 \\ 0 & 5 & 4 & 2 & 0 & 1 \end{pmatrix}.$$

Multiplying row by 3 by 3. we obtain

$$\begin{pmatrix} 1 & 1 & 2 & 1 & 0 & 0 \\ 0 & -3 & -3 & -3 & 1 & 0 \\ 0 & 15 & 12 & 6 & 0 & 3 \end{pmatrix}.$$

Adding 5 times row 2 to row 3, we obtain

$$\begin{pmatrix} 1 & 1 & 2 & 1 & 0 & 0 \\ 0 & -3 & -3 & -3 & 1 & 0 \\ 0 & 0 & -3 & -9 & 5 & 3 \end{pmatrix}.$$

Multiplying row 1 by 3, we obtain

$$\begin{pmatrix} 3 & 3 & 6 & 3 & 0 & 0 \\ 0 & -3 & -3 & -3 & 1 & 0 \\ 0 & 0 & -3 & -9 & 5 & 3 \end{pmatrix}.$$

Adding 2 times row 3 to row 1, we obtain

$$\begin{pmatrix} 3 & 3 & 0 & -15 & 10 & 6 \\ 0 & -3 & -3 & -3 & 1 & 0 \\ 0 & 0 & -3 & -9 & 5 & 3 \end{pmatrix}.$$

Adding –1 times row 3 to row 2, we obtain

$$\begin{pmatrix} 3 & 3 & 0 & -15 & 10 & 6 \\ 0 & -3 & 0 & 6 & -4 & -3 \\ 0 & 0 & -3 & -9 & 5 & 3 \end{pmatrix}.$$

Adding 1 times row 2 to row 1, we obtain

$$\begin{pmatrix} 3 & 3 & 0 & -9 & 6 & 3 \\ 0 & -3 & 0 & 6 & -4 & -3 \\ 0 & 0 & -3 & -9 & 5 & 3 \end{pmatrix}.$$

Multiplying row 1 by 1/3, we obtain

$$\begin{pmatrix} 1 & 0 & 0 & -3 & 2 & 1 \\ 0 & -3 & 0 & 6 & -4 & -3 \\ 0 & 0 & -3 & -9 & 5 & 3 \end{pmatrix}.$$

Multiplying row 2 by –1/3, we obtain

$$\begin{pmatrix} 1 & 0 & 0 & -3 & 2 & 1 \\ 0 & 1 & 0 & -2 & 4/3 & 1 \\ 0 & 0 & -3 & -9 & 5 & 3 \end{pmatrix}.$$

Multiplying row 3 by –1/3, we obtain

$$\begin{pmatrix} 1 & 0 & 0 & -3 & 2 & 1 \\ 0 & 1 & 0 & -2 & 4/3 & 1 \\ 0 & 0 & 1 & 3 & -5/3 & -1 \end{pmatrix}.$$

Note now that the array is in reduced row echelon form, and that the left hand half is the identity matrix I_3. It follows that the right hand half of the array represents the inverse A^{-1}. Hence

$$A^{-1} = \begin{pmatrix} -3 & 2 & 1 \\ -2 & 4/3 & 1 \\ 3 & -5/3 & -1 \end{pmatrix}.$$

Example. Consider the matrix

$$A = \begin{pmatrix} 1 & 1 & 2 & 3 \\ 2 & 2 & 4 & 5 \\ 0 & 3 & 0 & 0 \\ 0 & 0 & 0 & 1 \end{pmatrix}.$$

To find A^{-1}, we consider the array

$$(A \mid I_4) = \begin{pmatrix} 1 & 1 & 2 & 3 & 1 & 0 & 0 & 0 \\ 2 & 2 & 4 & 5 & 0 & 1 & 0 & 0 \\ 0 & 3 & 0 & 0 & 0 & 0 & 1 & 0 \\ 0 & 0 & 0 & 1 & 0 & 0 & 0 & 1 \end{pmatrix}.$$

We now perform elementary row operations on this array and try to reduce the left hand half to the matrix I_4. Adding –2 times row 1 to row 2, we obtain

$$\begin{pmatrix} 1 & 1 & 2 & 3 & 1 & 0 & 0 & 0 \\ 0 & 0 & 0 & -1 & -2 & 1 & 0 & 0 \\ 0 & 3 & 0 & 0 & 0 & 0 & 1 & 0 \\ 0 & 0 & 0 & 1 & 0 & 0 & 0 & 1 \end{pmatrix}.$$

Adding 1 times row 2 to row 4, we obtain

$$\begin{pmatrix} 1 & 1 & 2 & 3 & 1 & 0 & 0 & 0 \\ 0 & 0 & 0 & -1 & -2 & 1 & 0 & 0 \\ 0 & 3 & 0 & 0 & 0 & 0 & 1 & 0 \\ 0 & 0 & 0 & 0 & -2 & 1 & 0 & 1 \end{pmatrix}.$$

Interchanging rows 2 and 3, we obtain

$$\begin{pmatrix} 1 & 1 & 2 & 3 & 1 & 0 & 0 & 0 \\ 0 & 3 & 0 & 0 & 0 & 0 & 1 & 0 \\ 0 & 0 & 0 & -1 & -2 & 1 & 0 & 0 \\ 0 & 0 & 0 & 0 & -2 & 1 & 0 & 1 \end{pmatrix}.$$

At this point, we observe that it is impossible to reduce the left hand half of the array to I_4. For those who remain unconvinced, let us continue. Adding 3 times row 3 to row 1, we obtain

$$\begin{pmatrix} 1 & 1 & 2 & 0 & -5 & 3 & 0 & 0 \\ 0 & 3 & 0 & 0 & 0 & 0 & 1 & 0 \\ 0 & 0 & 0 & -1 & -2 & 1 & 0 & 0 \\ 0 & 0 & 0 & 0 & -2 & 1 & 0 & 1 \end{pmatrix}.$$

Adding –1 times row 4 to row 3, we obtain

$$\begin{pmatrix} 1 & 1 & 2 & 0 & -5 & 3 & 0 & 0 \\ 0 & 3 & 0 & 0 & 0 & 0 & 1 & 0 \\ 0 & 0 & 0 & -1 & -2 & 1 & 0 & 0 \\ 0 & 0 & 0 & 0 & -2 & 1 & 0 & 1 \end{pmatrix}.$$

Multiplying row 1 by 6 (here we want to avoid fractions in the next two steps), we obtain

$$\begin{pmatrix} 6 & 6 & 12 & 0 & -30 & 18 & 0 & 0 \\ 0 & 3 & 0 & 0 & 0 & 0 & 1 & 0 \\ 0 & 0 & 0 & -1 & 0 & 0 & 0 & -1 \\ 0 & 0 & 0 & 0 & -2 & 1 & 0 & 1 \end{pmatrix}.$$

Adding – 15 times row 4 to row 1, we obtain

$$\begin{pmatrix} 6 & 6 & 12 & 0 & 0 & 3 & 0 & -15 \\ 0 & 3 & 0 & 0 & 0 & 0 & 1 & 0 \\ 0 & 0 & 0 & -1 & 0 & 0 & 0 & -1 \\ 0 & 0 & 0 & 0 & -2 & 1 & 0 & 1 \end{pmatrix}.$$

Adding –2 times row 2 to row 1, we obtain

$$\begin{pmatrix} 6 & 0 & 12 & 0 & 0 & 3 & -2 & -15 \\ 0 & 3 & 0 & 0 & 0 & 0 & 1 & 0 \\ 0 & 0 & 0 & -1 & 0 & 0 & 0 & -1 \\ 0 & 0 & 0 & 0 & -2 & 1 & 0 & 1 \end{pmatrix}.$$

Multiplying row 1 by 1/6, multiplying row 2 by 1/3, multiplying row 3 by –1 and multiplying row 4 by –1/2, we obtain

$$\begin{pmatrix} 1 & 0 & 2 & 0 & 0 & 1/2 & -1/3 & -5/2 \\ 0 & 1 & 0 & 0 & 0 & 0 & 1/3 & 0 \\ 0 & 0 & 0 & 1 & 0 & 0 & 0 & 1 \\ 0 & 0 & 0 & 0 & 1 & -1/2 & 0 & -1/2 \end{pmatrix}.$$

Note now that the array is in reduced row echelon form, and that the left hand half is not the identity matrix I_4. Our technique has failed. In fact, the matrix A is not invertible.

Criteria for Invertibility

Examples. In this section, we shall obtain some partial answers to this question. Our first step here is the following simple observation.

Proposition. Every elementary matrix is invertible.

Proof. Let us consider elementary row operations. Recall that these are: (1) interchanging two rows; (2) adding a multiple of one row to another row; and (3) multiplying one row by a non-zero constant.

These elementary row operations can clearly be reversed by elementary row operations. For (1), weinterchange the two rows again. For (2), if we have originally added c times row i to row j, then we can reverse this by adding $-c$ times row i to row j. For (3), if we have multiplied any row by a non-zero constant c, we can reverse this by multiplying the same row by the constant $1/c$. Note now that each elementary matrix is obtained from I_n by an elementary row operation. The inverse of this elementary matrix is clearly the elementary matrix obtained from I_n by the elementary row operation that reverses the original elementary row operation.

Suppose that an $n \times n$ matrix B can be obtained from an $n \times n$ matrix A by a finite sequence of elementary row operations. Then since these elementary row operations can be reversed, the matrix A can be obtained from the matrix B by a finite sequence of elementary row operations.

Definition. An $n \times n$ matrix A is said to be row equivalent to an $n \times n$ matrix B if there exist a finite number of elementary $n \times n$ matrices $E_1 \dots E_k$ such that $B = E_k \dots E_1 A$.

Remark. Note that $B = E_k \dots\dots E_1 A$ implies that $A = E_1^{-1} \cdots E_k^{-1} B$. It follows that if A is row equivalent to B, then B is row equivalent to A. We usually say that A and B are row equivalent. The following result gives conditions equivalent to the invertibility of an $n \times n$ matrix A.

Proposition. Suppose that

$$A = \begin{pmatrix} a_{11} & \cdots & a_{1n} \\ \vdots & & \vdots \\ a_{n1} & \cdots & a_{nn} \end{pmatrix},$$

and that

$$x = \begin{pmatrix} x_1 \\ \vdots \\ x_n \end{pmatrix} \text{ and } 0 = \begin{pmatrix} 0 \\ \vdots \\ 0 \end{pmatrix}$$

are $n \times 1$ matrices, where $x_1, ..., x_n$ are variables.

(a) Suppose that the matrix A is invertible. Then the system $Ax = 0$ of linear equations has only the trivial solution.

(b) Suppose that the system $Ax = 0$ of linear equations has only the trivial solution. Then the matrices A and I_n are row equivalent.

(c) Suppose that the matrices A and I_n are row equivalent. Then A is invertible.

Proof. (a) Suppose that x_0 is a solution of the system $Ax = 0$. Then since A is invertible, we have

$$x_0 = I_n x_0 = (A^{-1}A)x_0 = A^{-1}(Ax_0) = A^{-1}0 = 0.$$

It follows that the trivial solution is the only solution.

(b) Note that if the system $Ax = 0$ of linear equations has only the trivial solution, then it can be reduced by elementary row operations to the system

$$x_1 = 0, ..., x_n = 0.$$

This is equivalent to saying that the array

$$\left(\begin{array}{ccc|c} a_{11} & \cdots & a_{1n} & 0 \\ \vdots & & \vdots & \vdots \\ a_{n1} & \cdots & a_{nn} & 0 \end{array}\right)$$

can be reduced by elementary row operations to the reduced row echelon form

$$\left(\begin{array}{ccc|c} 1 & \cdots & 0 & 0 \\ \vdots & & \vdots & \vdots \\ 0 & \cdots & 1 & 0 \end{array}\right).$$

Hence the matrices A and I_n are row equivalent.

(c) Suppose that the matrices A and I_n are row equivalent. Then there exist elementary nn matrices $E_1, ..., E_k$ such that

$$I_n = E_k, ..., E_1 A.$$

By Proposition, the matrices $E_1, ..., E_k$ are all invertible, so that

$$A = E_1^{-1} \cdots E_k^{-1} I_n = E_1^{-1} \ldots E_k^{-1}$$

is a product of invertible matrices, and is therefore itself invertible.

Consequences of Invertibility

Suppose that the matrix

$$A = \begin{pmatrix} a_{11} & \cdots & a_{1n} \\ \vdots & & \vdots \\ a_{n1} & \cdots & a_{nn} \end{pmatrix}$$

is invertible. Consider the system $Ax = b$, where

$$x = \begin{pmatrix} x_1 \\ \vdots \\ x_n \end{pmatrix} \text{ and } b = \begin{pmatrix} b_1 \\ \vdots \\ b_n \end{pmatrix}$$

are $n \times 1$ matrices, where $x_1, \ldots, x_n$ are variables and

$$b_1, \ldots, b_n \in \mathbb{R}$$

are arbitrary. Since A is invertible let us consider $x = A^{-1}b$. Clearly

$$Ax = A(A_{-1}\, b) = (A^{-1}\, b) = (AA^{-1})b = I_n b = b,$$

is invertible, Consider

so that $x = A^{-1}b$ is a solution of the system. On the other hand , let x_0 be any solution of the system. Then $Ax_0 = b$, so that

$$x_0 = I_n x_0 = (A^{-1}A)x_0 = A^{-1}(Ax_0) = A^{-1}b.$$

It follows that the system has unique solution. We have proved the following important result.

Proposition. Suppose that

$$A = \begin{pmatrix} a_{11} & \cdots & a_{1n} \\ \vdots & & \vdots \\ a_{n1} & \cdots & a_{nn} \end{pmatrix},$$

and that

$$x = \begin{pmatrix} x_1 \\ \vdots \\ x_n \end{pmatrix} \text{ and } b = \begin{pmatrix} b_1 \\ \vdots \\ b_n \end{pmatrix}$$

are $n \times 1$ matrices, where $x_1, \ldots, x_n$ are variables and

$$b_1, \ldots, b_n \in \mathbb{R}$$

are arbitrary. Suppose further that the matrix A is invertible. Then the system $Ax = b$ of linear equations has the unique solution $x = A^{-1}b$.

We next attempt to study the question in the opposite direction.

Proposition. Suppose that

$$A = \begin{pmatrix} a_{11} & \cdots & a_{1n} \\ \vdots & & \vdots \\ a_{n1} & \cdots & a_{nn} \end{pmatrix},$$

and that

$$x = \begin{pmatrix} x_1 \\ \vdots \\ x_n \end{pmatrix} \text{ and } b = \begin{pmatrix} b_1 \\ \vdots \\ b_n \end{pmatrix}$$

are $n \times 1$ matrices, where $x_1, ..., x_n$ are variables. Suppose further that for every $b_1, ..., b_n \in \mathbb{R}$, the system $Ax = b$ of linear equations is soluble. Then the matrix A is invertible.

Proof. Suppose that

$$b_1 = \begin{pmatrix} 1 \\ 0 \\ \vdots \\ 0 \\ 0 \end{pmatrix}, b_2 = \begin{pmatrix} 0 \\ 1 \\ 0 \\ \vdots \\ 0 \end{pmatrix}, ..., b_n = \begin{pmatrix} 0 \\ 0 \\ \vdots \\ 0 \\ 1 \end{pmatrix}.$$

In other words, for every $j = 1, ..., n$, b_j is an $n \times 1$ matrix with entry 1 on row and entry 0 elsewhere.

Now let

$$x_1 = \begin{pmatrix} x_{11} \\ \vdots \\ x_{n1} \end{pmatrix} \quad ..., x_n = \begin{pmatrix} x_{1n} \\ \vdots \\ x_{nn} \end{pmatrix}$$

denote respectively solutions of the systems of linear equations

$$Ax = b_1, ..., \; Ax = b_n$$

It is easy to check that

$$A(x_1, ..., x_n) = (b_1, ..., b_n);$$

in other words,

$$A \begin{pmatrix} a_{11} & \cdots & a_{1n} \\ \vdots & & \vdots \\ a_{n1} & \cdots & a_{nn} \end{pmatrix} = I_n,$$

so that A is invertible.

We can now summarize Propositions as follows.

Proposition. In the notation of Proposition, the following four statements are equivalent:

(a) The matrix A is invertible.

(b) The system $Ax = 0$ of linear equations has only the trivial solution.

(c) The matrices A and I_n are row equivalent.

(d) The system $Ax = b$ of linear equations is soluble for every $n \times 1$ matrix b.

Application to Economics

In this section, we describe briey the Leontief input-output model, where an economy is divided into n sectors.

For every $i = 1,..., n$, let x_i denote the monetary value of the total output of sector i over a fixed period, and let d_i denote the output of sector i needed to satisfy outside demand over the same xed period. Collecting together x_i and d_i for $i = 1,.... n$, we obtain the vectors

$$x = \begin{pmatrix} x_1 \\ \vdots \\ x_n \end{pmatrix} \in \mathbb{R}^n \text{ and } d = \begin{pmatrix} d_1 \\ \vdots \\ d_n \end{pmatrix} \in \mathbb{R},$$

known respectively as the production vector and demand vector of the economy.

On the other hand, each of the n sectors requires material from some or all of the sectors to produce its output. For $i, j = 1,... n$, let c_{ij} denote the monetary value of the output of sector i needed by sector j to produce one unit of monetary value of output. For every $j = 1, ... n$, the vector

$$c_j = \begin{pmatrix} c_{1j} \\ \vdots \\ c_{nj} \end{pmatrix} \in \mathbb{R}^n$$

is known as the unit consumption vector of sector j. Note that the column sum

$$c_{1j} + ... + c_{nj} \leq 1$$

in order to ensure that sector j does not make a loss. Collecting together the unit consumption vectors, we obtain the matrix

$$C = (c_1 \cdots c_n) = \begin{pmatrix} c_{11} & \cdots & c_{1n} \\ \vdots & & \vdots \\ c_{n1} & \cdots & c_{nn} \end{pmatrix},$$

is known as the consumption matrix of the economy.

Consider the matrix product

For every $i = 1, ..., n$, the entry $c_{i1}x_1 + ... + c_{in}x_n$ represents the monetary value of the output of sector

i needed by all the sectors to produce their output. This leads to the production equation

$$x = Cx + d$$

Here Cx represents the part of the total output that is required by the various sectors of the economy to produce the output in the first place, and d represents the part of the total output that is available to satisfy outside demand.

Clearly $(I - C)x = d$. If the matrix $(I - C) - d$ is invertible, then represents the perfect production level. We state without proof the following fundamental result.

Proposition. Suppose that the entries of the consumption matrix C and the demand vector d are non-negative. Suppose further that the inequality (5) holds for each column of C. Then the inverse matrix $(I - C)^{-1}$ exists, and the production vector $x = (I - C)^{-1} d$ has non-negative entries and is the unique solution of the production equation (6).

Let us indulge in some heuristics. Initially, we have demand d. To produce d, we need Cd as input. To produce this extra Cd, we need $C(Cd) = C^2d$ as input. To produce this extra C^2d, we need $C(C^2d) = C^3d$ as input. And so on. Hence we need to produce

$$d + Cd + C^2d + C^3d + ... = (I + C + C^2 + C^3 + ...)d$$

in total. Now it is not dicult to check that for every positive integer k, we have

$$(I - C)(I + C + C^2 + C^3 + ... + C^k) = \mathrm{I} - C^{k+1},$$

If the entries of C^{k+1} are all very small, then

$$(I - C)(I + C + C^2 + C^3 + ... + C^k) \approx \mathrm{I},$$

so that

$$(I - C)^{-1} \approx \mathrm{I} + C + C^2 + C^3 + ... + C^k.$$

This gives a practical way of approximating $(\mathrm{I} - C)^{-1}$, and |also suggests that

$$(I - C)^{-1} = I + C + C^2 + C^3 + ...$$

Example. An economy consists of three sectors. Their dependence on each other is summarized in the table below:

	To produce one unit of monetary value of output in sector		
	1	2	3
monetary value of output required from sector 1	0:3	0:2	0:1
monetary value of output required from sector 2	0:4	0:5	0:2
monetary value of output required from sector 3	0:1	0:1	0:3

Suppose that the final demand from sectors 1, 2 and 3 are respectively 30, 50 and 20. Then the production vector and demand vector are respectively

$$x = \begin{pmatrix} x_1 \\ x_2 \\ x_3 \end{pmatrix} \text{and } d = \begin{pmatrix} d_1 \\ d_2 \\ d_3 \end{pmatrix} = \begin{pmatrix} 30 \\ 50 \\ 20 \end{pmatrix},$$

while the consumption matrix is given by

$$C = \begin{pmatrix} 0.3 & 0.2 & 0.1 \\ 0.4 & 0.5 & 0.2 \\ 0.1 & 0.1 & 0.3 \end{pmatrix}, \text{ so that } I - C = \begin{pmatrix} 0.7 & -0.2 & -0.1 \\ -0.4 & 0.5 & -0.2 \\ -0.1 & -0.1 & 0.7 \end{pmatrix}.$$

The production equation $(I - C)\, x = d$ has augmented matrix

$$\left(\begin{array}{ccc|c} 0.7 & -0.2 & -0.1 & 30 \\ -0.4 & 0.5 & -0.2 & 50 \\ -0.1 & -0.1 & 0.7 & 20 \end{array}\right), \text{ equivalent to } \left(\begin{array}{ccc|c} 7 & -2 & -1 & 300 \\ -4 & 5 & -2 & 500 \\ -1 & -1 & 7 & 200 \end{array}\right),$$

and which can be converted to reduced row echelon form

$$\left(\begin{array}{ccc|c} 0 & 0 & 0 & 3200/27 \\ 0 & 0 & 0 & 6100/27 \\ 0 & 0 & 1 & 700/9 \end{array}\right).$$

This gives $x_1 \approx 119$, $x_2 \approx 226$ and $x_3 \approx 78$, to the nearest integers.

Matrix Transformation on the Plane

Let A be a 2×2 matrix with real entries. A matrix transformation $T : \mathbb{R}^2 \to \mathbb{R}^2$ can be dened as follows: For every

$$x = (x_1, x_2) \in \mathbb{R},$$

we write

$$T(x) = y,$$

where

$$y = (y_1, y_2) \in \mathbb{R}^2$$

satifies

$$\begin{pmatrix} y_1 \\ y_2 \end{pmatrix} = A \begin{pmatrix} x_1 \\ x_2 \end{pmatrix}.$$

Such a transformation is linear, in the sense that $T(x' + x') = T(x') + T(x')$ for every x' $x' \in \mathbb{R}^2$ and $T(cx) = cT(x)$ for every $x \in \mathbb{R}^2$ and every $c \in \mathbb{R}$. To see this, simply observe that

$$A = \begin{pmatrix} x_1' + x_1'' \\ x_2' + x_2'' \end{pmatrix} = A \begin{pmatrix} x_2' \\ x_2' \end{pmatrix} + A \begin{pmatrix} x_1'' \\ x_2'' \end{pmatrix} \text{ and } A \begin{pmatrix} cx_1 \\ cx_2 \end{pmatrix} = cA \begin{pmatrix} x_1 \\ x_2 \end{pmatrix}.$$

Here we conne ourselves to looking at a few simple matrix transformations on the plane.

Example. The matrix

$$A = \begin{pmatrix} 1 & 0 \\ 0 & -1 \end{pmatrix} \quad \text{satisfies} \quad A\begin{pmatrix} x_1 \\ x_2 \end{pmatrix} = \begin{pmatrix} 1 & 0 \\ 0 & -1 \end{pmatrix}\begin{pmatrix} x_1 \\ x_2 \end{pmatrix} = \begin{pmatrix} x_1 \\ -x_2 \end{pmatrix}$$

for every $(x_1, x_2) \in \mathbb{R}^2$, and so represents reflection across the x_1-axis, whereas the matrix

$$A = \begin{pmatrix} -1 & 0 \\ 0 & 1 \end{pmatrix} \quad \text{satisfies} \quad A\begin{pmatrix} x_1 \\ x_2 \end{pmatrix} = \begin{pmatrix} -1 & 0 \\ 0 & 1 \end{pmatrix}\begin{pmatrix} x_1 \\ x_2 \end{pmatrix} = \begin{pmatrix} -x_1 \\ x_2 \end{pmatrix}$$

for every $(x_1, x_2) \in \mathbb{R}^2$, and so represents reflection across the x_2-axis. On the other hand, the matrix

$$A = \begin{pmatrix} -1 & 0 \\ 0 & -1 \end{pmatrix} \quad \text{satisfies} \quad A\begin{pmatrix} x_1 \\ x_2 \end{pmatrix} = \begin{pmatrix} -1 & 0 \\ 0 & -1 \end{pmatrix}\begin{pmatrix} x_1 \\ x_2 \end{pmatrix} = \begin{pmatrix} -x_1 \\ -x_2 \end{pmatrix}$$

for every $(x_1, x_2) \in \mathbb{R}^2$, and so represents reection across the origin, whereas the matrix

$$A = \begin{pmatrix} 0 & 1 \\ 1 & 0 \end{pmatrix} \quad \text{satisfies} \quad A\begin{pmatrix} x_1 \\ x_2 \end{pmatrix} = \begin{pmatrix} 0 & 1 \\ 1 & 0 \end{pmatrix}\begin{pmatrix} x_1 \\ x_2 \end{pmatrix} = \begin{pmatrix} x_2 \\ x_1 \end{pmatrix}$$

for every $(x_1, x_2) \in \mathbb{R}^2$, and so represents reection across the line $x_1 = x_2$. We give a summary in the table below:

Transformation	Equations	matrix
Reflection across x_1 − axis	$\begin{cases} y_1 = x_1 \\ y_2 = -x_2 \end{cases}$	$\begin{pmatrix} 1 & 0 \\ 0 & -1 \end{pmatrix}$
Reflection across x_2 − axis	$\begin{cases} y_1 = -x_1 \\ y_2 = x_2 \end{cases}$	$\begin{pmatrix} -1 & 0 \\ 0 & 1 \end{pmatrix}$
Reflection across origin	$\begin{cases} y_1 = -x_1 \\ y_2 = -x_2 \end{cases}$	$\begin{pmatrix} -1 & 0 \\ 0 & -1 \end{pmatrix}$
Reflection across $x_1 = x_2$	$\begin{cases} y_1 = x_1 \\ y_2 = x_2 \end{cases}$	$\begin{pmatrix} 1 & 0 \\ 0 & 1 \end{pmatrix}$

Example. Let k be a fixed positive real number. The matrix

$$A = \begin{pmatrix} k & 0 \\ 0 & k \end{pmatrix} \quad \text{satisfies} \quad A\begin{pmatrix} x_1 \\ x_2 \end{pmatrix} = \begin{pmatrix} k & 0 \\ 0 & k \end{pmatrix}\begin{pmatrix} x_1 \\ x_2 \end{pmatrix} = \begin{pmatrix} kx_1 \\ kx_2 \end{pmatrix}$$

for every $(x_1, x_2) \in \mathbb{R}^2$, and so represents dilation if $k > 1$ and contraction if $0 < 1$. $k < 1$. On the other hand, the matrix

$$A=\begin{pmatrix} k & 0 \\ 0 & 1 \end{pmatrix} \quad \text{satisfies} \quad A\begin{pmatrix} x_1 \\ x_2 \end{pmatrix}=\begin{pmatrix} k & 0 \\ 0 & 1 \end{pmatrix}\begin{pmatrix} x_1 \\ x_2 \end{pmatrix}=\begin{pmatrix} kx_1 \\ x_2 \end{pmatrix}$$

for every $(x_1, x_2) \in \mathbb{R}^2$, and so represents an expansionn in the x_1 -direction if $k >$ 1and compression in the x_1-direction if $0 < k < 1$. whereas the matrix

$$A=\begin{pmatrix} 1 & 0 \\ 0 & k \end{pmatrix} \quad \text{satisfies} \quad A\begin{pmatrix} x_1 \\ x_2 \end{pmatrix}=\begin{pmatrix} 1 & 0 \\ 0 & k \end{pmatrix}\begin{pmatrix} x_1 \\ x_2 \end{pmatrix}=\begin{pmatrix} x_1 \\ kx_2 \end{pmatrix}$$

for every $(x_1, x_2) \in \mathbb{R}^2$, and so represents expansion in the x_2-direction if $k >1$ and compression in the x_2-direction if $0 < k < 1$. We give summary in the table below:

Transformation	Equations	matrix
Dilation or contract ion by factor $k > 0$	$\begin{cases} y_1 = kx_1 \\ y_2 = kx_2 \end{cases}$	$\begin{pmatrix} k & 0 \\ 0 & k \end{pmatrix}$
Expansion or compression in x_1-direction by factor $k > 0$	$\begin{cases} y_1 = kx_1 \\ y_2 = x_2 \end{cases}$	$\begin{pmatrix} k & 0 \\ 0 & 1 \end{pmatrix}$
Expansion or compression in x_2-direction by factor $k > 0$	$\begin{cases} y_1 = x_1 \\ y_2 = kx_2 \end{cases}$	$\begin{pmatrix} 1 & 0 \\ 0 & k \end{pmatrix}$

Example. Let k be a fixed real number. The matrix

$$A=\begin{pmatrix} 1 & k \\ 0 & 1 \end{pmatrix} \quad \text{satisfies} \quad A\begin{pmatrix} x_1 \\ x_2 \end{pmatrix}=\begin{pmatrix} 1 & k \\ 0 & 1 \end{pmatrix}\begin{pmatrix} x_1 \\ x_2 \end{pmatrix}=\begin{pmatrix} x_1 + kx_2 \\ x_2 \end{pmatrix}$$

for every $(x_1, x_2) \in \mathbb{R}^2$, and so represents a shear in the x_1-direction. For the case $k =$ 1, we have the following:

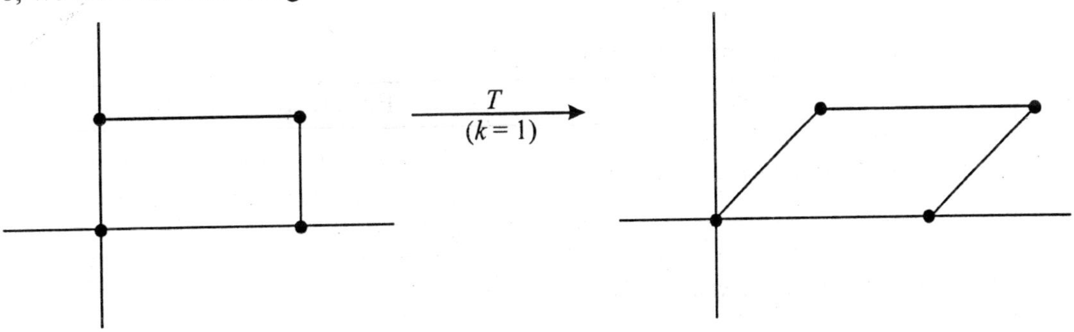

For the case $k = -1$, we have the following:

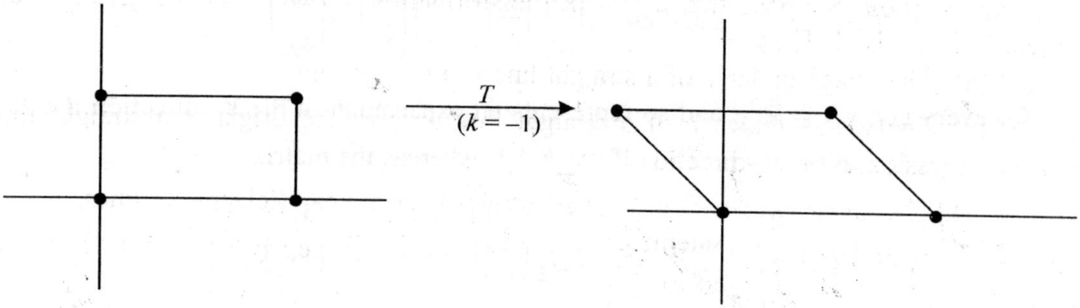

Similarly, the matrix

$$A = \begin{pmatrix} 1 & 0 \\ 0 & 1 \end{pmatrix} \quad \text{satisfies} \quad A\begin{pmatrix} x_1 \\ x_2 \end{pmatrix} = \begin{pmatrix} 1 & 0 \\ k & 1 \end{pmatrix}\begin{pmatrix} x_1 \\ x_2 \end{pmatrix} = \begin{pmatrix} x_1 \\ kx_1 + x_2 \end{pmatrix}$$

for every $(x_1, x_2) \in \mathbb{R}^2$, and so represents a shear in the x_2-direction. We give a summary in the table below:

Transformation	Equations	Matrix
Shear in x_1 – direction	$\begin{cases} y_2 = x_1 + kx_2 \\ y_2 = x_2 \end{cases}$	$\begin{pmatrix} 1 & k \\ 0 & 1 \end{pmatrix}$
Shear in x_1 – direction	$\begin{cases} y_1 = x_1 \\ y_2 = kx_1 + x_2 \end{cases}$	$\begin{pmatrix} 1 & 0 \\ k & 1 \end{pmatrix}$

Example. For anticlockwise rotation by an angle θ, we have $T(x_1, x_2) = (y_1, y_2)$, where

$$y_1 + iy_2 = (x_1 + ix_2)(\cos\theta + i\sin\theta),$$

and so

$$\begin{pmatrix} y_1 \\ y_2 \end{pmatrix} = \begin{pmatrix} \cos\theta & -\sin\theta \\ \sin\theta & \cos\theta \end{pmatrix}\begin{pmatrix} y_1 \\ y_2 \end{pmatrix}.$$

It follows that the matrix in question is given by

$$A = \begin{pmatrix} \cos\theta & -\sin\theta \\ \sin\theta & \cos\theta \end{pmatrix}.$$

We give a summary in the table below:

Transformation	Equations	Matrix
Anticlockwise rotation by angle θ	$\begin{cases} y_1 = x_1\cos\theta - x_2\sin\theta \\ y_1 = x_1\sin\theta + x_2\cos\theta \end{cases}$	$\begin{pmatrix} \cos\theta & -\sin\theta \\ \sin\theta & \cos\theta \end{pmatrix}$

We conclude this section by establishing the following result which reinforces the linearity of matrix transformations on the plane.

Proposition. Suppose that a matrix transformation $T:\mathbb{R}^2 \to \mathbb{R}^2$ is given by an invertible matrix A. Then

(a) The image under T of a straight line is a straight line;

(b) The image under T of a straight line through the origin is a straight line through the origin; and

(c) The images under T of parallel straight lines are parallel straight lines.

Proof. Suppose that $T(x_1, x_2) = (y_1, y_2)$. Since A is invertible, we have $x = A^{-1}y$, where

$$x = \begin{pmatrix} x_1 \\ x_2 \end{pmatrix} \quad \text{and} \quad y = \begin{pmatrix} y_1 \\ y_2 \end{pmatrix}.$$

The equation of a straight line is given by $\alpha x_1 + \beta x_2 = \gamma$ or, in matrix form, by

$$(\alpha \beta) = \begin{pmatrix} x_1 \\ x_2 \end{pmatrix} (\gamma).$$

Hence

$$(\alpha \beta) A^{-1} \begin{pmatrix} y_1 \\ y_2 \end{pmatrix} (\gamma).$$

Let

$$(\alpha' \beta') = (\alpha\ \beta)\, A^{-1}.$$

In other words, the image under T of the straight line

$$\alpha\beta_1 + \beta\alpha_2 = \gamma \text{ is } \alpha' y_1 + \beta' y_2 = \gamma,$$

clearly another straight line. This proves (a). To prove (b), note that straight lines through the origin correspond to $\gamma = 0$. To prove (c), note that parallel straight lines correspond to different values of γ for the same values of α and β.

Application to Computer Graphics

Example. Consider the letter M in the diagram below:

Following the boundary in the anticlockwise direction starting at the origin, the 12

vertices can be represented by the coordinates

$$\begin{pmatrix}0\\0\end{pmatrix}, \begin{pmatrix}1\\0\end{pmatrix}, \begin{pmatrix}1\\6\end{pmatrix}, \begin{pmatrix}4\\0\end{pmatrix}, \begin{pmatrix}7\\6\end{pmatrix}, \begin{pmatrix}7\\0\end{pmatrix},$$

$$\begin{pmatrix}8\\0\end{pmatrix}, \begin{pmatrix}8\\8\end{pmatrix}, \begin{pmatrix}7\\8\end{pmatrix}, \begin{pmatrix}4\\2\end{pmatrix}, \begin{pmatrix}1\\8\end{pmatrix}, \begin{pmatrix}0\\8\end{pmatrix},$$

Let us apply a matrix transformation to these vertices, using the matrix

$$A = \begin{pmatrix}1 & \frac{1}{2}\\ 0 & 1\end{pmatrix},$$

representing a shear in the x_1-direction with factor 0:5, so that

$$A\begin{pmatrix}x_1\\x_2\end{pmatrix} = \begin{pmatrix}x_1 + \frac{1}{2}x_2\\ x_2\end{pmatrix} \quad \text{for every } (x_1, x_2) \in \mathbb{R}^2.$$

Then the images of the 12 vertices are respectively

$$\begin{pmatrix}0\\0\end{pmatrix}, \begin{pmatrix}1\\0\end{pmatrix}, \begin{pmatrix}4\\6\end{pmatrix}, \begin{pmatrix}4\\0\end{pmatrix}, \begin{pmatrix}10\\6\end{pmatrix}, \begin{pmatrix}7\\0\end{pmatrix},$$

$$\begin{pmatrix}8\\0\end{pmatrix}, \begin{pmatrix}12\\8\end{pmatrix}, \begin{pmatrix}11\\8\end{pmatrix}, \begin{pmatrix}5\\2\end{pmatrix}, \begin{pmatrix}5\\8\end{pmatrix}, \begin{pmatrix}4\\8\end{pmatrix},$$

noting that

$$\begin{pmatrix}1 & \frac{1}{2}\\ 0 & 1\end{pmatrix}\begin{pmatrix}0 & 1 & 1 & 4 & 7 & 7 & 8 & 8 & 7 & 4 & 1 & 0\\ 0 & 0 & 6 & 0 & 6 & 0 & 0 & 8 & 8 & 2 & 8 & 8\end{pmatrix}$$

$$= \begin{pmatrix}0 & 1 & 4 & 4 & 10 & 7 & 8 & 12 & 11 & 5 & 5 & 4\\ 0 & 0 & 6 & 0 & 6 & 0 & 0 & 8 & 8 & 2 & 8 & 8\end{pmatrix}.$$

In view of Proposition, the image of any line segment that joins two vertices is a line segment that joins the images of the two vertices. Hence the image of the letter *M* under the shear looks like the following:

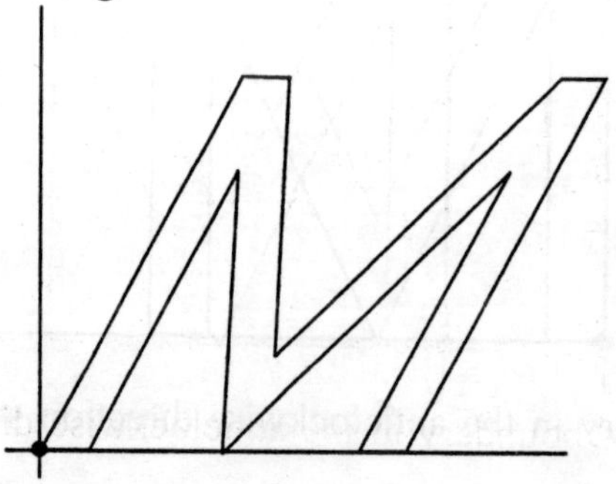

Next, we may wish to translate this image. However, translation is transformation by vector

$$h = (h_1, h_2) \in \mathbb{R}^2$$

is of the form

$$\begin{pmatrix} y_1 \\ y_2 \end{pmatrix} = \begin{pmatrix} x_1 \\ x_2 \end{pmatrix} + \begin{pmatrix} h_1 \\ h_2 \end{pmatrix} \text{ for every } (x_1, x_2) \in \mathbb{R}^2.$$

and this cannot be described by a matrix transformation on the plane. To overcome this deficiency, we introduce homogeneous coordinates. For every point

$$(x_1, x_2)\ (x_1, x_2) \in \mathbb{R}^2$$

we identify it with the point $(x_1, x_2, 1) \in \mathbb{R}^3$. Now we wish to translate a point (x_1, x_2) to

$$(x_1, x_2) + (h_1, h_2) = (x_1 + h_1, x_2 + h_2),$$

so we attempt to find a 3×3 matrix A* such that

$$\begin{pmatrix} x_1 + h_1 \\ x_2 + h_2 \\ 1 \end{pmatrix} = A^* \begin{pmatrix} x_1 \\ x_2 \\ 1 \end{pmatrix} \text{ for every } (x_1, x_2) \in \mathbb{R}^2.$$

It is easy to check that

$$\begin{pmatrix} x_1 + h_1 \\ x_2 + h_2 \\ 1 \end{pmatrix} = \begin{pmatrix} 1 & 0 & h_1 \\ 0 & 1 & h_2 \\ 0 & 0 & 1 \end{pmatrix} \begin{pmatrix} x_1 \\ x_2 \\ 1 \end{pmatrix} \text{ for every } (x_1, x_2) \in \mathbb{R}^2.$$

It follows that using homogeneous coordinates, translation by vector $h = (h_1, h_2) \in \mathbb{R}^2$. can be described by the matrix

$$A^* = \begin{pmatrix} 1 & 0 & h_1 \\ 0 & 1 & h_2 \\ 0 & 0 & 1 \end{pmatrix}.$$

Remark. Consider a matrix transformation $T : \mathbb{R}^2 \to \mathbb{R}^2$ on the plane given by a matrix

$$A = \begin{pmatrix} a_{11} & a_{12} \\ a_{21} & a_{22} \end{pmatrix}.$$

Suppose that $T(x_1, x_2) = (y_1, y_2)$. Then

$$\begin{pmatrix} y_1 \\ y_2 \end{pmatrix} A = \begin{pmatrix} x_1 \\ x_2 \end{pmatrix} = \begin{pmatrix} a_{11} & a_{12} \\ a_{21} & a_{22} \end{pmatrix} \begin{pmatrix} x_1 \\ x_2 \end{pmatrix}.$$

Under homogeneous coordinates, the image of the point $(x_1, x_2, 1)$ is now $(y_1, y_2, 1)$. Note that

$$\begin{pmatrix} y_1 \\ y_2 \\ 1 \end{pmatrix} = \begin{pmatrix} a_{11} & a_{12} & 0 \\ a_{21} & a_{22} & 0 \\ 0 & 0 & 1 \end{pmatrix} \begin{pmatrix} x_1 \\ x_2 \\ 1 \end{pmatrix}.$$

It follows that homogeneous coordinates can also be used to study all the matrix transformations we have discussed. By moving over to homogeneous coordinates, we simply replace the 2 × 2 matrix A by the 3 × 3 matrix

$$A^* = \begin{pmatrix} A & 0 \\ 0 & 1 \end{pmatrix}.$$

Example. The letter *M*, the 12 vertices are now represented by homogeneous coordinates, put in an array in the form

$$\begin{pmatrix} 0 & 1 & 1 & 4 & 7 & 7 & 8 & 8 & 7 & 4 & 1 & 0 \\ 0 & 0 & 6 & 0 & 6 & 0 & 0 & 8 & 8 & 2 & 8 & 8 \\ 1 & 1 & 1 & 1 & 1 & 1 & 1 & 1 & 1 & 1 & 1 & 1 \end{pmatrix}.$$

Then the 2 × 2 matrix

$$A = \begin{pmatrix} 1 & \frac{1}{2} \\ 0 & 1 \end{pmatrix}$$

is now replaced by the 3 × 3 matrix

$$A^* \begin{pmatrix} 1 & \frac{1}{2} & 0 \\ 0 & 1 & 0 \\ 0 & 0 & 1 \end{pmatrix}.$$

Note that

$$A^* \begin{pmatrix} 0 & 1 & 1 & 4 & 7 & 7 & 8 & 8 & 7 & 4 & 1 & 0 \\ 0 & 0 & 6 & 0 & 6 & 0 & 0 & 8 & 8 & 2 & 8 & 8 \\ 1 & 1 & 1 & 1 & 1 & 1 & 1 & 1 & 1 & 1 & 1 & 1 \end{pmatrix}$$

$$= \begin{pmatrix} 1 & \frac{1}{2} & 0 \\ 0 & 1 & 0 \\ 0 & 0 & 1 \end{pmatrix} \begin{pmatrix} 0 & 1 & 1 & 4 & 7 & 7 & 8 & 8 & 7 & 4 & 1 & 0 \\ 0 & 0 & 6 & 0 & 6 & 0 & 0 & 8 & 8 & 2 & 8 & 8 \\ 1 & 1 & 1 & 1 & 1 & 1 & 1 & 1 & 1 & 1 & 1 & 1 \end{pmatrix}$$

$$=\begin{pmatrix} 0 & 1 & 4 & 4 & 10 & 7 & 8 & 12 & 11 & 5 & 5 & 4 \\ 0 & 0 & 6 & 0 & 6 & 0 & 0 & 8 & 8 & 2 & 8 & 8 \\ 1 & 1 & 1 & 1 & 1 & 1 & 1 & 1 & 1 & 1 & 1 & 1 \end{pmatrix}.$$

Next, let us consider a translation by the vector (2; 3). The matrix under homogeneous coordinates for this translation is given by

$$B^*\begin{pmatrix} 1 & 0 & 2 \\ 0 & 1 & 3 \\ 0 & 0 & 1 \end{pmatrix}.$$

Note that

$$B^*A^*\begin{pmatrix} 0 & 1 & 4 & 4 & 7 & 7 & 8 & 8 & 7 & 4 & 1 & 0 \\ 0 & 0 & 6 & 0 & 6 & 0 & 0 & 8 & 8 & 2 & 8 & 8 \\ 1 & 1 & 1 & 1 & 1 & 1 & 1 & 1 & 1 & 1 & 1 & 1 \end{pmatrix}$$

$$=\begin{pmatrix} 1 & 0 & 2 \\ 0 & 1 & 3 \\ 0 & 0 & 1 \end{pmatrix}\begin{pmatrix} 0 & 1 & 4 & 4 & 10 & 7 & 8 & 12 & 11 & 5 & 5 & 4 \\ 0 & 0 & 6 & 0 & 6 & 0 & 0 & 8 & 8 & 2 & 8 & 8 \\ 1 & 1 & 1 & 1 & 1 & 1 & 1 & 1 & 1 & 1 & 1 & 1 \end{pmatrix}$$

$$=\begin{pmatrix} 2 & 3 & 6 & 6 & 12 & 9 & 10 & 14 & 13 & 7 & 7 & 6 \\ 3 & 3 & 9 & 3 & 9 & 3 & 9 & 11 & 11 & 5 & 11 & 11 \\ 1 & 1 & 1 & 1 & 1 & 1 & 1 & 1 & 1 & 1 & 1 & 1 \end{pmatrix},$$

giving rise to coordinates in $\mathbb{R}^2$, displayed as an array

$$\begin{pmatrix} 2 & 3 & 6 & 6 & 12 & 9 & 10 & 14 & 13 & 7 & 7 & 6 \\ 3 & 3 & 9 & 3 & 9 & 3 & 3 & 11 & 11 & 5 & 11 & 11 \end{pmatrix}$$

Hence the image of the letter *M* under the shear followed by translation looks like the following:

Example. Under homogeneous coordinates, the transformation representing a reflection across the x_1-axis, followed by a shear by factor 2 in the x_1-direction, followed by anticlockwise rotation by 90°, and followed by translation by vector (2, –1), has matrix

$$\begin{pmatrix}1 & 0 & 2\\0 & 1 & -1\\0 & 0 & 1\end{pmatrix}\begin{pmatrix}0 & -1 & 0\\1 & 0 & 0\\0 & 0 & 1\end{pmatrix}\begin{pmatrix}1 & 2 & 0\\0 & 1 & 0\\0 & 0 & 1\end{pmatrix}\begin{pmatrix}1 & 0 & 0\\0 & -1 & 0\\0 & 0 & 1\end{pmatrix}=\begin{pmatrix}1 & 1 & 2\\1 & -2 & -1\\0 & 0 & 1\end{pmatrix}.$$

Complexity of a Non-Homogeneous System

Consider the problem of solving a system of linear equations of the form $Ax = b$, where A is an $n \times n$ invertible matrix. We are interested in the number of operations required to solve such a system. By an operation, we mean interchanging, adding or multiplying two real numbers.

One way of solving the system $Ax = b$ is to write down the augmented matrix

$$\left(\begin{array}{ccc|c}a_{11} & \cdots & a_{1n} & b_1\\ \vdots & & \vdots & \vdots\\ a_{n1} & \cdots & a_{nn} & b_n\end{array}\right),$$

and then convert it to reduced row echelon form by elementary row operations.

The first step is to reduce it to row echelon form:

(I) First of all, we may need to interchange two rows in order to ensure that the top left entry in the array is non-zero. This requires $n + 1$ operations.

(II) Next, we need to multiply the new first row by a constant in order to make the top left pivot entry equal to 1. This requires $n + 1$ operations, and the array now looks like

$$\left(\begin{array}{cccc|c}1 & a_{12} & \cdots & a_{1n} & b_1\\ a_{21} & a_{22} & \cdots & a_{2n} & b_2\\ \vdots & \vdots & & \vdots & \vdots\\ a_{n1} & a_{n2} & \cdots & a_{nn} & b_n\end{array}\right).$$

Note that we are abusing notation somewhat, as the entry a_{12} here, for example, may well be different from the entry a_{12} in the augmented matrix.

(III) For each row $i = 2, \ldots, n$, we now multiply the first row by $-a_{i1}$ and then add to row i. This requires $2(n-1)(n+1)$ operations, and the array now looks like

$$\left(\begin{array}{cccc|c}1 & a_{12} & \cdots & a_{1n} & b_1\\ 0 & a_{22} & \cdots & a_{2n} & b_2\\ \vdots & \vdots & & \vdots & \vdots\\ 0 & a_{n2} & \cdots & a_{nn} & b_n\end{array}\right)$$

(IV) In summary, to proceed from the form II to the form III, the number of operations required is at most $2(n + 1) + 2(n - 1)(n + 1) = 2n(n + 1)$.

(V) Our next task is to convert the smaller array

$$\left(\begin{array}{ccc|c} a_{22} & \cdots & a_{2n} & b_2 \\ \vdots & & \vdots & \vdots \\ a_{n2} & \cdots & a_{nn} & b_n \end{array}\right),$$

to an array that looks like

$$\left(\begin{array}{cccc|c} 1 & a_{23} & \cdots & a_{2n} & b_2 \\ 0 & a_{33} & \cdots & a_{3n} & b_3 \\ \vdots & \vdots & & \vdots & \vdots \\ 0 & a_{n3} & \cdots & a_{nn} & b_n \end{array}\right).$$

These have one row and one column fewer than the arrays (II) and (III), and the number of operations required is at most $2m(m + 1)$, where $m = n - 1$. We continue in this way systematically to reach row echelon form, and conclude that the number of operations required to convert the augmented matrix (II) to row echelon form is at most

$$\sum_{m=1}^{n2} 2m(m+1) \approx \frac{2}{3}n^3.$$

The next step is to convert the row echelon form to reduced row echelon form. This is simpler, as many entries are now zero. It can be shown that the number of operations required is bounded by something like $2n^2$ indeed, by something like n^2 if one analyzes the problem more carefully. In any case, these estimates are insignicant compared to the estimate $\frac{2}{3}n^3$ earlier.

We therefore conclude that the number of operations required to solve the system $Ax = b$ by reducing the augmented matrix to reduced row echelon form is bounded by something like $\frac{2}{3}n^3$ when n is large.

Another way of solving thc system $Ax = b$ is to first nd the inverse matrix A^{-1}. This may involve converting the array

$$\left(\begin{array}{ccc|ccc} a_{11} & \cdots & a_{1n} & 1 & & \\ \vdots & & \vdots & & \ddots & \\ a_{n1} & \cdots & a_{nn} & & & 1 \end{array}\right)$$

to reduced row echelon form by elementary row operations. It can be shown that the number of operations required is something like $2n^3$, so this is less ecient than our first method.

Matrix Factorization

In some situations, we may need to solve systems of linear equations of the form $Ax = b$, with the same coefficient matrix A but for many different vectors b. If A is an invertible square matrix, then we can and its inverse A^{-1} and then compute $A^{-1}b$ for each vector b. However, the matrix A may not be a square matrix, and we may have to convert the augmented matrix to reduced row echelon form.

In this section, we describe a way for solving this problem in a more efficient way. To describe this, we first need a deffinition.

Definition. A rectangular array of numbers is said to be in quasi row echelon form if the following conditions are satised:

(1) The left-most non-zero entry of any non-zero row is called a pivot entry. It is not necessary for its value to be equal to 1.

(2) All zero rows are grouped together at the bottom of the array.

(3) The pivot entry of a non-zero row occurring lower in the array is to the right of the pivot entry of a non-zero row occurring higher in the array.

In other words, the array looks like row echelon form in shape, except that the pivot entries do not have to be equal to 1.

We consider first of all a special case.

Proposition. Suppose that an $m \times n$ matrix A can be converted to quasi row echelon form by elementary row operations but without interchanging any two rows. Then $A = LU$, where L is an $m \times m$ lower triangular matrix with diagonal entries all equal to 1 and U is a quasi row echelon form of A.

Proof. Recall that applying an elementary row operation to an $m \times n$ matrix corresponds to multiplying the matrix on the left by an elementary $m \times m$ matrix. On the other hand, if we are aiming for quasi row echelon form and not row echelon form, then there is no need to multiply any row of the array by a non-zero constant. Hence the only elementary row operation we need to perform is to add a multiple of one row to another row. In fact, it is sucient even to restrict this to adding a multiple of a row higher in the array to another row lower in the array, and it is easy to see that the corresponding elementary matrix is lower triangular, with diagonal entries all equal to 1. Let us call such elementary matrices unit lower triangular. If an $m \times n$ matrix A can be reduced in this way to quasi row echelon form U, then

$$U = E_k, \ldots, E_2E_1A,$$

where the elementary matrices $E_1, E_2, \ldots, E_k$ are all unit lower triangular. Let

$$L = (E_k, \ldots, E_2E_1)^{-1}.$$

Then

$$A = LU.$$

It can be shown that products and inverses of unit lower triangular matrices are also unit lower triangular. Hence L is a unit lower triangular matrix as required.

If $Ax = b$ and $A = LU$, then

$$L(Ux) = b.$$

Writing

$$y = Ux,$$

we have

$$Ly = b \text{ and } Ux = y.$$

It follows that the problem of solving the system $Ax = b$ corresponds to first solving the system $Ly = b$ and then solving the system $Ux = y$. Both of these systems are easy to solve since both L and U have many zero entries. It remains to and L and U.

If we reduce the matrix A to quasi row echelon form by only performing the elementary row operation of adding a multiple of a row higher in the array to another row lower in the array, then U can be taken as the quasi row echelon form resulting from this. It remains to nd L. However, note that $L = (E_k, \ldots, E_2E_1)^{-1}$, where $U = E_k, \ldots, E_2E_1A$, and so

$$I = E_k, \ldots, E_2E_1L$$

This means that the very elementary row operations that convert A to U will convert L to I. We therefore wish to create a matrix L such that this is satised. It is simplest to illustrate the technique by an example.

Example. **Consider the Matrix**

$$A = \begin{pmatrix} 2 & -1 & 2 & -2 & 3 \\ 4 & 1 & 6 & -5 & 8 \\ 2 & -10 & -4 & 8 & -5 \\ 2 & -13 & -6 & 16 & -5 \end{pmatrix}.$$

The entry 2 in row 1 and column 1 is a pivot entry, and column 1 is a pivot column. Adding –2 times row 1 to row 2, adding –1 times row 1 to row 3, and adding –1 times row 1 to row 4, we obtain

$$\begin{pmatrix} 2 & -1 & 2 & -2 & 3 \\ 0 & 3 & 2 & -1 & 2 \\ 0 & -9 & -6 & 10 & -8 \\ 0 & -12 & -8 & 18 & -8 \end{pmatrix}.$$

Note that the same three elementary row operations convert

$$\begin{pmatrix} 1 & 0 & 0 & 0 \\ 2 & 1 & 0 & 0 \\ 1 & * & 1 & 0 \\ 1 & * & * & 1 \end{pmatrix} \text{ to } \begin{pmatrix} 1 & 0 & 0 & 0 \\ 0 & 1 & 0 & 0 \\ 0 & * & 1 & 0 \\ 0 & * & * & 1 \end{pmatrix}.$$

Next, the entry 3 in row 2 and column 2 is a pivot entry, and column 2 is a pivot column. Adding 3 times row 2 to row 3, and adding 4 times row 2 to row 4, we obtain

$$\begin{pmatrix} 2 & -1 & 2 & -2 & 3 \\ 0 & 3 & 2 & -1 & 2 \\ 0 & 0 & 0 & 7 & -2 \\ 0 & 0 & 0 & 14 & 0 \end{pmatrix}.$$

Note that the same two elementary row operations convert

$$\begin{pmatrix} 1 & 0 & 0 & 0 \\ 0 & 1 & 0 & 0 \\ 0 & -3 & 1 & 0 \\ 0 & -4 & * & 1 \end{pmatrix} \text{ to } \begin{pmatrix} 1 & 0 & 0 & 0 \\ 0 & 1 & 0 & 0 \\ 0 & 0 & 1 & 0 \\ 0 & 0 & * & 1 \end{pmatrix}.$$

Next, the entry 7 in row 3 and column 4 is a pivot entry, and column 4 is a pivot column. Adding –2 times row 3 to row 4, we obtain the quasi row echelon form

$$U = \begin{pmatrix} 2 & -1 & 2 & -2 & 3 \\ 0 & 3 & 2 & -1 & 2 \\ 0 & 0 & 0 & 7 & -2 \\ 0 & 0 & 0 & 0 & 4 \end{pmatrix},$$

where the entry 4 in row 4 and column 5 is a pivot entry, and column 5 is a pivot column. Note that the same elementary row operation converts

$$\begin{pmatrix} 1 & 0 & 0 & 0 \\ 0 & 1 & 0 & 0 \\ 0 & 0 & 1 & 0 \\ 0 & 0 & 2 & 1 \end{pmatrix} \text{ to } \begin{pmatrix} 1 & 0 & 0 & 0 \\ 0 & 1 & 0 & 0 \\ 0 & 0 & 1 & 0 \\ 0 & 0 & 0 & 1 \end{pmatrix}.$$

Now observe that if we take

$$L = \begin{pmatrix} 1 & 0 & 0 & 0 \\ 2 & 1 & 0 & 0 \\ 1 & -3 & 1 & 0 \\ 1 & -4 & 2 & 1 \end{pmatrix},$$

then L can be converted to I_4 by the same elementary operations that convert A to U.

The strategy is now clear. Every time we nd a new pivot, we note its value and the entries below it. The lower triangular entries of L are formed by these columns with each column divided by the value of the pivot entry in that column.

Example. Let us examine our last example again. The pivot columns at the time of establishing the pivot entries are respectively

$$\begin{pmatrix}2\\4\\2\\2\end{pmatrix}, \begin{pmatrix}*\\3\\-9\\-12\end{pmatrix}, \begin{pmatrix}*\\*\\7\\14\end{pmatrix}, \begin{pmatrix}*\\*\\*\\4\end{pmatrix},$$

Dividing them respectively by the pivot entries 2, 3, 7 and 4, we obtain respectively the columns

$$\begin{pmatrix}1\\2\\1\\1\end{pmatrix}, \begin{pmatrix}*\\1\\-3\\-4\end{pmatrix}, \begin{pmatrix}*\\*\\1\\2\end{pmatrix}, \begin{pmatrix}1\\*\\*\\1\end{pmatrix},$$

Note that the lower triangular entries of the matrix

$$L = \begin{pmatrix}1 & 0 & 0 & 0\\2 & 1 & 0 & 0\\1 & -3 & 1 & 0\\1 & -4 & 2 & 1\end{pmatrix}$$

correspond precisely to the entries in these columns.

LU FACTORIZATION ALGORITHM.

(1) Reduce the matrix A to quasi row echelon form by only performing the elementary row operation of adding a multiple of a row higher in the array to another row lower in the array. Let U be the quasi row echelon form obtained.

(2) Record any new pivot column at the time of its first recognition, and modify it by replacing any entry above the pivot entry by zero and dividing every other entry by the value of the pivot entry.

(3) Let L denote the square matrix obtained by letting the columns be the pivot columns as modied in step (2).

Example. We wish to solve the system of linear equations $Ax = b$, where

$$A = \begin{pmatrix}3 & -1 & 2 & -4 & 1\\-3 & 3 & -5 & 5 & -2\\6 & -4 & 11 & -10 & 6\\-6 & 8 & -21 & 13 & -9\end{pmatrix} \text{ and } b = \begin{pmatrix}1\\-2\\9\\-15\end{pmatrix}.$$

Let us first apply LU factorization to the matrix A. The first pivot column is column 1, with modied version

$$\begin{pmatrix} 1 \\ -1 \\ 2 \\ -2 \end{pmatrix}.$$

Adding row 1 to row 2, adding – 2 times row 1 to row 3, and adding 2 times row 1 to row 4, we obtain

$$\begin{pmatrix} 3 & -1 & 2 & -4 & 1 \\ 0 & 2 & -3 & 1 & -1 \\ 0 & -2 & 7 & -2 & 4 \\ 0 & 6 & -17 & 5 & -7 \end{pmatrix}.$$

The second pivot column is column 2, with modied version

$$\begin{pmatrix} 0 \\ 1 \\ -1 \\ 3 \end{pmatrix}.$$

Adding row 2 to row 3, and adding –3 times row 2 to row 4, we obtain

$$\begin{pmatrix} 3 & -1 & 2 & -4 & 1 \\ 0 & 2 & -3 & 1 & -1 \\ 0 & 0 & 4 & -1 & 3 \\ 0 & 0 & -8 & 2 & -4 \end{pmatrix}$$

The third pivot column is column 3, with modied version

$$\begin{pmatrix} 0 \\ 0 \\ 1 \\ -2 \end{pmatrix}.$$

Adding 2 times row 3 to row 4, we obtain the quasi row echelon form

$$\begin{pmatrix} 3 & -1 & 2 & -4 & 1 \\ 0 & 2 & -3 & 1 & -1 \\ 0 & 0 & 4 & -1 & 3 \\ 0 & 0 & 0 & 0 & 2 \end{pmatrix}.$$

The last pivot column is column 5, with modied version

$$\begin{pmatrix} 0 \\ 0 \\ 0 \\ 1 \end{pmatrix}.$$

It follows that

$$L = \begin{pmatrix} 1 & 0 & 0 & 0 \\ -1 & 1 & 0 & 0 \\ 2 & -1 & 1 & 0 \\ -2 & 3 & -2 & 1 \end{pmatrix} \text{ and } U = \begin{pmatrix} 3 & -1 & 2 & -4 & 1 \\ 0 & 2 & -3 & 1 & -1 \\ 0 & 0 & 4 & -1 & 3 \\ 0 & 0 & 0 & 0 & 2 \end{pmatrix}$$

We now consider the system $Ly = b$, with augmented matrix

$$\left(\begin{array}{cccc|c} 1 & 0 & 0 & 0 & 1 \\ -1 & 1 & 0 & 0 & -2 \\ 2 & -1 & 1 & 0 & 9 \\ -2 & 3 & -2 & 1 & -15 \end{array}\right).$$

Using row 1, we obtain $y_1 = 1$. Using row 2, we obtain $y_2 - y_1 = -2$, so that $y_2 = -1$. Using row 3, we obtain $y_3 + 2y_1 - y_2 = 9$, so that $y_3 = 6$. Using row 4, we obtain

$$y_4 - 2y_1 + 3y_2 - 2y_3 = -15,$$

so that $y_4 = 2$.

Hence
$$y = \begin{pmatrix} 1 \\ -1 \\ 6 \\ 2 \end{pmatrix}.$$

We next consider the system $Ux = y$, with augmented matrix

$$\left(\begin{array}{ccccc|c} 3 & -1 & 2 & -4 & 1 & 1 \\ 0 & 2 & -3 & 1 & -1 & -1 \\ 0 & 0 & 4 & -1 & 3 & 6 \\ 0 & 0 & 0 & 0 & 2 & 2 \end{array}\right).$$

Here the free variable is x_4. Let $x_4 = t$. Using row 4, we obtain $2x_5 = 2$, so that $x_5 = 1$. Using row 3, we obtain $4x_3 = 6 + x_4 - 3x_5 = 3 + t$, so that $x_3 = \frac{3}{4} + \frac{1}{4}t$. Using row 2, we obtain

$$2x_2 = -1 + 3x_3 - x_4 + x_5 \; x_5 = \frac{9}{4} - \frac{1}{4}t,$$

so that $x_2 = \frac{9}{8} - \frac{1}{8}t$. Using row 1, we obtain $3x_1 = 1+x_2-2x_3+4x_4-x_5 = \frac{27}{8}t - \frac{3}{8}$, so that $x_1 = \frac{9}{8}t - \frac{1}{8}$. Hence

$$(x_1, x_2, x_3, x_4, x_5) = \left(\frac{9t-1}{8}, \frac{9-t}{8}, t\frac{3+t}{4}, t, 1\right), \text{ where } t \in \mathbb{R}.$$

Remarks. (1) In practical situations, interchanging rows is usually necessary to convert a matrix A to quasi row echelon form. The technique here can be modied to produce a matrix L which is not unit lower triangular, but which can be made unit lower triangular by interchanging rows.

(2) Computing an LU factorization of an $n \times n$ matrix takes approximately $\frac{2}{3}n^3$ operations. Solving the systems $Ly = b$ and $Ux = y$ requires approximately $2n^2$ operations.

(3) LU factorization is particularly ecient when the matrix A has many zero entries, in which case the matrices L and U may also have many zero entries.

Application to Games of Strategy

Consider a game with two players. Player R, usually known as the row player, has m possible moves, denoted by $i = 1, 2, 3, ..., m$, while player C, usually known as the column player, has n possible moves, denoted by $j = 1, 2, 3, ..., n$, For every $i = 1, 2, 3, ..., m$, and $j = 1, 2, 3, ..., n$, let a_{ij} denote the payo that player C has to make to player R if player R makes move i and player C makes move j. These numbers give rise to the payo matrix

$$A = \begin{pmatrix} a_{11} & \cdots & a_{1n} \\ \vdots & & \vdots \\ a_{m1} & \cdots & a_{mn} \end{pmatrix}.$$

The entries can be positive, negative or zero.

Suppose that for every $i = 1, 2, 3, ..., m$, player R makes move i with probability p_i, and that for every $j = 1, 2, 3, ..., n$, player C makes move j with probability q_j. Then

$$p_1 + ... + p_m = 1 \text{ and } q_1 + ... + q_n = 1.$$

Assume that the players make moves independently of each other. Then for every $i = 1, 2, 3, ..., m$, and $j = 1, 2, 3, ..., n$, the number $p_i q_j$ represents the probability that player R makes move i and player C makes move j. Then the double sum

$$E_A(p,q) = \sum_{i=1}^{m}\sum_{j=1}^{n} a_{ij} p_i q_j$$

represents the expected payo that player C has to make to player R.

The matrices

$$p = (p_1 \cdots p_m) \text{ and } q = \begin{pmatrix} q_1 \\ \vdots \\ q_n \end{pmatrix}$$

are known as the strategies of player *R* and player *C* respectively. Clearly the expected payo

$$E_A(p,q) = \sum_{i=1}^{m}\sum_{j=1}^{n} a_{ij}p_iq_j(p_1 \cdots p_m)\begin{pmatrix} a_{11} & \cdots & a_{1n} \\ \vdots & & \vdots \\ a_{m1} & \cdots & a_{mn} \end{pmatrix}\begin{pmatrix} q_1 \\ \vdots \\ q_n \end{pmatrix} = p.Aq.$$

Here we have slightly abused notation. The right hand side is a 1 × 1 matrix!

We now consider the following problem: Suppose that *A* is xed. Is it possible for player *R* to choose a strategy *p* to try to maximize the expected payo $E_A(p, q)$? Is it possible for player *C* to choose a strategy *q* to try to minimize the expected payo *EA*(*p*, *q*)?

Fundemental Theorem of Zero Sum Games. There exist strategies p^* and q^* such that

$$E_A(p^*, q) \geq E_A(p^*, q^*) \geq EA(p, q^*)$$

for every strategy p^* of player *R*, and every strategy q^* of player *C*. Remark. The strategy *p* is known as an optimal strategy for player *R*, and the strategy *q* is known as an optimal strategy for player *C*. The quantity $EA(p^*, q^*)$ is known as the value of the game. Optimal strategies are not necessarily unique. However, if p^{**} and q^{**} are another pair of optimal strategies, then $EA(p^*, q^*) = E_A(p^{**}, q^{**})$.

Zero sum games which are strictly determined are very easy to analyse. Here the payo matrix *A* contains saddle points. An entry a_{ij} in the payo matrix *A* is called a saddle point if it is a least entry in its row and a greatest entry in its column. In this case, the strategies

$$p^* = (0 \cdots 010 \cdots 0) \text{ and } q = \begin{pmatrix} 0 \\ \vdots \\ 0 \\ 1 \\ 0 \\ \vdots \\ 0 \end{pmatrix},$$

where the 1's occur in position *i* in p^* and position *j* in q^*, are optimal strategies, so that the value of the game is *aij* .

Remark. It is very easy to show that different saddle points in the payo matrix have the same value.

Example. In some sports mad school, the teachers require 100 students to each choose between rowing (*R*) and cricket (*C*). However, the students cannot make up their mind, and will only decide when the identities of the rowing coach and cricket coach are known. There are 3 possible rowing coaches and 4 possible cricket coaches the school can hire. The number of students who will choose rowing ahead of cricket in each scenario is as follows, where $R1$, $R2$ and $R3$ denote the 3 possible rowing coaches, and $C1$, $C2$, $C3$ and $C4$ denote the 4 possible cricket coaches:

	$C1$	$C2$	$C3$	$C5$
$R1$	75	50	45	60
$R2$	20	60	30	55
$R3$	45	70	35	30

[For example, if coaches $R2$ and $C1$ are hired, then 20 students will choose rowing, and so 80 students will choose cricket.] We first reset the problem by subtracting 50 from each entry and create a payo matrix

$$A = \begin{pmatrix} 25 & 0 & -5 & 10 \\ -30 & 10 & 20 & 5 \\ -5 & 20 & -15 & -20 \end{pmatrix}.$$

[For example, the top left entry denotes that if each sport starts with 50 students, then 25 is the number cricket concedes to rowing.] Here the entry –5 in row 1 and column 3 is a saddle point, so the optimal strategy for rowing is to use coach $R1$ and the optimal strategy for cricket is to use coach $C3$.

In general, saddle points may not exist, so that the problem is not strictly determined. Then the solution for these optimal problems are solved by linear programming techniques which we do not discuss here. However, in the case of 2×2 payo matrices

$$A = \begin{pmatrix} a_{11} & a_{12} \\ a_{21} & a_{22} \end{pmatrix}$$

which do not contain saddle points, we can write $p_2 = 1 - p_1$ and $q_2 = 1 - q_1$. Then

$$E_A(p, q) = a_{11}p_1q_1 + a_12p_1(1 - q_1) + a_{21}(1 - p_1)q_1 + a_{22}(1 - p_1)(1 - q_1)$$
$$= ((a_{11} - a_{12} - a_{21} + a_{22})p_1 - (a_{22} - a_{21}))q_1 + (a_{12} - a_{22})p_1 + a_{22}.$$

Let

$$p_1 = p_1^* = \frac{a_{22} - a_{21}}{a_{11} - a_{12} - a_{21} + a_{22}}.$$

Then

$$E_A(p^*, q) = \frac{(a_{12} - a_{22})(a_{22} - a_{21})}{a_{11} - a_{12} - a_{21} + a_{22}} + a_{22} = \frac{a_{11}a_{22} - a_{12}a_{21}}{a_{11} - a_{12} - a_{21} + a_{22}},$$

which is independent of q. Similarly, if

$$q_1 = q_1^* = \frac{a_{22} - a_{12}}{a_{11} - a_{12} - a_{21} + a_{22}},$$

then

$$E_A(p, q^*) = \frac{a_{11}a_{22} - a_{12}a_{21}}{a_{11} - a_{12} - a_{21} + a_{22}},$$

which is independent of p. Hence
$E_A(p^* q) = E_A(p^*, q^*) = E_A(p, q^*)$ for all strategies p and q:
Note that

$$p^* = \left(\frac{a_{22} - a_{21}}{a_{11} - a_{12} - a_{21} + a_{22}} \; \frac{a_{11} - a_{12}}{a_{11} - a_{12} - a_{21} + a_{22}} \right)$$

and

$$q^* = \begin{pmatrix} \dfrac{a_{22} - a_{12}}{a_{11} - a_{12} - a_{21} + a_{22}} \\ \dfrac{a_{11} - a_{21}}{a_{11} - a_{12} - a_{21} + a_{22}} \end{pmatrix},$$

with value

$$EA(p^*, q^*) = \frac{a_{11}a_{22} - a_{12}a_{21}}{a_{11} - a_{12} - a_{21} + a_{22}}.$$

ORTHOGONAL MATRICES

Definition. A square matrix A with real entries and satisfying the condition $A^{-1} = A^t$ is called an orthogonal matrix.

Example. Consider the euclidean space $\mathbb{R}^2$ with the euclidean inner product. The vectors $u_1 = (1, 0)$ and $u_2 = (0, 1)$ form an orthonormal basis $B = \{u_1, u_2\}$. Let us now rotate u_1 and u_2 anticlockwise by an angle to obtain $v1 = (\cos\theta \sin \theta)$ and $v_2 = (-\sin\theta , \cos\theta)$. Then $C = \{v_1, v_2\}$ is also an orthonormal basis.

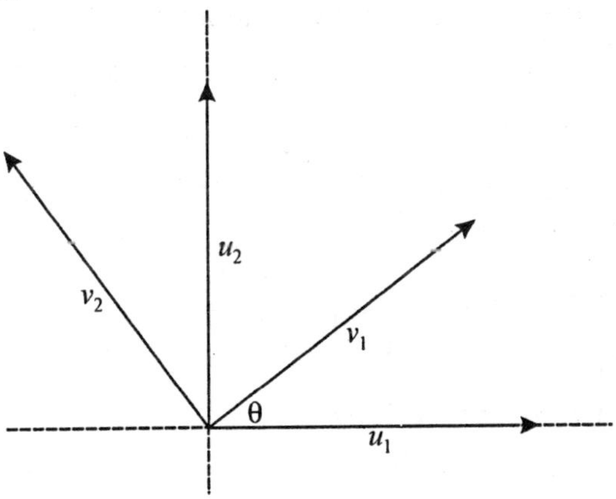

The transition matrix from the basis C to the basis B is given by

$$p=([v_1]_B[v_2]_B=\begin{pmatrix}\cos & -\sin\theta\\ \sin\theta & \cos\theta\end{pmatrix}.$$

Clearly

$$p^{-1}=p^t=\begin{pmatrix}\cos & \sin\theta\\ -\sin\theta & \cos\theta\end{pmatrix}.$$

In fact, our example is a special case of the following general result.

Proposition. Suppose that $B = \{u_1,... un\}$ and $C = \{v_1, ..., v_n\}$ are two orthonormal bases of a real inner product space V. Then the transition matrix P from the basis C to the basis B is an orthogonal matrix.

Example. The matrix

$$A=\begin{pmatrix}1/3 & -2/3 & 2/3\\ 2/3 & -1/3 & -2/3\\ 2/3 & 2/3 & 1/3\end{pmatrix}$$

is orthogonal, since

$$A^tA=\begin{pmatrix}1/3 & -2/3 & 2/3\\ -2/3 & -1/3 & 2/3\\ 2/3 & -2/3 & 1/3\end{pmatrix}\begin{pmatrix}1/3 & -2/3 & 2/3\\ 2/3 & -1/3 & -2/3\\ 2/3 & 2/3 & 1/3\end{pmatrix}=\begin{pmatrix}1 & 0 & 0\\ 0 & 1 & 0\\ 0 & 0 & 1\end{pmatrix}.$$

Note also that the row vectors of A, namely (1/3,–2/3, 2/3), (2/3,–1/3,–2/3) and (2/3, 2/3, 1/3) are orthonormal. So are the column vectors of A.

In fact, our last observation is not a coincidence.

Proposition. Suppose that A is an $n \times n$ matrix with real entries. Then

(a) A is orthogonal if and only if the row vectors of A form an orthonormal basis of $\mathbb{R}^n$ under the euclidean inner product; and

(b) A is orthogonal if and only if the column vectors of A form an orthonormal basis of $\mathbb{R}^n$ under the euclidean inner product.

Proof. We shall only prove (a), since the proof of (b) is almost identical. Let $r_1,....., r_n$ denote the row vectors of A. Then

$$AA^t=\begin{pmatrix}r_1\cdot r_1 & \cdots & r_1\cdot r_n\\ \vdots & & \vdots\\ r_n\cdot r_1 & \cdots & r_n\cdot r_n\end{pmatrix}.$$

It follows that $AA^t = I$ if and only if for every $i, j = 1,....., n$, we have

$$r_i . r_j \begin{cases} 1 & \text{if } i = j, \\ 0 & \text{if } i \neq j, \end{cases}$$

if and only if $r_1, ..., r_n$ are orthonormal.

Proposition. Suppose that A is an $n \times n$ matrix with real entries. Suppose further that the inner product in $\mathbb{R}^n$ is the euclidean inner product. Then the following are equivalent:

(a) A is orthogonal.

(b) For every $x \in \mathbb{R}^n$, we have $\| Ax \| = \| x \|$.

(c) For every $u, v \in \mathbb{R}^n$, we have $Au . Av = u . v$.

Proof. ((a)) $\Rightarrow$ (b)) Suppose that A is orthogonal, so that $A^tA = I$. It follows that for every $x \in \mathbb{R}^n$, we have

$$\| Ax \|^2 = Ax . Ax = x^tA^tAx = x^tIx = x^tx = x . x = \| x \|^2.$$

((b)) $\Rightarrow$ (c)) Suppose that $\| Ax \| = \| x \|$ for every $x \in \mathbb{R}^n$. Then for every $u, v \in \mathbb{R}^n$, we have

$$Au.Av = \frac{1}{4}\| Au + Av \|^2 - \frac{1}{4}\| Au - Av \|^2 = \frac{1}{4}\| A(u+v) \|^2 - \frac{1}{4}\| A(u - v \|^2$$

$$= \frac{1}{4}\| u + v \|^2 - \frac{1}{4}\| u - v \|^2 = u.v.$$

Suppose that $Au . Av = u . v$ for every $u, v \in \mathbb{R}^n$. Then

$$Iu . v = u . v = Au . Av = v^tA^tAu = A^tAu . v,$$

so that

$$(A^tA - I)u . v = 0.$$

In particular, this holds when $v = (A^tA - I)u$, so that

$$(A^tA - I)u . (A^tA - I) u = 0,$$

whence

$$(A^tA - I)u = 0, \qquad ...(i)$$

But then (1) is a system of n homogeneous linear equations in n unknowns satised by every $u \in \mathbb{R}^n$. Hence the coefficient matrix $A^t A - I$ must be the zero matrix, and so $A^tA = I$.

Proof. For every $u \in V$, we can write

$$u = \beta_1u_1 + ... + \beta_nu_n = \gamma_1v_1 + ... + \gamma_nv_n,$$

where

$$\beta_1,....,\beta_n, \gamma_1,..... \gamma_n \in \mathbb{R},$$

and where $B = \{u_1,....., u_n\}$ and $C = \{v_1,....., v_n\}$ are two orthonormal bases of V. Then

$$\| u \|^2 = \langle u, u \rangle = \langle \beta_1u_1 + ... + \beta_nu_n, \beta_1u_1 + ... + \beta_nu_n \rangle$$

$$\sum_{i=1}^{n}\sum_{j=1}^{n}\beta_i\beta_j\langle u_iu_j\rangle = \sum_{i=1}^{n}\beta_i^2 = (\beta_1,\dots, \beta_n)\ (\beta_1,\dots, \beta_n).$$

Similarly,

$$\|u\|^2 = \langle u, u\rangle = \langle \gamma_1 v_1 + \dots + \gamma_n v_n, \gamma_1 v_1 + \dots + \gamma_n v_n\rangle$$

$$= \sum_{i=1}^{n}\sum_{j=1}^{n}\gamma_i\gamma_j\langle v_iv_j\rangle = \sum_{i=1}^{n}\gamma_i^2$$

It follows that in $\mathbb{R}^n$ with the euclidean norm, we have $\| [u]_B \| = \| [u]_C \|$, and so

$$\| P[u]_C \| = \| [u]_C \|$$

for every $u \in V$. Hence $\| Px \| = \| x \|$ holds for every $x \in \mathbb{R}^n$. It now follows from Proposition that P is orthogonal.

Eigenvalues and Eigenvectors

We give a brief review on eigenvalues and eigenvectors :

Suppose that

$$A = \begin{pmatrix} a_{11} & \cdots & a_{1n} \\ \vdots & & \vdots \\ a_{n1} & \cdots & a_{nn} \end{pmatrix}$$

is an $n \times n$ matrix with real entries. Suppose further that there exist a number $\lambda \in \mathbb{R}$ and a non-zero vector $v \in \mathbb{R}^n$ such that $Av = v$. Then we say that λ is an eigenvalue of the matrix A, and that v is an eigenvector corresponding to the eigenvalue λ. In this case, we have $Av = \lambda v = \lambda Iv$, where I is the $n \times n$ identity matrix, so that $(A - \lambda I)v = 0$. Since $v \in \mathbb{R}^n$ is non-zero, it follows that we must have

$$\det(A - \lambda I) = 0. \qquad \dots(ii)$$

In other words, we must have

$$\det\begin{pmatrix} a_{11} & a_{12} & \cdots & a_{1n} \\ a_{21} & a_{22} - \lambda & & a_{2n} \\ \vdots & & \ddots & \vdots \\ a_{n1} & a_{n2} & \cdots & a_{nn} - \lambda \end{pmatrix} = 0$$

Note that (ii) is a polynomial equation. The polynomial $\det(A - \lambda I)$ is called the characteristic polynomial of the matrix A. Solving this equation (2) gives the eigenvalues of the matrix A.

On the other hand, for any eigenvalue of the matrix A, the set

$$\{ v \in \mathbb{R}^n : (A - \lambda I)v = 0\} \qquad \dots(iii)$$

is the nullspace of the matrix $A - \lambda I$, and forms a subspace of $\mathbb{R}^n$. This space (iii) is called the eigenspace corresponding to the eigenvalue λ. Suppose now that A has eigenvalues $\lambda_1, \ldots. \lambda_n \in \mathbb{R}$, not necessarily distinct, with corresponding eigenvectors $v_1, \ldots, v_n \in \mathbb{R}^n$, and that $v_1, \ldots. v_n$ are linearly independent. Then it can be shown that

$$P^{-1}AP = D,$$

where

$$P = (\, v_1 \ldots v_n \,) \text{ and } D = \begin{pmatrix} \lambda_1 & & \\ & \ddots & \\ & & \lambda_n \end{pmatrix}.$$

In fact, we say that A is diagonalizable if there exists an invertible matrix P with real entries such that $P^{-1}AP$ is a diagonal matrix with real entries. It follows that A is diagonalizable if its eigenvectors form a basis of $\mathbb{R}^n$. In the opposite direction, one can show that if A is diagonalizable, then it has n linearly independent eigenvectors in $\mathbb{R}^n$. It therefore follows that the question of diagonalizing a matrix A with real entries is reduced to one of linear independence of its eigenvectors.

We now summarize our discussion so far.

Diagonalization Process. Suppose that A is an $n \times n$ matrix with real entries.

(1) Determine whether the n roots of the characteristic polynomial $\det(A - \lambda I)$ are real.

(2) If' not, then A is not diagonalizable. If so, then nd the eigenvectors corresponding to these eigen-values. Determine whether we can nd n linearly independent eigenvectors.

(3) If not, then A is not diagonalizable. If so, then write

$$P = (\, v_1 \ldots. v_n \,) \text{ and } D = \begin{pmatrix} \lambda_1 & & \\ & \ddots & \\ & & \lambda_n \end{pmatrix},$$

where $\lambda_1, \ldots, \lambda_n \in \mathbb{R}$ are the eigenvalues of A and where $v_1, \ldots. v_n \in \mathbb{R}^n$ are respectively their corresponding eigenvectors. Then $P^{-1}AP = D$.

In particular, it can be shown that if A has distinct eigenvalues $\lambda_1, \ldots. \lambda_n \in \mathbb{R}$, with corresponding eigenvectors $v_1, \ldots, v_n \in \mathbb{R}^n$, then $v_1, \ldots. v_n$ are linearly independent. It follows that all such matrices A are diagonalizable.

Orthonormal Diagonalization

We now consider the euclidean space $\mathbb{R}^n$ an as inner product space with the euclidean inner product. Given any $n \times n$ matrix A with real entries, we wish to nd out whether there exists an orthonormal basis of $\mathbb{R}^n$ consisting of eigenvectors of A. Recall that in the Diagonalization process discussed in the last section, the columns of the matrix P are eigenvectors of A, and these vectors form a basis of $\mathbb{R}^n$. It follows from Proposition basis is orthonormal if and only if the matrix P is orthogonal.

Definition. An nn matrix A with real entries is said to be orthogonally dagonalizable if there exists an orthogonal matrix P with real entries such that $P^{-1}AP = P^tAP$ is a diagonal matrix with real entries. First of all, we would like to determine which matrices are orthogonally diagonalizable. For those that are, we then need to discuss how we may find an orthogonal matrix P to carry out the diagonalization. To study the first question, we have the following result which gives a restriction on those matrices that are orthogonally diagonalizable.

Proposition. Suppose that A is a orthogonally diagonalizable matrix with real entries. Then A is symmetric.

Proof. Suppose that A is orthogonally diagonalizable. Then there exists an orthogonal matrix P and a diagonal matrix D, both with real entries and such that $P^tAP = D$. Since $PP^t = P^tP = I$ and $Dt = D$, we have

$$A = PDP^t = PD^tP^t,$$

so that

$$A^t = (PD^tP^t)^t = (P^t)^t(D^t)^tP^t = PDP^t = A,$$

whence A is symmetric. Our first question is in fact answered by the following result which we state without proof.

Proposition. Suppose that A is an $n \times n$ matrix with real entries. Then it is orthogonally diagonalizable if and only if it is symmetric. The remainder of this section is devoted to nding a way to orthogonally diagonalize a symmetric matrix with real entries. We begin by stating without proof the following result. The proof requires results from the theory of complex vector spaces.

Proposition. Suppose that A is a symmetric matrix with real entries. Then all the eigenvalues of A are real.

Our idea here is to follow the Diagonalization process discussed in the last section, knowing that since A is diagonalizable, we shall find a basis of $\mathbb{R}^n$ consisting of eigenvectors of A. We may then wish to orthogonalize this basis by the Gram-Schmidt process. This last step is considerably simplied in view of the following result.

Proposition. Suppose that u_1 and u_2 are eigenvectors of a symmetric matrix A with real entries, corresponding to distinct eigenvalues 1 and 2 respectively. Then $u_1 . u_2 = 0$. In other words, eigenvectors of a symmetric real matrix corresponding to distinct eigenvalues are orthogonal.

Proof. Note that if we write u_1 and u_2 as column matrices, then since A is symmetric, we have

$$Au_1.u_2 = u_2^t Au_1 = u_2^t A^t u_1 = (Au_2)^t u_1 = u_1.Au_2.$$

It follows that

$$\lambda_1 u^1 u_2 = Au_1 . u_2 = u_1 \; Au_2 = u_1 . 2u_2,$$

so that $(\lambda_1 - \lambda 2)(u_1 . u_2) = 0$. Since $\lambda_1 \neq \lambda_2$, we must have $u_1 . u_2 = 0$.

We can now follow the procedure below.

Orthogonal Diagonalization Process. Suppose that A is a symmetric $n \times n$ matrix with real entries.

(1) Determine the n real roots $\lambda 1, \ldots . \lambda_n$ of the characteristic polynomial det $(A - \lambda I)$, and find n linearly independent eigenvectors $u_1, \ldots ., u_n$ of A corresponding to these eigenvalues as in the Diagonalization process.

(2) Aply the Gram-Schmidt orthogonalization process to the eigenvectors $u_1, \ldots, u_n$ to obtain orthogonal eigenvectors $v_1 m, \ldots, v_n$ of A, noting that eigenvectors corresponding to distinct eigenvalues are already orthogonal.

(3) Normalize the orthogonal eigenvectors $v_1, \ldots, v_n$ to obtain orthonormal eigenvectors $w_1, \ldots, w_n$ of A. These form an orthonormal basis of $\mathbb{R}^n$. Furthermore, write

$$P = (w_1 \ldots w_n) \text{ and } D = \begin{pmatrix} \lambda_1 & & \\ & \ddots & \\ & & \lambda_n \end{pmatrix},$$

where $\lambda_1, \ldots, \lambda_n \in \mathbb{R}$ are the eigenvalues of A and where $w_1, \ldots, w_n \in \mathbb{R}^n$ are respectively their orthogonalized and normalized eigenvectors. Then $P^t AP = D$.

Remark. Note that if we apply the Gram-Schmidt orthogonalization process to eigenvectors corresponding to the same eigenvalue, then the new vectors that result from this process are also eigenvectors corresponding to this eigenvalue. Why?

Example. Consider the matrix

$$A = \begin{pmatrix} 2 & 2 & 1 \\ 2 & 5 & 2 \\ 1 & 2 & 2 \end{pmatrix}.$$

To find the eigenvalues of A, we need to nd the roots of

$$\det \begin{pmatrix} 2-\lambda & 2 & 1 \\ 2 & 5-\lambda & 2 \\ 1 & 2 & 2-\lambda \end{pmatrix} = 0;$$

in other words, $(\lambda - 7)(\lambda - 1)^2 = 0$. The eigenvalues are therefore $\lambda_1 = 7$ and (double root)

$$\lambda_2 = \lambda_3 = 1.$$

An eigenvector corresponding to $\lambda_1 = 7$ is a solution of the system

$$(A-7I)u=\begin{pmatrix} -5 & 2 & 1 \\ 2 & -2 & 2 \\ 1 & 2 & -5 \end{pmatrix}u=0 \text{ with root } u_1=\begin{pmatrix}1\\2\\1\end{pmatrix}.$$

Eigenvectors corresponding to $\lambda_2 = \lambda_3 = 1$ are solutions of the system

$$(A-7I)u=\begin{pmatrix} 1 & 2 & 1 \\ 2 & 4 & 2 \\ 1 & 2 & 1 \end{pmatrix}u=0 \text{ with root } u_2=\begin{pmatrix}1\\0\\-1\end{pmatrix} \text{ and } u_3=\begin{pmatrix}2\\-1\\0\end{pmatrix}$$

which are linearly independent. Next, we apply the Gram-Schmidt orthogonalization process to u_2 and u_3, and obtain

$$v_2=\begin{pmatrix}1\\0\\-1\end{pmatrix} \text{ and } v_3=\begin{pmatrix}1\\-1\\1\end{pmatrix}$$

which are now orthogonal to each other. Note that we do not have to do anything to u1 at this stage, in view of Proposition. We now conclude that

$$v_1=\begin{pmatrix}1\\2\\1\end{pmatrix},\ v_2=\begin{pmatrix}1\\0\\-1\end{pmatrix}, \text{ and } v_3=\begin{pmatrix}1\\-1\\1\end{pmatrix}$$

form an orthogonal basis of $\mathbb{R}^3$. Normalizing each of these, we obtain respectively

$$w_1=\begin{pmatrix}1/\sqrt{6}\\2/\sqrt{6}\\1/\sqrt{6}\end{pmatrix},\ w_2=\begin{pmatrix}1/\sqrt{2}\\0\\-1/\sqrt{2}\end{pmatrix},\ w_3=\begin{pmatrix}1/\sqrt{3}\\-1/\sqrt{3}\\1/\sqrt{3}\end{pmatrix}.$$

We now take

$$P=(w_1 \quad w_2 \quad w_3)=\begin{pmatrix} 1/\sqrt{6} & 1/\sqrt{2} & 1/\sqrt{3} \\ 2/\sqrt{6} & 0 & -1/\sqrt{3} \\ 1/\sqrt{6} & -1/\sqrt{2} & 1/\sqrt{3} \end{pmatrix}.$$

Then

$$P=P^t=\begin{pmatrix} 1/\sqrt{6} & 2/\sqrt{6} & 1/\sqrt{6} \\ 1/\sqrt{2} & 0 & -1/\sqrt{2} \\ 1/\sqrt{3} & -1/\sqrt{3} & 1/\sqrt{3} \end{pmatrix} \text{ and } p^tAP=\begin{pmatrix} 7 & 0 & 0 \\ 0 & 1 & 0 \\ 0 & 0 & 1 \end{pmatrix}.$$

Example. Consider the matrix

$$A = \begin{pmatrix} -1 & 6 & -12 \\ 0 & -13 & 30 \\ 0 & -9 & 20 \end{pmatrix}.$$

To find the eigenvalues of A, we need to nd the roots of

$$\det\begin{pmatrix} -1-\lambda & 6 & -12 \\ 0 & -13-\lambda & 30 \\ 0 & -9 & 20-\lambda \end{pmatrix} 0;$$

in other words, $(\lambda + 1)(\lambda - 2)(\lambda - 5) = 0$. The eigenvalues are therefore

$$\lambda_1 = -1, \lambda_2 = 2 \text{ and } \lambda_3 = 5.$$

An eigenvector corresponding $\lambda_1 = -1$ is a solution of the system

$$(A+I)u = \begin{pmatrix} 0 & 6 & -12 \\ 0 & -12 & 30 \\ 0 & -9 & 21 \end{pmatrix} u = 0, \text{ with root } u_1 = \begin{pmatrix} 1 \\ 0 \\ 0 \end{pmatrix}.$$

An eigenvector corresponding to $\lambda_2 = 2$ is a solution of the system

$$(A+5I)u = \begin{pmatrix} -6 & 6 & -12 \\ 0 & -18 & 30 \\ 0 & -9 & 15 \end{pmatrix} u = 0, \text{ with root } u_3 = \begin{pmatrix} 1 \\ -5 \\ -3 \end{pmatrix}.$$

An eigenvector corresponding to $\lambda_3 = 5$ is a solution of the system

$$(A+5I)u = \begin{pmatrix} -6 & 6 & -12 \\ 0 & -18 & 30 \\ 0 & -9 & 15 \end{pmatrix} u = 0, \text{ with root } u_3 = \begin{pmatrix} 1 \\ -5 \\ -3 \end{pmatrix}.$$

Note that while u_1, u_2, u_3 correspond to distinct eigenvalues of A, they are not orthogonal. The matrix A is not symmetric, and so Proposition 10G does not apply in this case.

Example. Consider the matrix

$$A = \begin{pmatrix} 5 & -2 & 0 \\ -2 & 6 & 2 \\ 0 & 2 & 7 \end{pmatrix}.$$

To find the eigenvalues of A, we need to nd the roots of

$$\det\begin{pmatrix} 5-\lambda & -2 & 0 \\ -2 & 6-\lambda & 2 \\ 0 & 2 & 7-\lambda \end{pmatrix} = 0;$$

in other words, $(\lambda - 3)(\lambda - 6)(\lambda - 9) = 0$. The eigenvalues are therefore

$$\lambda_1 = 3, \lambda_2 = 6$$

and $\lambda_3 = 9$. An eigenvector corresponding $\lambda_1 = 3$ is a solution of the system

$$(A-3I)u = \begin{pmatrix} 2 & -2 & 0 \\ -2 & 3 & 2 \\ 0 & 2 & 4 \end{pmatrix} u = 0, \text{ with root } u_1 = \begin{pmatrix} 2 \\ 2 \\ -1 \end{pmatrix}.$$

An eigenvector corresponding to $\lambda_2 = 6$ is a solution of the system

$$(A-6I)u = \begin{pmatrix} -1 & -2 & 0 \\ -2 & 0 & 2 \\ 0 & 2 & 1 \end{pmatrix} u = 0, \text{ with root } u_2 = \begin{pmatrix} 2 \\ -1 \\ 2 \end{pmatrix}.$$

An eigenvector corresponding to 3 = 9 is a solution of the system

$$(A-9I)u = \begin{pmatrix} -4 & -2 & 0 \\ -2 & -3 & 2 \\ 0 & 2 & -2 \end{pmatrix} u = 0, \text{ with root } u_3 = \begin{pmatrix} -1 \\ 2 \\ 2 \end{pmatrix}.$$

Note now that the eigenvalues are distinct, so it follows from Proposition that u_1, u_2, u_3 are orthogonal, so we do not have to apply Step (2) of the Orthogonal diagonalization process. Normalizing each of these vectors, we obtain respectively

$$w_1 = \begin{pmatrix} 2/3 \\ 2/3 \\ -1/3 \end{pmatrix}, \quad w_2 = \begin{pmatrix} 2/3 \\ -1/3 \\ 2/3 \end{pmatrix}, \quad w_3 \begin{pmatrix} -1/3 \\ 2/3 \\ 2/3 \end{pmatrix}.$$

We now take

$$P = (w_1 \quad w_2 \quad w_3) = \begin{pmatrix} 2/3 & 2/3 & -1/3 \\ 2/3 & -1/3 & 2/3 \\ -1/3 & 2/3 & 2/3 \end{pmatrix}.$$

Then

$$P^{-1} = P^t = \begin{pmatrix} 2/3 & 2/3 & -1/3 \\ 2/3 & -1/3 & 2/3 \\ -1/3 & 2/3 & 2/3 \end{pmatrix} \text{ and } P^t AP = \begin{pmatrix} 3 & 0 & 0 \\ 0 & 6 & 0 \\ 0 & 0 & 9 \end{pmatrix}.$$

Chapter 4

Determinants

Introduction

The reader probably already met determinants in calculus or algebra, at least the determinants of 2×2 and 3×3 matriccs. For a 2×2 matrix

$$\begin{pmatrix} a & b \\ c & d \end{pmatrix}$$

the determinant is simply $ad - bc$; the determinant of a 3×3 matrix can be found by the "Star of David" rule.

In this chapter we would like to introduce determinants for $n \times n$ matrices. I don't want just to give a formal definition. First *I* want to give some motivation, and then derive some properties the determinant should have. Then if we want to have these properties, we do not have any choice, and arrive to several equivalent definitions of the determinant.

It is more convenient to start not with the determinant of a matrix, but with determinant of a system of vectors. There is no real difference here, since we always can join vectors together (say as columns) to form a matrix.

Let us have n vectors $v_1, v_2, ..., v_n$ in $\mathbb{R}^n$ (notice that the number of vectors coincide with dimension), and we want to find the n–dimensional volume of the parallelepiped determined by these vectors.

The parallelepiped determined by the vectors $v_1, v_2, ..., v_n$ can be defined as the collection of all vectors $v \in \mathbb{R}^n$ that can be represented as

$$v = t_1v_1 + t_2v_2 + ...+ t_nv_n, 0 \leq t_k \leq 1 \ \forall k = 1, 2, ..., n.$$

It can be easily visualized when $n = 2$ (parallelogram) and $n = 3$ (parallelepiped).

So, what is the n–dimensional volume?

If $n = 2$ it is area; if $n = 3$ it is indeed the volume. In dimension 1 is it just the length.

Finally, let us introduce some notation. For a system of vectors (columns) $v_1, v_2, ..., v_n$ we will denote its determinant (that we are going to construct) as $D(v_1, v_2, ..., vn)$. If we join these vectors in a matrix A (column number k of a is v_k), then we will use the notation det A,

$$\det A = D(v_1, v_2, ..., vn)$$

Also, for a matrix

$$A = \begin{pmatrix} a_{1,1} & a_{1,2} & \cdots & a_{1,n} \\ a_{2,1} & a_{2,2} & \cdots & a_{2,n} \\ \vdots & \vdots & & \vdots \\ an,_1 & an,_2 & \cdots & a_{n,n} \end{pmatrix}$$

its determinant is often is denoted by

$$\begin{vmatrix} a_{1.1} & a_{1,2} & \cdots & a_{1,n} \\ a_{2,1} & a_{2,2} & \cdots & a_{2,n} \\ \vdots & \vdots & & \vdots \\ a_{n,1} & a_{n,2} & \cdots & a_{n,n} \end{vmatrix}.$$

WHAT PROPERTIES DETERMINANT SHOULD HAVE

We know, that for dimensions 2 and 3 "volume" of a parallelepiped is determined by the *base times height* rule: if we pick one vector, then height is the distance from this vector to the subspace spanned by the remaining vectors, and the base is the $(n - 1)$ – dimensional volume of the parallelepiped determined by the remaining vectors.

Now let us generalize this idea to higher dimensions. For a moment we do not care about how exactly to determine height and base. We will show, that if we assume that the base and the height satisfy some natural properties, then we do not have any choice, and the volume (determinant) is uniquely defined.

Linearity in Each Argument

First of all, if we multiply vector v_1 by a positive number a, then the height (*i.e.*, the distance to the linear span £ $(v_2, ..., v_n)$) is multiplied by a. If we admit negative heights (and negative volumes), then this property holds for all scalars a, and so the determinant $D(v_1, v_2, ..., v_n)$ of the system $v_1, v_2, ..., v_n$ should satisfy

$$D(\alpha v_1, v_2, ..., v_n) = \alpha D(v_1, v_2, ..., vn).$$

Of course, there is nothing special about vector v_1, so for any index k

$$D(v_1,\ldots,\underset{k}{\alpha v_k},\ldots,v_n) = \alpha D(v_1, ...,\underset{k}{v_k}, ..., v_n)$$

To get the next property, let us notice that if we add 2 vectors, then the "height" of the result should be equal the sum of the "heights" of summands, *i.e.*, that

$$D(v_1,\ldots,\underbrace{u_k+v_k}_{k},\ldots,v_n) = D(v_1\ldots,\underset{k}{u_k},\ldots,v_n)+D(v_1,\ldots,\underset{k}{v_k},\ldots,v_n)$$

In other words, the above two properties say that the determinant of n vectors is *linear in each argument* (*vector*), meaning that if we fix $n-1$ vectors and interpret the remaining vector as a variable (argument), we get a linear function.

Remark. We already know that *linearity* is a very nice property, that helps in many situations. So, admitting negative heights (and therefore negative volumes) is a very small price to pay to get linearity, since we can always put on the absolute value afterwards.

In fact, by admitting negative heights, we did not sacrifice anything! To the contrary, we even gained something, because the sign of the determinant contains some information about the system of vectors (orientation).

Preservation Under "Column Replacement"

The next property also seems natural. Namely, if we take a vector, say v_j, and add to it a multiple of another vector v_k, the "height" does not change, so

$$D(v_1, \ldots \underbrace{v_j + \alpha v_k}_{j}, \ldots, \underset{k}{v_k}, \ldots, v_n) = D\left(v_1, \ldots, \underset{j}{v_j}, \ldots, \underset{k}{v_k}, \ldots, v_n\right)$$

In other words, if we apply the *column operation* of the third type, the determinant does not change.

Remark. Although it is not essential here, let us notice that the second part of linearity is not independent: it can be deduced from properties.

Antisymmetry

The next property the determinant should have, is Functions of several variables that change sign

$$D\left(v_1, \ldots, \underset{j}{v_k}, \ldots, \underset{k}{v_j}, \ldots, v_n\right) = -D(v_1, \ldots, \underset{j}{v_j}, \ldots \underset{k}{v_k}, \ldots, v_n).$$

At first sight this property does not look natural, but it can be deduced from the previous ones. Namely, applying property three times, and then using we get

$$\begin{aligned}
& D(v_1, \ldots, \underset{j}{v_j}, \ldots, \underset{k}{v_k}, \ldots, v_n) = \\
& = D(v_1, \ldots, \underset{j}{v_j}, \ldots, \underbrace{v_k - v_j}_{k}, \ldots, v_n) \\
& = D\left(v_1, \ldots, \underbrace{v_j + (v_k - v_j)}_{j}, \ldots, \underbrace{v_k - v_j}_{k}, \ldots, v_n\right) \\
& = D\left(v_1, \ldots, \underset{j}{v_k}, \ldots, \underbrace{v_k - v_j}_{k}, \ldots, v_n\right) \\
& = D\left(v_1, \ldots, \underset{j}{v_k}, \ldots, \underbrace{(v_k - v_j) - v_k}_{k}, \ldots, v_n\right)
\end{aligned}$$

$$= D\left(v_1,\ldots,\underset{j}{v_k},\ldots,\underset{k}{-v_j},\ldots,v_n\right)$$

$$= -D\left(v_1,\ldots,\underset{j}{v_k},\ldots,\underset{k}{v_j},\ldots,v_n\right).$$

Normalization

The last property is the easiest one. For the $D(e_1, e_2, \ldots, e_n) = 1$.

In matrix notation this can be written as

$$\det(I) = 1$$

Constructing the Determinant

The plan of the game is now as follows: using the properties that the determinant should have, we derive other properties of the determinant, some of them highly non–trivial. We will show how to use these properties to compute the determinant using our old friend—row reduction.

We will show that the determinant, *i.e.*, a function with the desired properties exists and unique. After all we have to be sure that the object we are computing and studying exists.

Basic Properties. We will use the following basic properties of the determinant:

1. Determinant is linear in each column, *i.e.*, in vector notation for every index k

$$D(v_1,\ldots,\underbrace{\alpha u_k + \beta v_k}_{k},\ldots,v_n) =$$

$$\alpha D(v_1,\ldots,\underset{k}{u_k},\ldots,v_n) + \beta D(v_1,\ldots \underset{k}{v_k},\ldots v_n)$$

for all scalars α, β.

2. *Determinant is antisymmetric, i.e.,* if one interchanges two columns, the determinant changes sign.
3. Normalization property: det $I = 1$.

The first propertyis just the combined. The second one and the last one is the normalization property. Note, that we did not use property: it can be deduced from the above three. These three properties completely define determinant.

Properties of Determinant Deduced from the Basic Properties

Proposition. For a square matrix a the following statements hold:

1. *If a has a zero column, then* det$A$ = *0.*
2. *If a has two equal columns, then* det A = *0;*
3. *If one column of a is a multiple of another, then* det$A$ = *0;*
4. *If columns of a are linearly dependent, i.e., if the matrix is not invertible, then* det $A = 0$.

Proof. Statement 1 follows immediately from linearity. If we multiply the zero column by zero, we do not change the matrix and its determinant. But by the property 1 above, we should get 0. The fact that determinant is antisymmetric, implies statement 2. Indeed, if we interchange two equal columns, we change nothing, so the determinant remains the same. On the other hand, interchanging two columns changes sign of determinant, so

$$\det A = -\det A,$$

which is possible only if $\det A = 0$. Statement 3 is immediate corollary of statement 2 and linearity.

To prove the last statement, let us first suppose that the first vector v_1 is a linear combination of the other vectors,

$$v_1 = \alpha_2 v_2 + \alpha_3 v_3 + ... + \alpha_n v_n = \sum_{k=2}^{n} a_k v_k.$$

Then by linearity we have (in vector notation)

$$D(v_1, v_2, ..., v_n) = D\left(\left(\sum_{k=2}^{n} \alpha_k v_k\right), v_2, v_3, ..., v_n\right)$$

$$= \sum_{k=2}^{n} \alpha_k D(v_k, v_2, v_3, ..., v_n)$$

and each determinant in the sum is zero because of two equal columns.

Let us now consider general case, *i.e.*, let us assume that the system $v_1, v_2, ..., v_n$ is linearly dependent. Then one of the vectors, say v_k can be represented as a linear combination of the others. Interchanging this vector with v_1 we arrive to the situation we just treated, so

$$D(v_1, ..., \underset{k}{V_k}, ..., v_n) = -D(v_k, ..., \underset{k}{V_1}, ..., v_n) = -0 = 0,$$

so the determinant in this case is also 0.

The next proposition generalizes property. As we already have said above, this property can be deduced from the three "basic" properties of the determinant, we are using in this section.

Proposition. The determinant does not change if we add to a column a linear combination of the other columns (leaving the other columns intact). In particular, the determinant is preserved under "column replacement" (column operation of third type).

Proof. Fix a vector v_k, and let u be a linear combination of the other vectors,

$$u = \sum_{j \neq k} \alpha_j v_j.$$

Then by linearity

$$D(v_1 ..., \underbrace{v_k + u}_{k}, ..., v_n) = D(v_1, ..., \underset{k}{v_k}, ..., v_n) + D(v_1, ..., \underset{k}{u}, ..., v_n),$$

and by Proposition the last term is zero.

Determinants of Diagonal and Triangular Matrices

Now we are ready to compute determinant for some important special classes of matrices.

The first class is the so–called *diagonal* matrices. Let us recall that a square matrix $A = \{aj, k\}_{j,j=1}^{n}$ is called *diagonal* if all entries *off the main diagonal* are zero, *i.e.*, if $a_{j,k} = 0$ for all $j \neq k$. We will often use the notation diag$\{a_1, a_2, ..., an\}$ for the diagonal matrix

$$\begin{pmatrix} a_1 & 0 & \cdots & 0 \\ 0 & a_2 & \cdots & 0 \\ \vdots & \vdots & \ddots & 0 \\ 0 & 0 & \cdots & a_n \end{pmatrix}.$$

Since a diagonal matrix diag$\{a_1, a_2, ..., a_n\}$ can be obtained from the identity matrix I by multiplying column number k by a_k,

Determinant of a diagonal matrix equal the product of the diagonal entries,det(diag$\{a_1, a_2, ..., a_n\}$) = $a_1 a_2 .... a_n$.

The next important class is the class of so–called *triangular* matrices. *A* square matrix $A = \{a_{j,k}\}_{j,j=1}^{n}$ is called *upper triangular* if all entries below the main diagonal are 0, *i.e.*, if $a_{j,k} = 0$ for all $k < j$. *A* square matrix is called *lower triangular* if all entries *above* the main are 0, i.e if $a_{j,k} = 0$ for all $j < k$. We call a matrix *triangular*, if it is either lower or upper triangular matrix.

It is easy to see that

Determinant of a *triangular* matrix equals to the product of the diagonal entries,

$$\det A = a_{1,1} a_{2,2} \cdots a_{n,} n.$$

Indeed, if a triangular matrix has zero on the main diagonal, it is not invertible (this can easily be checked by column operations) and therefore both sides equal zero. If all diagonal entries are non–zero, then using column replacement (column operations of third type) one can transform the matrix into a diagonal one with the same diagonal entries: For upper triangular matrix one should first subtract appropriate multiples of the first column from the columns number 2, 3, ..., *n*, "killing" all entries in the first row, then subtract appropriate multiples of the second column from columns number 3, ..., *n*, and so on.

To treat the case of lower triangular matrices one has to do "column reduction" from the left to the right, *i.e.*, first subtract appropriate multiples of the last column from columns number $n - 1$, ..., 2, 1, and so on.

Computing the Determinant

Now we know how to compute determinants, using their properties: one just need to do column reduction (*i.e.*, row reduction for A^T) keeping track of column operations changing the determinant. Fortunately, the most often used operation — row replacement, *i.e.,* operation of third type does not change the determinant. So we only need to keep track of interchanging of columns and of multiplication of column by a scalar.

If an echelon form of A^T does not have pivots in every column (and row), then a is not invertible, so det $A = 0$. If a is invertible, we arrive at a triangular matrix, and det A is the product of diagonal entries times the correction from column interchanges and multiplications.

The above algorithm implies that det A can be zero only if a matrix A is not invertible. Combining this with the last statement of Proposition we get Proposition. det $A = 0$ if and only if a is not invertible. An equivalent statement: det $A \neq 0$ *if and only if A is invertible.*

Note, that although we now know how to compute determinants, the determinant is still not defined. One can ask: why don't we define it as the result we get from the above algorithm? The problem is that formallythis result is not well defined: that means we did not prove that different sequences of column operations yield the same answer.

Determinants of Transpose and Product

In this section we prove two important theorems

Theorem. (Determinant of a transpose). *For a square matrix A,*

$$\det A = \det(A^T).$$

This theorem implies that for all statement about columns the corresponding statements about rows are also true. In particular, determinants behave under *row operations* the same way they behave under *column operations*. So, we can use row operations to compute determinants.

Theorem. (Determinant of a product). *For n × n matrices A and B*

$$\det(AB) = (\det A)(\det B)$$

In other words

Determinant of a product equals product of deteramamts

To prove both theorem we need the following lemma

Lemma. For a square matrix A and an elementary matrix E (of the same size)

$$\det(AE) = (\det A)\ (\det E)$$

Proof. The proof can be done just by direct checking: determinants of special matrices are easy to compute; right multiplication by an elementary matrix is a column operation, and eect of column operations on the determinant is well known.

This can look like a lucky coincidence, that the determinants of elementary matrices agree with the corresponding column operations, but it is not a coincidence at all.

Namely, for a column operation the corresponding elementary matrix can be obtained from the identity matrix I by this column operation. So, its determinant is 1 (determinant of I) times the eect of the column operation. And that is all! It may be hard to realise at first, but the above paragraph is a *complete and rigorous* proof of the lemma!

Applying N times Lemma we get the following corollary.

Corollary. For any matrix A and any sequence of elementary matrices $E_1, E_2, ..., E_N$ (all matrices are $n \times n$)

$$\det(AE_1E_2, ..., E_N) = (\det A)(\det E_1)(\det E_2), ..., (\det E_N)$$

Lemma. Any invertible matrix is a product of elementary matrices.

Proof. We know that any invertible matrix is row equivalent to the identity matrix, which is its reduced echelon form. So

$$I = ENEN^{-1}, ..., E_2E_1A,$$

and therefore any invertible matrix can be represented as a product of elementary matrices,

$$A = E_1^{-1}E_2^{-1}...E_{N-1}^{-1}E_N^{-1}I = E_1^{-1}E_2^{-1}...E_{N-1}^{-1}E_N^{-1}$$

(inverse of an elementary matrix is an elementary matrix).

Proof. First of all, it can be easily checked, that for an elementary matrix E we have $\det E = \det(E^T)$. Notice, that it is sucient to prove the theorem only for invertible matrices A, since if A is not invertible then A^T is also not invertible, and both determinants are zero.

By Lemma matrix A can be represented as a product of elementary matrices,

$$A = E_1E_2, ..., E_N,$$

and by Corollary the determinant of A is the product of determinants of the elementary matrices. Since taking the transpose just transposes each elementary matrix and reverses their order, Corollary implies that

$$\det A = \det A^T.$$

Proof. Let us first suppose that the matrix B is invertible. Then Lemma implies that B can be represented as a product of elementary matrices

$$B = E_1E_2, ..., E_N,$$

and so by Corollary

$$\det(AB) = (\det A)[(\det E_1)(\det E_2), ..., (\det E_N)] = (\det A)(\det B).$$

If B is not invertible, then the product AB is also not invertible, and the theorem just says that $0 = 0$.

To check that the product $AB = C$ is not invertible, let us assume that it is invertible. Then multiplying the identity $AB = C$ by C^{-1} from the left, we get $C^{-1}AB = I$, so $C^{-1}A$ is a left inverse of B. So B is left invertible, and since it is square, it is invertible. We got a contradiction.

Properties of Determinant

First of all, let us say once more, that determinant is defined only for square matrices Since we now know that $\det A = \det(A^T)$, the statements that we knew about columns are true for rows too.

1. Determinant is linear in each row (column) when the other rows (columns) are fixed.
2. If one interchanges two rows (columns) of a matrix A, the determinant changes sign.

3. For a triangular (in particular, for a diagonal) matrix its determinant is the product of the diagonal entries. In particular, det $I = 1$.
4. If a matrix A has a zero row (or column), det $A = 0$.
5. If a matrix A has two equal rows (columns), det $A = 0$.
6. If one of the rows (columns) of A is a linear combination of the other rows (columns), *i.e.*, if the matrix is not invertible, then det $A = 0$; More generally,
7. det $A = 0$ if and only if A is not invertible, or equivalently
8. det $A \neq 0$ if and only if A is invertible.
9. det A does not change if we add to a row (column) a linear combination of the other rows (columns). In particular, the determinant is preserved under the row (column) replacement, *i.e.*, under the row (column) operation of the third kind.
10. det A^T = det A.
11. det(AB) = (det A)(det B). And finally,
12. If a is an $n \times n$ matrix, then det(aA) = an det A.

The last property follows from the linearity of the determinant, if we recall that to multiply a matrix A by a we have to multiply each row by a, and that each multiplication multiplies the determinant by a.

Existence and Uniqueness of the Determinant

In this section we arrive to the formal definition of the determinant. We show that a function, satisfying the basic properties 1, 2, 3 from Section 3 exists, and moreover, such function is unique, *i.e.*, we do not have any choice in constructing the determinant.

Consider an $n \times n$ matrix $A = \{a_{j,k}\} nj, k = 1$, and let $v_1, v_2, \ldots, v_n$ be its columns, *i.e.*,

$$v_k = \begin{pmatrix} a_{1,k} \\ a_{2,k} \\ \vdots \\ a_{n,k} \end{pmatrix} = a_{1,k}\, e_1 + a_{2,k}\, e_2 + \ldots + a_{n,k}\, e_n = \sum_{j=1}^{n} a_{j,k}\, e_j.$$

Using linearity of the determinant we expand it in the first column

$$v_1\text{: } D(v_1, v_2, \ldots, v_n) = D(\sum_{j=1}^{n} a_{j,1}e_j, v_2 \ldots, e_n = \sum_{j=1}^{n} a_{j,1}\, D(e_j, v_2, \ldots v_n).$$

Then we expand it in the second column, then in the third, and so on. We get

$$D(v_1, v_2, \ldots, v_n) = \sum_{j_1=1}^{n} \sum_{j_2=1}^{n} \cdots \sum_{j_n=1}^{n} a_{j_1,1} a_{j_2,2} \ldots a_{j_n,n} D(e_{j_1} . e_{j_2}, \ldots e_{j_n}).$$

Notice, that we have to use a different index of summation for each column:we call them $j_1, j_2, ..., j_n$; the index j_1 here is the same as the index j.

It is a huge sum, it contains n^n terms. Fortunately, some of the terms are zero. Namely, if any 2 of the indices $j_1, j_2, ..., j^n$ coincide, the determinant $D(ej_1 \,.\, ej_2, ... ej^n)$ is zero, because there are two equal rows here.

So, let us rewrite the sum, omitting all zero terms. The most convenient way to do that is using the notion of a permutation. a *permutation* of an ordered set $\{1, 2, ..., n\}$ is a rearrangement of its elements. a convenient way to represent a permutation is by using a function

$$\sigma : \{1, 2, ..., n\} \to \{1, 2, ..., n\},$$

where $\sigma(1), \sigma(2), ..., (n)$ gives the new order of the set $1, 2, ..., n$. In other words, the permutation rearranges the ordered set $1, 2, ..., n$ into $\sigma(1), \sigma(2), ..., (n)$.

Such function σ has to be one–to–one (different values for different arguments) and onto (assumes all possible values from the target space). Such functions (one–to–one and onto) are called bijections, and they give one–toone correspondence between two sets.

Although it is not directly relevant here, let us notice, that it is wellknown in combinatorics, that the number of different perturbations of the set $\{1, 2, ..., n\}$ is exactly $n!$. The set of all permutations of the set $\{1, 2, ..., n\}$ will be denoted Perm(n).

Using the notion of a permutation, we can rewrite the determinant as

$$D(v_1, v_2, ..., v_n) = \sum_{\sigma \in \text{Perm}(n)} a_{\sigma(1),1} a_{\sigma(2),2} \cdots a_{\sigma(n),n} D(e_{\sigma(1)}, e_{\sigma(2)}, \ldots, e_{\sigma(n)}).$$

The matrix with columns $e_{\sigma(1)}, e_{\sigma(2)}, ..., e_{\sigma(n)}$ can be obtained from the identity matrix by finitely many column interchanges, so the determinant

$$D(e_{\sigma(1)}, e_{\sigma(2)}, ..., e_{\sigma(n)})$$

is 1 or –1 depending on the number of column interchanges.

To formalize that, we define *sign* (denoted sign σ) of a permutation to be 1 if even number of interchanges is necessary to rearrange the ntuple $1, 2, ..., n$ into $\sigma(1), \sigma(2), ..., \sigma(n)$, and $\text{sign}(\sigma) = -1$ if the number of interchanges is odd.

It is a well–known fact from the combinatorics, that the sign of permutation is well defined, *i.e.*, that although there are infinitely many ways to get the n–tuple (1), (2), ..., (n) from $1, 2, ..., n$, the number of interchanges is either always odd or always even.

One of the ways to show that is to count the number K of pairs $j, k, j < k$ such that $\sigma(j) > \sigma(k)$, and see if the number is even or odd. We call the permutation *odd* if K is odd and even if K is even. Then define signum of to be $(-1)K$. We want to show that signum and sign coincide, so sign is well defined.

If $(k) = k\ \forall\ k$, then the number of such pairs is 0, so signum of such *identity* permutation is 1. Note also, that any elementary transpose, which interchange two neighbors, changes the signum of a permutation, because it changes (increases or decreases) the number of the pairs exactly by 1. So, to get from a permutation to another one always needs an even number of elementary transposes if the permutation have the same signum, and an oddnumber if the signums are different.

Finally, any interchange of two entries can be achieved by an odd number of elementary transposes. This implies that signum changes under an interchange of two entries. So, to get from 1, 2, ..., n to an even permutation (positive signum) one always need even number of interchanges, and odd number of interchanges is needed to get an odd permutation (negative signum). That means *signum* and *sign* coincide, and so sign is well defined.

So, if we want determinant to satisfy basic properties 1–3 from Section 3, we must define it as

$$\det A = \sum_{\sigma \in \text{Perm}(n)} a_{\sigma(1),1}, a_{\sigma(2),2} \ldots, a_{\sigma(n),n} sign(\sigma),$$

where the sum is taken over all permutations of the set $\{1, 2, ..., n\}$.

If we define the determinant this way, it is easy to check that it satisfies the basic properties. Indeed, it is linear in each column, because for each column every term (product) in the sum contains exactly one entry from this column. Interchanging two columns of a just adds an extra interchange to the perturbation, so right side in changes sign. Finally, for the identitymatrix I, the right side is 1 (it has one non–zero term).

COFACTOR EXPANSION

For an $n \times n$ matrix $A = \{a_{j,k}\}_{j,k=1}^{n}$ let $A_{j,k}$ denotes the $(n-1) \times (n-1)$ matrix obtained from A by crossing out row number j and column number k.

Theorem. (Cofactor expansion of determinant). Let *A be an n × n matrix. For each j,* $1 \leq j \leq n$, *determinant of A can be expanded in the row number j as*

$$\det A = a_{j,1}(-1)^{j+1} \det A_{j,1} + a_{j,2}(-1)_{j+2} \det A_{j,2} + \ldots$$
$$+ a_{j,1}(-1)^{j+n} \det A_{j,n}$$
$$= \sum_{k=1}^{n} a_{j,k}(-1)^{j+k} \det A_{j,k}.$$

Similarly, for each k, $1 \leq k \leq n$, *the determinant can be expanded in the column number k,*

$$\det A = \sum_{k=1}^{n} a_{j,k}(-1)^{j+k} \det A_{j,k}.$$

Proof. Let us first prove the formula for the expansion in row number 1. The formula for expansion in row number k then can be obtained from it by interchanging rows number 1 and k. Since $\det A = \det A^T$, column expansion follows automatically.

Let us first consider a special case, when the first row has one non zero term $a_{1,1}$. Performing column operations on columns 2, 3, ..., n we transform a to the lower triangular form. The determinant of A then can be computed as the product of diagonal entries of the triangular matrix

the product of diagonal entries of the triangular matrix × correcting factor from the column operations.

But the product of all diagonal entries except the first one (*i.e.*, without $a_{1,1}$) times the correcting factor is exactly $\det A_{1,1}$, so in this particular case $\det A = a_{1,1} \det A_{1,1}$.

Let us now consider the case when all entries in the first row except $a_{1,2}$ are zeroes. This case can be reduced to the previous one by interchanging columns number 1 and 2, and,therefore in this case $\det A = (-1) \det A_{1,2}$.

The case when $a_{1,3}$ is the only non–zero entry in the first row, can be reduced to the previous one by interchanging rows 2 and 3, so in this case $\det A = a_{1,3} \det A_{1,3}$.

Repeating this procedure we get that in the case when $a_{1,k}$ is the only non–zero entry in the first row $\det A = (-1)^{1+k} a_{1,k} \det A_{1,k}$.

In the general case, linearity of the determinant implies that

$$\det A = \det A^{(1)} + \det A^{(2)} + \dots + \det A^{(n)} = \sum_{k=1}^{n} \det A^{(k)}$$

where the matrix $A^{(k)}$ is obtained from A by replacing all entries in the first row except $a_{1,k}$ by 0. As we just discussed above

$$\det A^{(k)} = (-1)^{1+k} a_{1,k} \det A_{1,k},$$

so

$$\det A = \sum_{k=1}^{n} (-1)^{1+k} a_{1,k} \det A_{1,k}.$$

To get the cofactor expansion in the second row, we can interchange the first and second rows and apply the above formula. The row exchange changes the sign, so we get

$$\det A = -\sum_{k=1}^{n} (-1)^{1+k} a_{2,k} \det A_{2,k} = \sum_{k=1}^{n} (-1)^{2+k} a_{2,k} \det A_{2,k}.$$

Exchanging rows 3 and 2 and expanding in the second row we get formula

$$\det A = \sum_{k=1}^{n} (-1)^{3+k} a_{3,k} \det A_{3,k},$$

and so on.

To expand the determinant $\det A$ in a column one need to apply the row expansion formula for A^T.

Definition. The numbers

$$C_{j,k} = (-1)^{j+k} \det A_{j,k}$$

are called *cofactors.*

Using this notation, the formula for expansion of the determinant in the row number j can be rewritten as

$$\det A = a_{j,1}C_{j,1} + a_{j,2}C_{j,2} + ... + a_{j,n}C_{j,n} = \sum_{k=1}^{n} a_{j,k}C_{j,k}.$$

Similarly, expansion in the row number k can be written as

$$\det A = a_{1,k}C_{1,k} + a_{2,k}C_{2,k} + ... + a_{n,k}C_{n,k} = \sum_{j=1}^{n} a_{j,k}C_{j,k}.$$

Remark. Very often the cofactor expansion formula is used as the definition of determinant. It is not dicult to show that the quantity given by this formula satisfies the basic properties of the determinant: the normalization property is trivial, the proof of antisymmetry is easy. However, the proof of linearity is a bit tedious (although not too dicult).

Remark. Although it looks very nice, the cofactor expansion formula is not suitable for computing determinant of matrices bigger than 3×3.

As one can count it requires $n!$ multiplications, and $n!$ grows very rapidly. For example, cofactor expansion of a 20×20 matrix require $20! \approx 2.4 \cdot 10^{18}$ multiplications: it would take a computer performing a *billion* multiplications per second over 77 years to perform the multiplications.

On the other hand, computing the determinant of an $n \times n$ matrix using *row reduction* requires $(n^3 + 2n - 3)/3$ multiplications (and about the same number of additions). It would take a computer performing a *million* operations per second (very slow, by today's standards) a fraction of a second to compute the determinant of a 100×100 matrix by row reduction.

It can only be practical to apply the cofactor expansion formula in higher dimensions if a row (or a column) has a lot of zero entries. However, the cofactor expansion formula is of great theoretical importance, as the next section shows.

Cofactor Formula for Inverse Matrix

The matrix $C = \{C_{j,k}\}_{j,k=1}^{n}$ whose entries are cofactors of A given matrix A is called the *cofactor matrix* of A.

Theorem. Let a be an invertible matrix and let C be its cofactor matrix.

Then

$$A^{-1} = \frac{1}{\det A} C^T.$$

Proof. Let us find the product AC^T. The diagonal entry number j is obtained by multiplying jth row of a by jth column of a (*i.e.*, jth row of C), so

$$(AC^T)_{j,j} = a_{j,1}C_{j,1} + a_{j,2}C_{j,2} + ...+ a_{j,n}C_{j,n} = \det A,$$

by the cofactor expansion formula.

To get the off diagonal terms we need to multiply jth row of A by kth column of C^T, $j \neq k$, to get

$$a_{j,1}C_{k,1} + a_{j,2}C_{k,2} + ...+ a_{j,n}C_{k,n}.$$

It follows from the cofactor expansions formula (expanding in kth row) that this is the determinant of the matrix obtained from a by replacing row number k by the row number j (and leaving all other rows as they were). But the rows j and k of this matrix coincide, so the determinant is 0. So, all off–diagonal entries of AC^T are zeroes (and all diagonal ones equal det A), thus

$$AC^T = (\det A)\, I.$$

That means that the matrix $\frac{1}{\det A}\, C^T$ is a right inverse of A, and since a is square, it is the inverse. Recalling that for an invertible matrix A the equation $Ax = b$ has a unique solution

$$x = A^{-1}\, b = \frac{1}{\det A}\, C^T b.$$

We get the following corollary of the above theorem.

Corollary. (Cramer's rule). For an invertible matrix a the entry number *k of the solution of the equation* $Ax = b$ *is given by the formula*

$$x_k = \frac{\det B_k}{\det A},$$

where the matrix B_k *is obtained from a by replacing column number k of A by the vector b.*

Some applications of the cofactor formula for the inverse. Example (Inverting 2 × 2 matrices). The cofactor formula really shines when one needs to invert a 2 × 2 matrix

$$A = \begin{pmatrix} a & b \\ c & d \end{pmatrix}.$$

The cofactors are just entries (1 × 1 matrices), the cofactor matrix is

$$\begin{pmatrix} d & -b \\ -c & a \end{pmatrix},$$

so the inverse matrix A^{-1} is given by the formula

$$A^{-1} = \frac{1}{\det A} \begin{pmatrix} d & -b \\ -c & a \end{pmatrix}.$$

While the cofactor formula for the inverse does not look practical for dimensions higher than 3, it has a great theoretical value, as the examples below illustrate.

Example. (Matrix with integer inverse). Suppose that we want to construct a matrix a with integer entries, such that its inverse also has integer entries (inverting such matrix would make a nice homework problem: no messing with fractions). If det $A = 1$ and its entries are integer, the cofactor formula for inverses implies that A^{-1} also have integer entries.

Note, that it is easy to construct an integer matrix A with det $A = 1$: one should start with a triangular matrix with 1 on the main diagonal, and then apply several row or column replacements (operations of the third type) to make the matrix look generic.

Example (Inverse of a polynomial matrix). Another example is to consider a *polynomial matrix* $A(x)$, *i.e.*, a matrix whose entries are not numbers but polynomials $a_{j,k}(x)$ of the variable x. If det $A(x)$ 1, then the inverse matrix $A^{-1}(x)$ is also a polynomial matrix.

If det $A(x) = p(x) \equiv 0$, it follows from the cofactor expansion that $p(x)$ is a polynomial, so $A^{-1}(x)$ is a has rational entries: moreover, $p(x)$ is a multiple of each denominator.

Minors and Rank

For a matrix A let us consider its $k \times k$ submatrix, obtained by taking k rows and k columns. The determinant of this matrix is called a minor of order k. Note, that an $m \times n$ matrix has $\binom{m}{k}\cdot\binom{n}{k}$ different $k \times k$ submatrices, and so it has $\binom{m}{k}\cdot\binom{n}{k}$ minors of order k.

Theorem. For a non–zero matrix a its rank equals to the maximal integer k such that there exists a non–zero minor of order k.

Proof. Let us first show, that if $k >$ rank A then all minors of order k are 0. Indeed, since the dimension of the column space Ran A is rank $A < k$, any k columns of A are linearly dependent. Therefore, for any $k \times k$ submatrix of A its column are linearly dependent, and so all minors of order k are 0.

To complete the proof we need to show that there exists a non–zero minor of order k = rank A. There can be many such minors, but probably the easiest way to get such a minor is to take pivot rows and pivot column (*i.e.*, rows and columns of the original matrix, containing a pivot). This $k \times k$ submatrix has the same pivots as the original matrix, so it is invertible (pivot in every column and every row) and its determinant is non–zero.

This theorem does not look very useful, because it is much easier to perform row reduction than to compute all minors. However, it is of greattheoretical importance, as the following corollary shows.

Corollary. Let $A = A(x)$ be an $m \times n$ polynomial matrix (i.e., a matrix whose entries are polynomials of x). Then rank $A(x)$ is constant everywhere, except maybe finitely many points.

Proof. Let r be the largest integer such that rank$A(x) = r$ for some x. To show that such r exists, we first try $r = \min\{m, n\}$. If there exists x such that rank $A(x) = r$, we found r. If not, we replace r by $r-1$ and try again. After finitely many steps we either stop or hit 0. So, r exists.

Let x_0 be a point such that rank$A(x_0) = r$, and let M be a minor of order k such that $M(x_0) \neq 0$. Since $M(x)$ is the determinant of a $k \times k$ polynomial matrix, $M(x)$ is a polynomial. Since $M(x_0) \neq 0$, it is not identically zero, so it can be zero only at finitely many points. So, everywhere except maybe finitely many points rank$A(x) \geq r$. But by the definition of r, rank$A(x) \leq r$ for all x.

DETERMINANTS

We have related the question of the invertibility of a square matrix to a question of solutions of systems of linear equations. In some sense, this is unsatisfactory, since it is not simple to nd an answer to either of these questions without a lot of work. We shall relate these two uestions to the question of the determinant of the matrix in question. The task is reduced to checking whether this determinant is zero or non-zero. So what is the determinant?

Let us start with 1 × 1 matrices, of the form

$$A = (a)$$

Note here that $I_1 = (1)$. If a 6 $\neq$ 0, then clearly the matrix A is invertible, with inverse matrix

$$A - 1 = (a - 1)$$

On the other hand, if $a = 0$, then clearly no matrix B can satisfy $AB = BA = I_1$, so that the matrix A is not invertible. We therefore conclude that the value a is a good "determinant" to determine whether the 1 × 1 matrix A is invertible, since the matrix A is invertible if and only if $a \neq 0$.

Let us then agree on the following definition.

Definition. Suppose that

$$A = (a).$$

is a 1 × 1 matrix. We write

$$\det (A) = a,$$

and call this the determinant of the matrix A.

Next, let us turn to 202 0 × 2 matrices, of the form

$$A = \begin{pmatrix} a & b \\ c & d \end{pmatrix}.$$

We shall use elementary row operations to nd out when the matrix A is invertible. So we consider the array

$$(A/I_2) = \begin{pmatrix} a & b & 1 & 0 \\ c & d & 0 & 1 \end{pmatrix},$$

and try to use elementary row operations to reduce the left hand half of the array to I_2. Suppose first of all that $a = c = 0$. Then the array becomes

$$\begin{pmatrix} 0 & b & 1 & 0 \\ 0 & d & 0 & 1 \end{pmatrix},$$

and so it is impossible to reduce the left hand half of the array by elementary row operations to the matrix I_2. Consider next the case a 6 ≠ 0. Multiplying row 2 of the array (1) by a, we obtain

Adding $-c$ times row 1 to row 2, we obtain

$$\begin{pmatrix} a & b & 1 & 0 \\ 0 & ad - bc & -c & a \end{pmatrix}.$$

If $D = ad - bc = 0$, then this becomes

$$\begin{pmatrix} a & b & 1 & 0 \\ 0 & 0 & -c & a \end{pmatrix},$$

and so it is impossible to reduce the left hand half of the array by elementary row operations to the matrix I_2. On the other hand, if $D = ad - bc \neq 0$, then the array (2) can be reduced by elementary row operations to

$$\begin{pmatrix} 1 & 0 & d/D & -b/D \\ 0 & 1 & -c/D & a/D \end{pmatrix},$$

so that

$$A^{-1} = \frac{1}{ad-bc}\begin{pmatrix} d & -b \\ -c & a \end{pmatrix}.$$

Consider nally the case $c \neq 0$. Interchanging rows 1 and 2 of the array (1), we obtain

$$\begin{pmatrix} c & d & 0 & 1 \\ a & b & 1 & 0 \end{pmatrix}.$$

Multiplying row 2 of the array by c, we obtain

$$\begin{pmatrix} c & d & 0 & 1 \\ ac & bc & c & 0 \end{pmatrix}.$$

Adding $-a$ times row 1 to row 2, we obtain

$$\begin{pmatrix} c & d & 0 & 1 \\ 0 & bc-ad & c & -a \end{pmatrix}.$$

Multiplying row 2 by -1, we obtain

$$\begin{pmatrix} c & d & 0 & 1 \\ 0 & ad-bc & -c & a \end{pmatrix}.$$

Again, if $D = ad - bc = 0$, then this becomes

$$\begin{pmatrix} c & d & 0 & 1 \\ 0 & 0 & -c & a \end{pmatrix},$$

and so it is impossible to reduce the left hand half of the array by elementary row operations to the matrix I_2. On the other hand, if $D = ad - bc = 0$, then the array (3) can be reduced by elementary row operations to

$$\begin{pmatrix} 1 & 0 & d/D & -b/D \\ 0 & 1 & -c/D & a/D \end{pmatrix},$$

so that

$$A^{-1} = \frac{1}{ad-bc}\begin{pmatrix} d & -b \\ -c & a \end{pmatrix}.$$

Finally, note that $a = c = 0$ is a special case of $ad - bc = 0$. We therefore conclude that the value $ad - bc$ is a good determinant" to determine whether the 2×2 matrix A is invertible, since the matrix A is invertible if and only if $ad - bc \neq 0$.

Let us then agree on the following definition.

Definition. Suppose that

$$A = \begin{pmatrix} a & b \\ c & d \end{pmatrix}$$

is a 2 × 2 matrix. We write

$$\det(A) = ad - bc,$$

and call this the determinant of the matrix A.

Determinants for Square Matrices of Higher Order

If we attempt to repeat the argument for 2 × 2 matrices to 3 × 3 matrices, then it is very likely that we shall end up in a mess with possibly no firm conclusion. Try the argument on 4 × 4 matrices if you must. Those who have their feet firmly on the ground will try a different approach. Our approach is inductive in nature. In other words, we shall dene the determinant of 2 × 2 matrices in terms of determinants of 1 × 1 matrices, dene the determinant of 3 × 3 matrices in terms of determinants of 2 × 2 matrices, dene the determinant of 4 × 4 matrices in terms of determinants of 3 × 3 matrices,

and so on. Suppose now that we have dened the determinant of $(n - 1) \times (n - 1)$ matrices. Let

$$A = \begin{pmatrix} a_{11} & \cdots & a_{1n} \\ \vdots & & \vdots \\ an_1 & \cdots & a_{nn} \end{pmatrix}$$

be an n matrix. For every $i, j = 1, \ldots, n$, let us delete row i and column j of **A** to obtain the $(n - 1)$ $(n - 1)$ matrix

$$A_{ij} = \begin{pmatrix} a_{11} & \cdots & a_{1(j-1)} & \bullet & a_{1(j+1)} & \cdots & a_{1n} \\ \vdots & & \vdots & \bullet & \vdots & & \vdots \\ a_{(i-1)1} & \cdots & a_{(i-1)(j-1)} & \bullet & a_{(i-1)(j+1)} & \cdots & a_{(i-1)^n} \\ \bullet & \cdots & \bullet & \bullet & \bullet & \cdots & \bullet \\ a_{(i+1)1} & \cdots & a_{(i+1)(j-1)} & \bullet & a_{(j+1)(j+1)} & \cdots & a_{(i+1)_n} \\ \vdots & & \vdots & \vdots & \vdots & & \vdots \\ a_{n1} & \cdots & a_{n(j-1)} & \bullet & a_{n(j+1)} & \cdots & a_{nn} \end{pmatrix}.$$

Here denotes that the entry has been deleted.

Definition. The number $C_{ij} = (-1)^{i+j} \det(A_{ij})$ is called the cofactor of the entry a_{ij} of A. In other words, the cofactor of the entry a_{ij} is obtained from A by first deleting the row and the column containing the entry a_{ij}, then calculating the determinant of the resulting $(n - 1) \times (n - 1)$ matrix, and nally multiplying by a sign $(-1)^{i+j}$.

Note that the entries of A in row i are given by

$$(a_{i1}, \ldots, a_{in}).$$

Definition. By the cofactor expansion of A by row i, we mean the expression

$$\sum_{j=1}^{n} a_{ij} C_{ij} = a_{ij} C_{i1} + \ldots + a_{in} C_{in.}$$

Note that the entries of A in column j are given by

$$\begin{pmatrix} a_{1j} \\ \vdots \\ a_{nj} \end{pmatrix}$$

Definition. By the cofactor expansion of A by column j, we mean the expression

$$\sum_{i=1}^{n} a_{ij}C_{ij} = a_{1j}C_{1j} + \ldots a_{nj}C_{nj}.$$

Proposition. Suppose that A is an $n \times n$ matrix. Then the expressions are all equal and independent of the row or column chosen.

Definition. Suppose that A is an $n \times n$ matrix. We call the common value in the determinant of the matrix A, denoted by $\det(A)$.

Let us check whether this agrees with our earlier definition of the determinant of a 2×2 matrix. Writing

$$A = \begin{pmatrix} a_{11} & a_{12} \\ a_{21} & a_{22} \end{pmatrix},$$

we have

$$C_{11} = a_{22},\ C_{12} = -a_{21},\ C_{21} = -a_{12},\ C_{22} = a_{11}:$$

It follows that

by row 1 : $a_{11}C_{11} + a_{12}C_{12} = a_{11}a_{22} - a_{12}a_{21}$,

by row 2 : $a_{21}C_{21} + a_{22}C_{22} = -a_{21}a_{12} + a_{22}a_{11}$,

by column 1 : $a_{11}C_{11} + a_{21}C_{21} = a_{11}a_{22} - a_{21}a_{12}$,

by column 2 : $a_{12}C_{12} + a_{22}C_{22} = -a_{12}a_{21} + a_{22}a_{11}$:

The four values are clearly equal, and of the form $ad - bc$ as before.

Example. Consider the matrix

$$A = \begin{pmatrix} 2 & 3 & 5 \\ 1 & 4 & 2 \\ 2 & 1 & 5 \end{pmatrix}.$$

Let us use cofactor expansion by row 1. Then

$$C_{11} = (-1)^{1+1} \det \begin{pmatrix} 4 & 2 \\ 1 & 5 \end{pmatrix} = (-1)^2\,(20-2) = 18,$$

$$C_{12} = (-1)^{1+2} \det 1 \begin{pmatrix} 1 & 2 \\ 2 & 5 \end{pmatrix} = (-1)^3\,(5-4) = -1,$$

$$C_{13} = (-1)^{1+3} \det 1 \begin{pmatrix} 1 & 4 \\ 2 & 1 \end{pmatrix} = (-1)^4\,(1-8) = -7,$$

so that

$$\det(A) = a_{11}C_{11} + a_{12}C_{12} + a_{13}C_{13} = 36 - 3 - 35 = -2:$$

Alternatively, let us use cofactor expansion by column 2. Then

$$C_{12} = (-1)^{1+2} \det \begin{pmatrix} 1 & 2 \\ 2 & 5 \end{pmatrix} = (-1)^3(5-4) = -1,$$

$$C_{22} = (-1)^{2+2} \det \begin{pmatrix} 2 & 5 \\ 2 & 5 \end{pmatrix} = (-1)^4(10-10) = 0,$$

$$C_{32} = (-1)^{3+2} \det \begin{pmatrix} 2 & 5 \\ 1 & 2 \end{pmatrix} = (-1)^5(4-5) = 1,$$

so that

$$\det(A) = a_{12}C_{12} + a_{22}C_{22} + a_{32}C_{32} = -3 + 0 + 1 = -2.$$

When using co-factor expansion, we should choose a row or column with as few non-zero entries as possible in order to minimize the calculations.

Example. Consider the matrix

$$A = \begin{pmatrix} 2 & 3 & 0 & 5 \\ 1 & 4 & 0 & 2 \\ 5 & 4 & 8 & 5 \\ 2 & 1 & 0 & 5 \end{pmatrix}.$$

Here it is convenient to use cofactor expansion by column 3, since then

$$\det(A) = a_{13}C_{13} + a_{23}C_{23} + a_{33}C_{33} + a_{43}C_{43} = 8C_{33} = 8(-1)^{3+3} \det \begin{pmatrix} 2 & 3 & 5 \\ 1 & 4 & 2 \\ 2 & 1 & 5 \end{pmatrix} = -16,$$

in view of Example.

Some Simple Observations

In this section, we shall describe two simple observations which follow immediately from the definition of the determinant by cofactor expansion.

Proposition. Suppose that a square matrix A has a zero row or has a zero column. Then det (A) = 0.

$$\det(A) = 0.$$

Proof. We simply use cofactor expansion by the zero row or zero column.

Definition. Consider an $n \times n$ matrix

$$A = \begin{pmatrix} a_{11} & \dots & a_{1n} \\ \vdots & & \vdots \\ an_1 & \dots & a_{nn} \end{pmatrix}.$$

If $a_{ij} = 0$ whenever $i > j$, then A is called an upper triangular matrix. If $a_{ij} = 0$ whenever $i < j$, then A is called a lower triangular matrix. We also say that A is a triangular matrix if it is upper triangular or lower triangular.

Example. The matrix

$$\begin{pmatrix} 1 & 2 & 3 \\ 0 & 4 & 5 \\ 0 & 0 & 6 \end{pmatrix}$$

is upper triangular.

Example. A diagonal matrix is both upper triangular and lower triangular.

Proposition. Suppose that the n × n matrix is triangular.

$$A = \begin{pmatrix} a_{11} & \cdots & a_{1n} \\ \vdots & & \vdots \\ an_1 & \cdots & a_{nn} \end{pmatrix}$$

Then $\det(A) = a_{11}a_{22}, \ldots, a_{nn}$, *the product of the diagonal entries.*

Proof. Let us assume that A is upper triangular for the case when A is lower triangular, change the term left-most column to the term top row in the proof. Using cofactor expansion by the left-most column at each step, we see that

$$\det(\mathrm{A}) = \mathrm{a}_{11} \det \begin{pmatrix} a_{22} & \cdots & a_{2n} \\ \vdots & & \vdots \\ a_{n2} & \cdots & a_{nn} \end{pmatrix} = a_{11}\, a_{22}$$

$$\det \begin{pmatrix} a_{33} & \cdots & a_{3n} \\ \vdots & & \vdots \\ a_{n3} & \cdots & a_{nn} \end{pmatrix} = \ldots\, a_{11}\, a_{22} \ldots a_{nn}$$

as required.

Elementary Row Operations

We now study the eect of elementary row operations on determinants. Recall that the elementary row operations that we consider are: (1) interchanging two rows, (2) adding a multiple of one row to another row, and (3) multiplying one row by a non-zero constant.

Proposition. (ELEMENTARY ROW OPERATIONS) Suppose that A is an $n \times n$ matrix.

(*a*) *Suppose that the matrix B is obtained from the matrix A by interchanging two rows of A. Then det(B) = –det(A).*

(*b*) *Suppose that the matrix B is obtained from the matrix A by adding a multiple of one row of A to another row. Then det(B) = det(A).*

(*c*) *Suppose that the matrix B is obtained from the matrix A by multiplying one row of A by a non-zero constant c. Then det(B) = c det(A).*

Proof. (*a*) The proof is by induction on n. It is easily checked that the result holds when $n = 2$. When $n > 2$, we use co-factor expansion by a third row, say row i. Then

$$\det(B) = \sum_{j=1}^{n} a_{ij}(-1)^{i+j} \det(B_{ij}).$$

Note that the $(n - 1) \times (n - 1)$ matrices B_{ij} are obtained from the matrices A_{ij} by interchanging two rows of A_{ij}, so that $\det(B_{ij}) = -\det(A_{ij})$. It follows that

$$\det(B) = -\sum_{j=1}^{n} a_{ij}(-1)^{i+j} \det(A_{ij}) = -\det(A)$$

as required.

(b) Again, the proof is by induction on n. It is easily checked that the result holds when $n = 2$. When $n > 2$, we use co-factor expansion by a third row, say row i. Then

$$\det(B) = \sum_{j=1}^{n} a_{ij}(-1)^{i+j}\det(B_{ij})$$

Note that the $(n-1) \times (n-1)$ matrices B_{ij} are obtained from the matrices A_{ij} by adding a multiple of one row of A_{ij} to another row, so that $\det(B_{ij}) = \det(A_{ij})$. It follows that

$$\det(B) = \sum_{j=1}^{n} a_{ij}(-1)^{i+j}\det(A_{ij}) = \det(A)$$

as required.

(c) This is simpler. Suppose that the matrix B is obtained from the matrix A by multiplying row i of A by a non-zero constant c. Then

$$\det(B) = \sum_{j=1}^{n} ca_{ij}(-1)^{i+j}\det(B_{ij})$$

Note now that

$$B_{ij} = A_{ij},$$

since row i has been removed respectively from B and A. It follows that

$$\det(B) = \sum_{j=1}^{n} ca_{ij}(-1)^{i+j}\det(A_{ij}) = c\det(A)$$

as required.

In fact, the above operations can also be carried out on the columns of A. More precisely, we have the following result.

Proposition. Suppose that A is an $n \times n$ matrix.

(*a*) *Suppose that the matrix B is obtained from the matrix A by interchanging two columns of A. Then det(B) = – det(A).*

(*b*) *Suppose that the matrix B is obtained from the matrix A by adding a multiple of one column of A to another column. Then det(B) = det(A).*

(*c*) *Suppose that the matrix B is obtained from the matrix A by multiplying one column of A by a non-zero constant c. Then det(B) = c det(A).*

Elementary row and column operations can be combined with cofactor expansion to calculate the determinant of a given matrix. We shall illustrate this point by the following examples.

Example. Consider the matrix

$$A = \begin{pmatrix} 2 & 3 & 2 & 5 \\ 1 & 4 & 1 & 2 \\ 5 & 4 & 4 & 5 \\ 2 & 2 & 0 & 4 \end{pmatrix}$$

Adding –1 times column 3 to column 1, we have

$$\det(A) = \det\begin{pmatrix} 0 & 3 & 2 & 5 \\ 0 & 4 & 1 & 2 \\ 1 & 4 & 4 & 5 \\ 2 & 2 & 0 & 4 \end{pmatrix}.$$

Adding –1/2 times row 4 to row 3, we have

$$\det(A) = \det\begin{pmatrix} 0 & 3 & 2 & 5 \\ 0 & 4 & 1 & 2 \\ 0 & 3 & 4 & 3 \\ 2 & 2 & 0 & 4 \end{pmatrix}.$$

Using cofactor expansion by column 1, we have

$$\det(A) = 2(-1)^{4+1} \det\begin{pmatrix} 3 & 2 & 5 \\ 4 & 1 & 2 \\ 3 & 4 & 3 \end{pmatrix} = -2 \det\begin{pmatrix} 3 & 2 & 5 \\ 4 & 1 & 2 \\ 3 & 4 & 3 \end{pmatrix}.$$

Adding –1 times row 1 to row 3, we have

$$\det(A) = -2 \det\begin{pmatrix} 3 & 2 & 5 \\ 4 & 1 & 2 \\ 0 & 2 & -2 \end{pmatrix}.$$

Adding 1 times column 2 to column 3, we have

$$\det(\underline{A}) = -2 \det\begin{pmatrix} 3 & 2 & 7 \\ 4 & 1 & 3 \\ 0 & 2 & 0 \end{pmatrix}.$$

Using cofactor expansion by row 3, we have

$$\det(A) = -2.2(-1)^{3+2} \det\begin{pmatrix} 3 & 7 \\ 4 & 3 \end{pmatrix} = \det\begin{pmatrix} 3 & 7 \\ 4 & 3 \end{pmatrix}.$$

Using the formula for the determinant of 2 × 2 matrices, we conclude that

$$\det(A) = 4(9 - 28) = -76.$$

Let us start again and try a different way. Dividing row 4 by 2, we have

$$\det(A) = 2 \det\begin{pmatrix} 2 & 3 & 2 & 5 \\ 1 & 4 & 1 & 2 \\ 5 & 4 & 4 & 5 \\ 1 & 1 & 0 & 2 \end{pmatrix}.$$

Adding –1 times row 4 to row 2, we have

$$\det(A) = 2 \det = \begin{pmatrix} 2 & 3 & 2 & 5 \\ 0 & 3 & 1 & 0 \\ 5 & 4 & 4 & 5 \\ 1 & 1 & 0 & 2 \end{pmatrix}$$

Adding –3 times column 3 to column 2, we have

$$\det(A) = 2\det\begin{pmatrix} 2 & -3 & 2 & 5 \\ 0 & 0 & 1 & 0 \\ 5 & -8 & 4 & 5 \\ 1 & 1 & 0 & 2 \end{pmatrix}.$$

Using cofactor expansion by row 2, we have

$$\det(A) = 2.1(-1)^{2+3}\det\begin{pmatrix} 2 & -3 & 5 \\ 5 & -8 & 5 \\ 1 & 1 & 2 \end{pmatrix} = -2\det\begin{pmatrix} 2 & -3 & 5 \\ 5 & -8 & 5 \\ 1 & 1 & 2 \end{pmatrix}.$$

Adding –2 times row 3 to row 1, we have

$$\det(A) = -2\det\begin{pmatrix} 0 & -5 & 1 \\ 5 & -8 & 5 \\ 1 & 1 & 2 \end{pmatrix}$$

Adding –5 times row 3 to row 2, we have

$$\det(A) = -2\det\begin{pmatrix} 0 & -5 & 1 \\ 0 & -13 & -5 \\ 1 & 1 & 2 \end{pmatrix}.$$

Using cofactor expansion by column 1, we have

$$\det(A) = -2\ .\ 1(-1)^{3+1}\det\begin{pmatrix} -5 & 1 \\ -13 & -5 \end{pmatrix} = -2\det\begin{pmatrix} -5 & 1 \\ -13 & -5 \end{pmatrix}.$$

Using the formula for the determinant of 22 matrices, we conclude that

$$\det(A) = -2(25 + 13) = -76.$$

Example. Consider the matrix

$$A = \begin{pmatrix} 2 & 1 & 0 & 1 & 3 \\ 2 & 3 & 1 & 2 & 5 \\ 4 & 7 & 2 & 3 & 7 \\ 1 & 0 & 1 & 1 & 3 \\ 2 & 1 & 0 & 2 & 0 \end{pmatrix}.$$

Here we have the least number of non-zero entries in column 3, so let us work to get more zeros into this column. Adding –1 times row 4 to row 2, we have

$$\det(A) = \det\begin{pmatrix} 2 & 1 & 0 & 1 & 3 \\ 2 & 3 & 1 & 2 & 5 \\ 4 & 7 & 2 & 3 & 7 \\ 1 & 0 & 1 & 1 & 3 \\ 2 & 1 & 0 & 2 & 0 \end{pmatrix}.$$

Adding –2 times row 4 to row 3, we have

$$\det(A) = \det\begin{pmatrix} 2 & 1 & 0 & 1 & 3 \\ 1 & 3 & 0 & 1 & 2 \\ 2 & 7 & 0 & 1 & 1 \\ 1 & 0 & 1 & 1 & 3 \\ 2 & 1 & 0 & 2 & 0 \end{pmatrix}.$$

Using cofactor expansion by column 3, we have

$$\det(A) = 1(-1)^{4+3}\det\begin{pmatrix} 2 & 1 & 1 & 3 \\ 1 & 3 & 1 & 2 \\ 2 & 7 & 1 & 1 \\ 2 & 1 & 2 & 0 \end{pmatrix} = -\det\begin{pmatrix} 2 & 1 & 1 & 3 \\ 1 & 3 & 1 & 2 \\ 2 & 7 & 1 & 1 \\ 2 & 1 & 2 & 0 \end{pmatrix}.$$

Adding −1 times column 3 to column 1, we have

$$\det(A) = -\det\begin{pmatrix}1 & 1 & 1 & 3\\ 0 & 3 & 1 & 2\\ 1 & 7 & 1 & 1\\ 0 & 1 & 2 & 0\end{pmatrix}.$$

Adding −1 times row 1 to row 3, we have

$$\det(A) = -\det\begin{pmatrix}1 & 1 & 1 & 3\\ 0 & 3 & 1 & 2\\ 0 & 6 & 0 & -2\\ 0 & 1 & 2 & 0\end{pmatrix}.$$

Using cofactor expansion by column 1, we have

$$\det(A) = -1(-1)^{1+1}\det\begin{pmatrix}3 & 1 & 2\\ 6 & 0 & -2\\ 1 & 2 & 0\end{pmatrix} = -\det\det(A) = -\det\begin{pmatrix}3 & 1 & 2\\ 6 & 0 & -2\\ 1 & 2 & 0\end{pmatrix}.$$

Adding 1 times row 1 to row 2, we have

$$\det(A) = -\det\begin{pmatrix}3 & 1 & 2\\ 9 & 1 & 0\\ 1 & 2 & 0\end{pmatrix}.$$

Using cofactor expansion by column 3, we have

$$\det(A) = -2(-1)^{1+3}\det\begin{pmatrix}9 & 1\\ 1 & 2\end{pmatrix} = -2\det\begin{pmatrix}9 & 1\\ 1 & 2\end{pmatrix}.$$

Using the formula for the determinant of 2 × 2 matrices, we conclude that $\det(A) = -2(18-1) = -34$.

Example. Consider the matrix

$$A = \begin{pmatrix}1 & 0 & 2 & 4 & 1 & 0\\ 2 & 4 & 5 & 7 & 6 & 2\\ 4 & 6 & 1 & 9 & 2 & 1\\ 3 & 5 & 0 & 1 & 2 & 5\\ 2 & 4 & 5 & 3 & 6 & 2\\ 1 & 0 & 2 & 5 & 1 & 0\end{pmatrix}.$$

Here note that rows 1 and 6 are almost identical. Adding −1 times row 1 to row 6, we have

$$\det(A) = \det\begin{pmatrix}1 & 0 & 2 & 4 & 1 & 0\\ 2 & 4 & 5 & 7 & 6 & 2\\ 4 & 6 & 1 & 9 & 2 & 1\\ 3 & 5 & 0 & 1 & 2 & 5\\ 2 & 4 & 5 & 3 & 6 & 2\\ 0 & 0 & 0 & 1 & 0 & 0\end{pmatrix}.$$

Adding −1 times row 5 to row 2, we have

$$\det(A) = \det\begin{pmatrix} 1 & 0 & 2 & 4 & 1 & 0 \\ 0 & 0 & 0 & 4 & 0 & 0 \\ 4 & 6 & 1 & 9 & 2 & 1 \\ 3 & 5 & 0 & 1 & 2 & 5 \\ 2 & 4 & 5 & 3 & 6 & 2 \\ 0 & 0 & 0 & 1 & 0 & 0 \end{pmatrix}.$$

Adding –4 times row 6 to row 2, we have

$$\det(A) = \det\begin{pmatrix} 1 & 0 & 2 & 4 & 1 & 0 \\ 0 & 0 & 0 & 0 & 0 & 0 \\ 4 & 6 & 1 & 9 & 2 & 1 \\ 3 & 5 & 0 & 1 & 2 & 5 \\ 2 & 4 & 5 & 3 & 6 & 2 \\ 0 & 0 & 0 & 1 & 0 & 0 \end{pmatrix}.$$

It follows from Proposition 3*B* that $\det(A) = 0$.

Further Properties of Determinants

Definition. Consider the $n \times n$ matrix

$$A = \begin{pmatrix} a_{11} & \dots & a_{1n} \\ \vdots & & \vdots \\ a_{n1} & \dots & a_{nn} \end{pmatrix}.$$

By the transpose A^t of A, we mean the matrix obtained from A by transposing rows and columns.

$$A^t = \begin{pmatrix} a_{11} & \dots & a_{n1} \\ \vdots & & \vdots \\ a_{1n} & \dots & a_{nn} \end{pmatrix}.$$

Example. Consider the matrix

$$A = \begin{pmatrix} 1 & 2 & 3 \\ 4 & 5 & 6 \\ 7 & 8 & 9 \end{pmatrix}.$$

Then

$$A^t = \begin{pmatrix} 1 & 4 & 7 \\ 2 & 5 & 8 \\ 3 & 6 & 9 \end{pmatrix}.$$

Recall that determinants of 2×2 matrices depend on determinants of 1×1 matrices, in turn, determinants of 3×3 matrices depend on determinants of 2×2 matrices, and so on.

It follows that determinants of $n \times n$ matrices ultimately depend on determinants of 1×1 matrices. Note now that transposing a 1×1 matrix does not aect its determinant (why?). The result below follows in view of Proposition. *For every* $n \times n$ *matrix A, we have* $\det(A_t) = \det(A)$.

Example. We have

$$\det = \begin{pmatrix} 2 & 2 & 4 & 1 & 2 \\ 1 & 3 & 7 & 0 & 1 \\ 0 & 1 & 2 & 1 & 0 \\ 1 & 2 & 3 & 1 & 2 \\ 3 & 5 & 7 & 3 & 0 \end{pmatrix} = \det \begin{pmatrix} 2 & 1 & 0 & 1 & 3 \\ 2 & 3 & 1 & 2 & 5 \\ 4 & 7 & 2 & 3 & 7 \\ 1 & 0 & 1 & 1 & 3 \\ 2 & 1 & 0 & 2 & 0 \end{pmatrix} = -34.$$

Next, we shall study the determinant of a product. We shall sketch a proof of the following important result

Proposition. For every $n \times n$ matrices A and B, we have det(AB) = det(A) det(*B*).

Proposition. Suppose that the $n \times n$ matrix A is invertible. Then

$$\det(A^{-1}) = \frac{1}{\det(A)}.$$

Proof. In view of Propositions 3*G* and 3*C*, we have $\det(A)\det(A^{-1}) = \det(I_n) = 1$. The result follows immediately. Finally, the main reason for studying determinants, as outlined in the introduction, is summarized by the following result.

Proposition. Suppose that A is an $n \times n$ matrix. Then A is invertible if and only if $\det(A) \neq 0$.

Proof. Suppose that A is invertible. Then $\det(A) \neq 0$ follows immediately from Proposition. Suppose now that $\det(A) \neq 0$. Let us now reduce A by elementary row operations to reduced row echelon form B. Then there exist a finite sequence $E_1, \ldots, E_k$ of elementary $n \times n$ matrices such that

$$B = E_k, \ldots, E_1 A$$

It follows from Proposition that

$$\det(B) = \det(E_k), \ldots, \det(E_1)\det(A)$$

Recall that all elementary matrices are invertible and so have non-zero determinants. It follows that $\det(B) \neq 0$, so that B has no zero rows by Proposition. Since B is an $n \times n$ matrix in reduced row echelon form, it must be In. We therefore conclude that A is row equivalent to *In*. It now follows from Proposition that A is invertible. Combining Propositions, we have the following result.

Proposition. In the notation of Proposition, the following statements are equivalent:

(a) The matrix A is invertible.

(b) The system $Ax = 0$ of linear equations has only the trivial solution.

(c) The matrices A and In are row equivalent.

(d) The system $Ax = b$ of linear equations is soluble for every n 1 matrix b.

(e) The determinant $\det(A) \neq 0$.

Application to Curves and Surfaces

A special case of Proposition states that a homogeneous system of n linear equations in n variables has a non-trivial solution if and only if the determinant if the coefficient matrix is equal to zero. In this section, we shall use this to solve some problems in geometry. We illustrate our ideas by a few simple examples.

Example. Suppose that we wish to determine the equation of the unique line on the xy-plane that passes through two distinct given points (x_1, y_1) and (x_2, y_2). The equation of a line on the xy-plane is of the form $ax + by + c = 0$. Since the two points lie on the line, we must have $ax_1 + by_1 + c = 0$ and $ax_2 + by_2 + c = 0$. Hence $xa + yb + c = 0$,

$$x_1a + y_1b + c = 0,$$
$$x_2a + y_2b + c = 0.$$

Written in matrix notation, we have

$$\begin{pmatrix} x & y & 1 \\ x_1 & y_1 & 1 \\ x_2 & y_2 & 1 \end{pmatrix}\begin{pmatrix} a \\ b \\ c \end{pmatrix} = \begin{pmatrix} 0 \\ 0 \\ 0 \end{pmatrix}.$$

Clearly there is a non-trivial solution (a, b, c) to this system of linear equations, and so we must have

$$\det\begin{pmatrix} x & y & 1 \\ x_1 & y_1 & 1 \\ x_2 & y_2 & 1 \end{pmatrix} = 0.$$

the equation of the line required.

Example. Suppose that we wish to determine the equation of the unique circle on the xy-plane that passes through three distinct given points (x_1, y_1), (x_2, y_2) and (x_3, y_3), not all lying on a straight line. The equation of a circle on the xy-plane is of the form

$$a(x_2 + y_2) + bx + cy + d = 0.$$

Since the three points lie on the circle, we must have

$$a(a_1^2 + y_1^2) + bx_1 + cy_1 + d = 0, \quad a(x_2^2 + y_2^2) + bx_2 + cy_2 + d = 0$$

and

$$a(x_3^2 + y_3^2) + bx_3 + cy_3 + d = 0$$

Hence

$$(x_2 + y_2)a + xb + yc + d = 0,$$
$$(x_1^2 + y_1^2)a + x_1b + y_1c + d = 0$$
$$(x_2^2 + y_2^2)a + x_2b + y_2c + d, 0$$
$$(x_3^2 + y_3^2)a + x_3b + y_3c + d = 0$$

Written in matrix notation, we have

$$\begin{pmatrix} x^2 + y^2 & x & y & 1 \\ y_1^2 + y_1^2 & x_1 & y_1 & 1 \\ x_2^2 + y_2^2 & x_2 & y_2 & 1 \\ x_3^2 + y_3^2 & x_3 & y_3 & 1 \end{pmatrix}\begin{pmatrix} a \\ b \\ c \\ d \end{pmatrix} = \begin{pmatrix} 0 \\ 0 \\ 0 \\ 0 \end{pmatrix}.$$

Clearly there is a non-trivial solution (a, b, c, d) to this system of linear equations, and so we must have

$$\det\begin{pmatrix} x^2+y^2 & x & y & 1 \\ x_1^2+y_1^2 & x_1 & y_1 & 1 \\ x_2^2+y_2^2 & x_2 & y_2 & 1 \\ x_3^2+y_3^2 & x_3 & y_3 & 1 \end{pmatrix} = 0,$$

the equation of the circle required.

Example. Suppose that we wish to determine the equation of the unique plane in 3-space that passes through three distinct given points (x_1, y_1, z_1), (x_2, y_2, z_2) and (x_3, y_3, z_3), not all lying on a straight line. The equation of a plane in 3-space is of the form $ax + by + cz + d = 0$. Since the three points lie on the plane, we must have $ax_1 + by_1 + cz_1 + d = 0$, $ax_2 + by_2 + cz_2 + d = 0$, and $ax_3 + by_3 + cz_3 + d = 0$. Hence

$$xa + yb + zc + d = 0,$$
$$x_1a + y_1b + z_1c + d = 0,$$
$$x_2a + y_2b + z_2c + d = 0,$$
$$x_3a + y_3b + z_3c + d = 0:$$

Written in matrix notation, we have

$$\begin{pmatrix} x^2+y^2 & x & y & 1 \\ x_1^2+y_1^2 & x_1 & y_1 & 1 \\ x_2^2+y_2^2 & x_2 & y_2 & 1 \\ x_3^2+y_3^2 & x_3 & y_3 & 1 \end{pmatrix}\begin{pmatrix} a \\ b \\ c \\ d \end{pmatrix} = \begin{pmatrix} 0 \\ 0 \\ 0 \\ 0 \end{pmatrix}.$$

Clearly there is a non-trivial solution (a, b, c, d) to this system of linear equations, and so we must have the equation of the plane required.

$$\det\begin{pmatrix} x & y & z & 1 \\ x_1 & y_1 & z_1 & 1 \\ z_2 & y_2 & z_2 & 1 \\ x_3 & y_3 & z_3 & 1 \end{pmatrix} = 0,$$

Example. Suppose that we wish to determine the equation of the unique sphere in 3-space that passes through four distinct given points (x_1, y_1, z_1), (x_2, y_2, z_2), (x_3, y_3, z_3) and (x_4, y_4, z_4), not all lying on a plane. The equation of a sphere in 3-space is of the form

$$a(x_2 + y_2 + z_2) + bx + cy + dz + e = 0.$$

Since the four points lie on the sphere, we must have

$$a\left(x_1^2 + y_1^2 + z_1^2\right) + bx_1 + cy_1 + dz_1 + e = 0,$$

$$a\left(x_2^2 + y_2^2 + z_2^2\right) + bx_2 + cy_2 + dz_2 + e = 0,$$

$$a\left(x_3^2 + y_3^2 + z_3^2\right) + bx_3 + cy_3 + dz_3 + e = 0,$$

$$a\left(x_4^2+y_4^2+z_4^2\right)+bx_4+cy_4+dz_4+e=0,$$

Hence

$$(x^2+y^2+z^2)a+xb+yc+zd+e=0,$$
$$\left(x_1^2+y_1^2+z_1^2\right)a+x_1b+y_1c+z_1d+e=0,$$
$$\left(x_2^2+y_2^2+z_2^2\right)a+x_2b+y_2c+z_2d+e=0,$$
$$\left(x_3^2+y_3^2+z_3^2\right)a+x_3b+y_3c+z_3d+e=0,$$
$$\left(x_4^2+y_4^2+z_4^2\right)a+x_4b+y_4c+z_4d+e=0.$$

Written in matrix notation, we have

$$\begin{pmatrix} x^2+y^2+z^2 & x & y & z & 1 \\ x_1^2+y_1^2+z_1^2 & x_1 & y_1 & z_1 & 1 \\ x_2^2+y_2^2+z_2^2 & x_2 & y_2 & z_2 & 1 \\ x_3^2+y_3^2+z_3^2 & x_3 & y_3 & z_3 & 1 \\ x_4^2+y_4^2+z_4^2 & x_4 & y_4 & z_4 & 1 \end{pmatrix}\begin{pmatrix} a \\ b \\ c \\ d \\ e \end{pmatrix}=\begin{pmatrix} 0 \\ 0 \\ 0 \\ 0 \\ 0 \end{pmatrix}.$$

Clearly there is a non-trivial solution (*a, b, c, d, e*) to this system of linear equations, and so we must have

$$\det\begin{pmatrix} x_2+y_2+z_2 & x & y & z & 1 \\ x_1^2+y_1^2+z_1^2 & x_1 & y_1 & z_1 & 1 \\ x_2^2+y_2^2+z_2^2 & x_2 & y_2 & z_2 & 1 \\ x_3^2+y_3^2+z_3^2 & x_3 & y_3 & z_3 & 1 \\ x_4^2+y_4^2+z_4^2 & x_4 & y_4 & z_4 & 1 \end{pmatrix}=0,$$

the equation of the sphere required.

Some Useful Formulas

In this section, we shall discuss two very useful formulas which involve determinants only. The rst one enables us to nd the inverse of a matrix, while the second one enables us to solve a system of linear equations. The interested reader is referred to Section 3.8 for proofs. Recall rst of all that for any $n \times n$ matrix

$$A=\begin{pmatrix} a_{11} & \dots & a_{1n} \\ \vdots & & \vdots \\ a_{n1} & \dots & a_{nn} \end{pmatrix},$$

the number $C_{ij}=(-1)^{i+j}\det(A_{ij})$ is called the cofactor of the entry a_{ij}, and the $(n-1)$ $(n-1)$ matrix

$$A_{ij} = \begin{pmatrix} a_{11} & \cdots & a_{1(j-1)} & \bullet & a_{1(j+1)} & \cdots & a_{1n} \\ \vdots & & \vdots & \vdots & \vdots & & \vdots \\ a_{(i-1)1} & \cdots & a_{(i-1)(j-1)} & \bullet & a_{(i-1)(j+1)} & \cdots & a(i-1)_n \\ \bullet & \cdots & \bullet & \bullet & \bullet & \cdots & a(i+1)_n \\ a_{(i+1)1} & \cdots & a_{(i-1)(i-1)} & \bullet & a_{(i+1)(i+1)} & \cdots & a(i+1)_n \\ \vdots & & \vdots & \vdots & \vdots & & \vdots \\ a_{n1} & \cdots & a_{n(j-1)} & \bullet & a_{n(j+1)} & \cdots & a_{nn} \end{pmatrix}$$

is obtained from A by deleting row i and column j, here denotes that the entry has been deleted.

Definition. The $n \times n$ matrix

$$adj\ (A) = \begin{pmatrix} C_{11} & \cdots & C_{n1} \\ \vdots & & \vdots \\ C_{1n} & \cdots & C_{nn} \end{pmatrix}$$

is called the adjoint of the matrix A.

Remark. Note that $adj(A)$ is obtained from the matrix A rst by replacing each entry of A by its cofactor and then by transposing the resulting matrix.

Proposition. Suppose that the $n \times n$ matrix A is invertible. Then

$$A^{-1} = \frac{1}{\det(A)} adj(A).$$

Example. Consider the matrix

$$A = \begin{pmatrix} 1 & -1 & 0 \\ 0 & 1 & 2 \\ 2 & 0 & 3 \end{pmatrix}.$$

Then

$$adj(A) = \begin{pmatrix} \det\begin{pmatrix} 1 & 2 \\ 0 & 3 \end{pmatrix} & -\det\begin{pmatrix} -1 & 0 \\ 0 & 3 \end{pmatrix} & \det\begin{pmatrix} -1 & 0 \\ 1 & 2 \end{pmatrix} \\ -\det\begin{pmatrix} 0 & 2 \\ 2 & 3 \end{pmatrix} & \det\begin{pmatrix} 1 & 0 \\ 2 & 3 \end{pmatrix} & -\det\begin{pmatrix} 1 & 0 \\ 0 & 2 \end{pmatrix} \\ \det\begin{pmatrix} 0 & 1 \\ 2 & 0 \end{pmatrix} & -\det\begin{pmatrix} 1 & -1 \\ 2 & 0 \end{pmatrix} & \det\begin{pmatrix} 1 & -1 \\ 0 & 1 \end{pmatrix} \end{pmatrix} = \begin{pmatrix} 3 & 3 & -2 \\ 4 & 3 & -2 \\ -2 & -2 & 1 \end{pmatrix}.$$

On the other hand, adding 1 times column 1 to column 2 and then using cofactor expansion on row 1, we have

$$\det(A) = \det\begin{pmatrix} 1 & -1 & 0 \\ 0 & 1 & 2 \\ 2 & 0 & 3 \end{pmatrix} = \det\begin{pmatrix} 1 & 0 & 0 \\ 0 & 1 & 2 \\ 2 & 2 & 3 \end{pmatrix} = \det\begin{pmatrix} 1 & 2 \\ 2 & 3 \end{pmatrix} = -1.$$

It follows that

$$A^{-1} = \begin{pmatrix} -3 & -3 & 2 \\ -4 & -3 & 2 \\ 2 & 2 & -1 \end{pmatrix}.$$

Next, we turn our attention to systems of n linear equations in n unknowns, of the form

$$an_1x_1 + \ldots a_{nn}x_n = b_n,$$

represented in matrix notation in the form

$$Ax = b,$$

where

$$A = \begin{pmatrix} C_{11} & \cdots & C_{1n} \\ \vdots & & \vdots \\ C_{n1} & \cdots & C_{nn} \end{pmatrix} \text{ and } b = \begin{pmatrix} b_1 \\ \vdots \\ b_n \end{pmatrix}$$

represent the coefficients and

$$x = \begin{pmatrix} x_1 \\ \vdots \\ x_n \end{pmatrix}$$

represents the variables. For every $j = 1, \ldots, k$, write in other words, we replace column j of the matrix A by the column b.

$$A_j(b) = \begin{pmatrix} a_{11} & \cdots & a_{1(j-1)} & b_1 & a_{1(j+1)} & \cdots & a_{1n} \\ \vdots & & \vdots & \vdots & \vdots & & \vdots \\ a_{n1} & \cdots & a_{n(j-1)} & b_n & a_{n(j+1)} & \cdots & a_{nn} \end{pmatrix};$$

Proposition. (Cramer's Rule) *Suppose that the matrix A is invertible. Then the unique solution of the system Ax = b, where A, x and b are given by equation, is given by*

$$x_1 = \frac{\det(A_1(b))}{\det(A)}, \ \ldots \ x_n = \frac{\det(A_n(b))}{\det(A)},$$

where the matrices $A_1(b)$, ..., $A_1(b)$.

Example. Consider the system $Ax = b$, where

$$A = \begin{pmatrix} 1 & -1 & 0 \\ 0 & 1 & 2 \\ 2 & 0 & 3 \end{pmatrix} \text{ and } b = \begin{pmatrix} 1 \\ 2 \\ 3 \end{pmatrix}.$$

Recall that $\det(A) = -1$. By Cramer's rule, we have

$$x_1 = \frac{\det\begin{pmatrix} 1 & -1 & 0 \\ 2 & 1 & 2 \\ 3 & 0 & 3 \end{pmatrix}}{\det(A)} = -3, \quad x_2 = \frac{\det\begin{pmatrix} 1 & 1 & 0 \\ 0 & 2 & 2 \\ 2 & 3 & 3 \end{pmatrix}}{\det(A)} = -4,$$

$$x_3 = \frac{\det\begin{pmatrix}1 & -1 & 1\\ 0 & 1 & 2\\ 2 & 0 & 3\end{pmatrix}}{\det(A)} = 3,$$

Let us check our calculations. Recall from Example that

$$A^{-1} = \begin{pmatrix}-3 & -3 & 2\\ -4 & -3 & 2\\ 2 & 2 & -1\end{pmatrix}.$$

We therefore have

$$\begin{pmatrix}x_1\\ x_2\\ x_3\end{pmatrix} = \begin{pmatrix}-3 & -3 & 2\\ -4 & -3 & 2\\ 2 & 2 & -1\end{pmatrix}\begin{pmatrix}1\\ 2\\ 3\end{pmatrix} = \begin{pmatrix}-3\\ -4\\ 3\end{pmatrix}.$$

Further Discussion

In this section, we shall first discuss a definition of the determinant in terms of permutations. In order to do so, we need to make a digression and discuss first the rudiments of permutations on non-empty finite sets.

Definition. Let X be a non-empty finite set. A permutation ϕ on X is a function : $X \to X$ which is one-to-one and onto. If $x \in X$, we denote by x the image of x under the permutation. It is not dicult to see that if : $\phi\, X \to X$ and : $X \to X$ are both permutations on X, then : $X \to X$, dened by $x\phi\psi = (x\phi)\psi$ for every $x \in X$ so that is followed by, is also a permutation on X.

Remark. Note that we use the notation x instead of our usual notation (x) to denote the image of x under . Note also that we write to denote the composition. We shall do this only for permutations. The reasons will become a little clearer later in the discussion.

Since the set X is non-empty and finite, we may assume, without loss of generality, that it is $\{1, 2, \ldots, n\}$, where $n \in \mathbb{N}$. We now let S_n denote the set of all permutations on the set $\{1, 2, \ldots, n\}$. In other words, Sn denotes the collection of all functions from $(1, 2, \ldots, n)$ to $\{1, 2, \ldots, n\}$ that are both one-to-one and onto.

Proposition. For every $n \in \mathbb{N}$, the set S_n has $n!$ elements.

Proof. There are n choices for 1ϕ. For each such choice, there are $(n - 1)$ choices left for 2ϕ. And so on.

To represent particular elements of S_n, there are various notations. For example, we can use the notation

$$\begin{pmatrix}1 & 2 & \ldots & n\\ 1\phi & 2\phi & \ldots & n\phi\end{pmatrix}$$

to denote the permutation ϕ.

Example. In S_4,

$$\begin{pmatrix}1 & 2 & 3 & 4\\ 2 & 4 & 1 & 3\end{pmatrix}$$

denotes the permutation , where 1 = 2, 2 = 4, 3 = 1 and 4 = 3. On the other hand, the reader can easily check that

$$\begin{pmatrix}1 & 2 & 3 & 4\\ 2 & 4 & 1 & 3\end{pmatrix}\begin{pmatrix}1 & 2 & 3 & 4\\ 3 & 2 & 4 & 1\end{pmatrix}=\begin{pmatrix}1 & 2 & 3 & 4\\ 2 & 1 & 3 & 4\end{pmatrix}.$$

A more convenient way is to use the cycle notation. The permutations

$$\begin{pmatrix}1 & 2 & 3 & 4\\ 2 & 4 & 1 & 3\end{pmatrix} \text{ and } \begin{pmatrix}1 & 2 & 3 & 4\\ 3 & 2 & 4 & 1\end{pmatrix}$$

can be represented respectively by the cycles (1 2 4 3) and (1 3 4). Here the cycle (1 2 4 3) gives the information $1\phi = 2$, $2\phi = 4$, $4\phi = 3$ and $3\phi = 1$. Note also that in the latter case, since the image of 2 is 2, it is not necessary to include this in the cycle. Furthermore, the information

$$\begin{pmatrix}1 & 2 & 3 & 4\\ 2 & 4 & 1 & 3\end{pmatrix}\begin{pmatrix}1 & 2 & 3 & 4\\ 3 & 2 & 4 & 1\end{pmatrix}=\begin{pmatrix}1 & 2 & 3 & 4\\ 2 & 1 & 3 & 4\end{pmatrix}.$$

can be represented in cycle notation by (1 2 4 3)(1 3 4) = (1 2). We also say that the cycles (1 2 4 3), (1 3 4) and (1 2) have lengths 4, 3 and 2 respectively.

Example. In S_6, the permutation

$$\begin{pmatrix}1 & 2 & 3 & 4 & 5 & 6\\ 2 & 4 & 1 & 3 & 6 & 5\end{pmatrix}$$

can be represented in cycle notation as (1 2 4 3)(5 6).

Example. In S_4 or S_6, we have (1 2 4 3) = (1 2)(1 4)(1 3).

The last example motivates the following important idea.

Definition. Suppose that $n \in \mathbb{N}$. A permutation in S_n that interchanges two numbers among the elements of {1, 2, ..., n} and leaves all the others unchanged is called a transposition. *Remark.* It is obvious that a transposition can be represented by a 2-cycle, and is its own inverse. Two cycles $(x_1, x_2, \ldots, x_k)$ and $(y_1, y_2, \ldots, y_l)$ in S_n are said to be disjoint if the elements $x_1, \ldots, x_k, y_1, \ldots, y_l$ are all different. The interested reader may try to prove the following result.

Proposition. Suppose that $n \in \mathbb{N}$.

(a) Every permutation in S_n can be written as a product of disjoint cycles.

(b) For every subset $(x_1, x_2, \ldots, x_k)$ of the set {1, 2, ..., n}, where the elements $x_1, x_2, \ldots, x_k$ are distinct, the cycle $(x_1, x_2, \ldots, x_k)$ satises

$$(x_1\, x_2, \ldots, x_k) = (x_1, x_2)(x_1, x_3), \ldots, (x_1, xk),$$

in other words, every cycle can be written as a product of transpositions.

(c) Consequently, every permutation in S_n can be written as a product of transpositions.

Example. In S_9, the permutation

$$\begin{pmatrix}1 & 2 & 3 & 4 & 5 & 6 & 7 & 8 & 9\\ 3 & 2 & 5 & 1 & 7 & 8 & 4 & 9 & 6\end{pmatrix}$$

can be written in cycle notation as (1 3 5 7 4)(6 8 9). By Theorem 3*P*(*b*), we have

$$(1\ 3\ 5\ 7\ 4) = (1\ 3)(1\ 5)(1\ 7)(1\ 4) \text{ and } (6\ 8\ 9) = (6\ 8)(6\ 9).$$

Hence the permutation can be represented by (1 3)(1 5)(1 7)(1 4)(6 8)(6 9).

Definition. Suppose that $n \in \mathbb{N}$. Then a permutation in S_n is said to be even if it is representable as the product of an even number of transpositions and odd if it is representable as the product of an odd number of transpositions. Furthermore, we write

$$\varepsilon(\phi) = \begin{cases} +1 \text{ if } \phi \text{ is even} \\ -1 \text{ if } f \text{ is odd.} \end{cases}$$

Remark. It can be shown that no permutation can be simultaneously odd and even.

We are now in a position to dene the determinant of a matrix. Suppose that

$$A = \begin{pmatrix} a_{11} & \cdots & a_{1n} \\ \vdots & & \vdots \\ a_{n1} & \cdots & a_{nn} \end{pmatrix}$$

is an $n \times n$ matrix.

Definition. By an elementary product from the matrix A, we mean the product of n entries of A, no two of which are from the same row or same column.

It follows that any such elementary product must be of the form

$$a_1(1\phi)a_2(2\phi) \ldots a_n(nf),$$

where ϕ is a permutation in S_n.

Definition. By the determinant of an $n \times n$ matrix A of the form (11), we mean the sum

$$\det(A) = \sum_{\phi \varepsilon S_n} e(\phi)a_1(1\phi)a_2(2\phi)\ldots a_n(n\phi),$$

where the summation is over all the $n!$ permutations in S_n.

It is be shown that the determinant dened in this way is the same as that dened earlier by row or column expansions. Indeed, one can use (12) to establish Proposition. The very interested reader may wish to make an attempt. Here we conne our study to the special cases when $n = 2$ and $n = 3$.

In the two examples below, we use e to denote the identity permutation.

Example. Suppose that $n = 2$. We have the following:

elementary product	permutation	sign	contribution
$a_{11}a_{22}$	e	+1	$+a_{11}a_{22}$
$a_{12}a_{21}$	(1 2)	–1	$-a_{12}a_{21}$

Hence det $(A) = a_{11}a_{22} - a_{12}a_{21}$ as shown before.

Example. Suppose that $n = 3$. We have the following:

elementary product	permutation	sign	contribution
$a_{11}a_{22}a_{33}$	e	+1	$+a_{11}a_{22}a_{33}$
$a_{12}a_{21}a_{33}$	(1 2 3)	+1	$+a_{12}a_{21}a_{31}$
$a_{13}a_{21}a_{32}$	(1 3 2)	+1	$+a_{13}a_{21}a_{32}$
$a_{13}a_{22}a_{31}$	(1 3)	–1	$-a_{13}a_{22}a_{31}$

$a_{11}a_{23}a_{32}$	(2 3)	–1	$-a_{11}a_{23}a_{32}$
$a_{12}a_{21}a_{33}$	(1 2)	–1	$-a_{12}a_{21}a_{33}$

Hence $\det(A) = a_{11}a_{22}a_{33} + a_{12}a_{23}a_{31} + a_{13}a_{21}a_{32} - a_{13}a_{22}a_{31} - a_{11}a_{23}a_{32} - a_{12}a_{21}a_{33}$. We have the picture below:

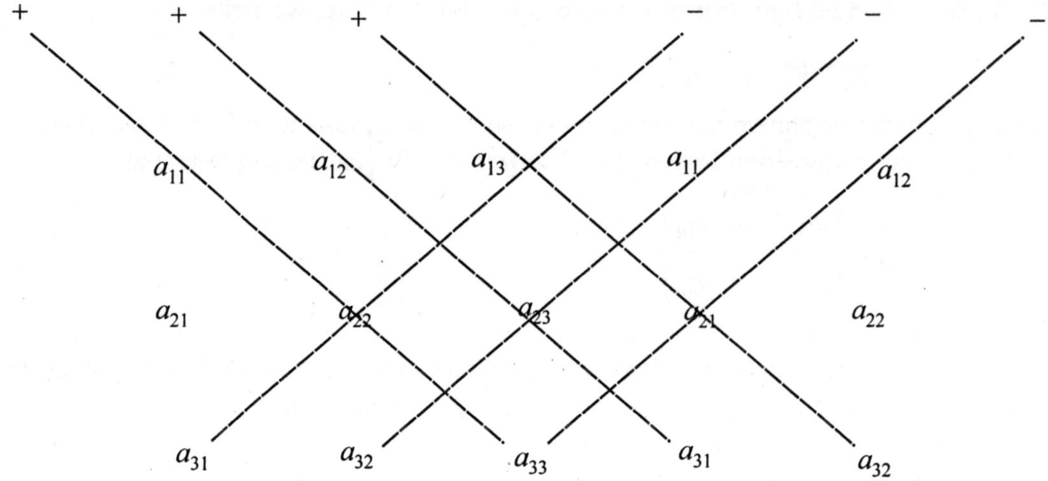

Next, we discuss briey how one may prove Proposition concerning the determinant of the product of two matrices. The idea is to use elementary matrices. Corresponding to Proposition, we can easily establish the following result.

Proposition. Suppose that E is an elementary matrix.

(a) *If E arises from interchanging two rows of In, then det(E)* = *–1.*

(b) *If E arises from adding one row of In to another row, then det(E)* = *1.*

(c) *If E arises from multiplying one row of In by a non-zero constant c, then det(E)* = *c.*

Combining Propositions 3D and 3Q, we can establish the following intermediate result.

Proposition. Suppose that E is an n × n elementary matrix. Then for any n × n matrix B, we have det(EB) = *det(E) det(B).*

Proof of Proposition. Let us reduce A by elementary row operations to reduced row echelon form A'. Then there exist a finite sequence $G_1, \ldots, G_k$ of elementary matrices such that $A' = G_k, \ldots, G_1A$.

Since elementary matrices are invertible with elementary inverse matrices, it follows that there exist a nite sequence $E_1, \ldots, E_k$ of elementary matrices such that

$$A = E_1 \ldots E_kA'$$

Suppose first of all that det(A) = 0. Then it follows from (13) that the matrix A_0 must have a zero row. Hence A' B must have a zero row, and so $\det(A' B) = 0$. But $AB = E_1, \ldots, E_k(A'B)$, so it follows from Proposition that $\det(AB) = 0$. Suppose next that $\det(A) \neq 0$. Then $A' = I_n$, and so it follows from Equation that $AB = E_1, \ldots, E_kB$.

Proof. It sucess to show that

$$Aadj(A) = \det(A)I_n,$$

as this clearly implies

$$A\left(\frac{1}{\det(A)}adj(A)\right) = I_n,$$

giving the result. To show, note that

$$Aadj(A) = \begin{pmatrix} a11 & \cdots & a_{1n} \\ \vdots & & \vdots \\ a_{n1} & \cdots & a_{nn} \end{pmatrix} \begin{pmatrix} C_{11} & \cdots & C_{n1} \\ \vdots & & \vdots \\ C_{1n} & \cdots & C_{nn} \end{pmatrix}.$$

Suppose that the right hand side of is equal to

$$(B) = \begin{pmatrix} b11 & \cdots & b_{1n} \\ \vdots & & \vdots \\ b_{n1} & \cdots & b_{nn} \end{pmatrix}.$$

Then for every $i, j = 1, \ldots, n$, we have

$$b_{ij} = a_{i1}C_{j1} + \ldots + a_{in}C_{jn}.$$

It follows that when $i = j$, we have

$$b_{ii} = a_{i1}C_{i1} + \ldots + a_{in}C_{in} = \det(A):$$

On the other hand, if $i \neq j$, then equation is equal to the determinant of the matrix obtained from A by replacing row j by row i. This matrix has therefore two identical rows, and so the determinant is 0 (why?).

Proof. Since A is invertible, we get

$$A^{-1} = \frac{1}{\det(A)}adj(A)$$

By Proposition, the unique solution of the system $Ax = b$ is given by

$$x = A^{-1} = \frac{1}{\det(A)}adj(A)b.$$

Written in full, this becomes

$$\begin{pmatrix} x_1 \\ \vdots \\ x_n \end{pmatrix} = \frac{1}{\det(A)} \begin{pmatrix} C_{11} & \cdots & C_{n1} \\ \vdots & & \vdots \\ C_{1n} & \cdots & C_{nn} \end{pmatrix} \begin{pmatrix} b_1 \\ \vdots \\ b_n \end{pmatrix} = \frac{1}{\det(A)} \begin{pmatrix} b_1C_{11} \cdots + b_nC_{n1} \\ \vdots \\ b_1C_{1n} + \ldots + b_nC_{nn} \end{pmatrix}.$$

Hence, for every $j = 1, \ldots, n$, we have

$$x_j = \frac{b_1C_{1j} + \ldots + b_nC_{nj}}{\det(A)}.$$

To complete the proof, it remains to show that

$$b_1C_{1j} + \ldots + b_nC_{nj} = \det(A_j(b)):$$

Note, on using cofactor expansion by column j, that

$$\det(A_j(b)) = \sum_{i=1}^{n} b_i(-1)^{i+j} \det\begin{pmatrix} a_{11} & \cdots & a_{1(j-1)} & \bullet & a_{1(j+1)} & \cdots & a_{1n} \\ \vdots & & \vdots & \vdots & \vdots & & \vdots \\ a_{(i-1)1} & \cdots & a_{(i-1)(j-1)} & \bullet & a_{(i-1)(j-1)} & \cdots & a_{(i-1)n} \\ \bullet & \cdots & \bullet & \bullet & \bullet & \cdots & \bullet \\ a_{(i+1)1} & \cdots & a_{(i+1)(j-1)} & \bullet & a_{(i+1)(j+1)} & \cdots & a_{(i+1)n} \\ \vdots & & \vdots & \vdots & \vdots & & \vdots \\ a_{n1} & \cdots & a_{n(j-1)} & \bullet & a_{n(j+1)} & \cdots & a_{nn} \end{pmatrix}$$

$$= \sum_{i=1}^{n} b_i(-1)^{i+j} \det(A_{ij}) = \sum_{i=1}^{n} b_i C_{ij} \text{ as required.}$$

Chapter 5

Introduction to Spectral Theory

EIGENVALUES AND EIGENVECTORS

Spectral theory is the main tool that helps us to understand the structure of a linear operator. In this chapter we consider only operators acting from a vector space to itself (or, equivalently, $n \times n$ matrices). If we have such a linear transformation $A : V \to V$, we can multiply it by itself, take any power of it, or any polynomial.

The main idea of spectral theory is to split the operator into simple blocks and analyse each block separately. To explain the main idea, let us consider difference equations. Many processes can be described by the equations of the following type

$$x_{n+1} = Ax_n, \ n = 0, 1, 2, ...,$$

where a : $V \to V$ is a linear transformation, and x_n is the state of the system at the time n. Given the initial state x_0 we would like to know the state x_n at the time n, analyse the long time behaviour of x_n, etc. 1

At the first glance the problem looks trivial: the solution x_n is given by the formula $xn = A^n x_0$. But what if n is huge: thousands, millions? Or what if we want to analyse the behaviour of x_n as $n \to \infty$?

Here the idea of *eigenvalues* and *eigenvectors* comes in. Suppose that $Ax_0 = \lambda x_0$, where λ is some scalar. Then $A^2 ax_0 = \lambda^2 x_0, \lambda^2 x_0 = \lambda^2 x_0, ..., A^n x_0 = \lambda^n x_0$, so the behaviour of the solution is very well understood

In this section we will consider only operators in finite–dimensional spaces. Spectral theory in infinitely many dimensions if significantly more complicated, and most of the results presented here fail in infinite–dimensional setting.

MAIN DEFINITIONS

Eigenvalues, Eigenvectors, Spectrum

A scalar λ is called an eigenvalue of an operator $A : V \to V$ if there exists a non–zero vector $v \in V$ such that

$$Av = \lambda v.$$

The vector v is called the eigenvector of a (corresponding to the eigenvalue λ).

If we know that λ is an eigenvalue, the eigenvectors are easy to find: one just has to solve the equation $Ax = \lambda x$, or, equivalently

$$(A - \lambda I)x = 0.$$

So, finding *all* eigenvectors, corresponding to an eigenvalue is simply finding the nullspace of $A - \lambda I$. The nullspace $Ker(A - \lambda I)$, *i.e.*, the set of all eigenvectors and 0 vector, is called the *eigenspace*.

The set of all eigenvalues of an operator A is called spectrum of A, and is usually denoted $\sigma(A)$.

Finding Eigenvalues: Characteristic Polynomials

A scalar λ is an eigenvalue if and only if the nullspace Ker$(A - \lambda I)$ is non–trivial (so the equation $(A - \lambda I)\,x = 0$ has a non–trivial solution).

Let a act on $\mathbb{R}^n$ (*i.e.*, a : $\mathbb{R}^n \to \mathbb{R}^n$). Since the matrix of A is square, $A - I$ has a non–trivial nullspace if and only if it is not invertible. We know that a square matrix is not invertible if and only if its determinant is 0. Therefore

$$\boxed{l \in \sigma(A), i.e. \lambda \text{ is an eigenvalue of } A \Leftrightarrow \det\ (A\ - \lambda I) = 0}$$

If A is an $n \times n$ matrix, the determinant $\det(A - \lambda I)$ is a polynomial of degree n of the variable λ. This polynomial is called the *characteristic polynomial* of A. So, to find all eigenvalues of A one just needs to compute the characteristic polynomial and find all its roots.

This method of finding the spectrum of an operator is not very practical in higher dimensions. Finding roots of a polynomial of high degree can be a very dicult problem, and it is impossible to solve the equation of degree higher than 4 in radicals. So, in higher dimensions different numerical methods of finding eigenvalues and eigenvectors are used.

Characteristic Polynomial of an Operator

So we know how to find the spectrum of a matrix. But how do we find eigenvalues of an operator acting in an abstract vector space? The recipe is simple:

Take an arbitrary basis, and compute eigenvalues of the matrix of the operator in this basis.

But how do we know that the result does not depend on a choice of the basis?

There can be several possible explanations. One is based on the notion of *similar matrices*. Let us recall that square matrices A and B are called similar if there exist an invertible matrix S such that

$$A = SBS^{-1}.$$

Note, that determinants of similar matrices coincide. Indeed

$$\det A = \det(SBS^{-1}) = \det S \det B \det S^{-1} = \det B$$

because $\det S^{-1} = 1/ \det S$. Note that if $A = SBS^{-1}$ then

$$A - \lambda I = SBS^{-1} - \lambda SIS^{-1} = S(BS^{-1} - \lambda IS^{-1}) = S(B - \lambda I)S^{-1},$$

so the matrices $A - lI$ and $B - lI$ are similar. Therefore

$$\det(A - \lambda I) = \det(B - \lambda I),$$

i.e.,

characteristic polynomials of similar matrices coincide.

If $T : V \rightarrow V$ is a linear transformation, and A and B are two bases in V, then

$$[T]_{AA} = [I]_{AB}[T]_{BB}[I]_{BA}$$

and since $[I]_{BA} = ([I]_{AB})^{-1}$ the matrices $[T]_{AA}$ and $[T]_{BB}$ are similar.

In other words, matrices of a linear transformation in different bases are similar.

Therefore, we can define the characteristic polynomial of an operator as the characteristic polynomial of its matrix in some basis. As we have discussed above, the result does not depend on the choice of the basis, so characteristic polynomial of an operator is well defined.

Multiplicities of Eigenvalues

Let us remind the reader, that if p is a polynomial, and λ is its root (*i.e.*, $p(\lambda) = 0$) then $z - \lambda$ divides $p(z)$, *i.e.*, p can be represented as $p(z) = (z - \lambda)q(z)$, where q is some polynomial. If $q(\lambda) = 0$, then q also can be divided by $z -$, so $(z -)^2$ divides p and so on.

The largest positive integer k such that $(z - \lambda)^k$ divides $p(z)$ is called the multiplicity of the root λ.

If λ is an eigenvalue of an operator (matrix) A, then it is a root of the characteristic polynomial $p(z) = \det(A - zI)$. The multiplicity of this root is called the (*algebraic*) *multiplicity* of the eigenvalue λ.

Any polynomial $p(z) = \sum_{k=0}^{n} a_k z^k$ of degree n has exactly n complex roots, counting multiplicity. The words *counting multiplicities* mean that if a root has multiplicity d we have to count it d times. In other words, p can be represented as

$$p(z) = a_n(z - \lambda_1)(z - \lambda_2) \ldots (z - \lambda_n).$$

where $\lambda_1, \lambda_2, \ldots, \lambda_n$ are its complex roots, counting multiplicities. There is another notion of multiplicity of an eigenvalue: the dimension of the eigenspace Ker(A–I) is called *geometric multiplicity* of the eigenvalue λ.

Geometric multiplicity is not as widely used as algebraic multiplicity. So, when people say simply "multiplicity" they usually mean *algebraic multiplicity*.

Let us mention, that algebraic and geometric multiplicities of an eigenvalue can differ.

Proposition. Geometric multiplicity of an eigenvalue cannot exceed its algebraic multiplicity.

Trace and Determinant

Theorem. Let A be $n \times n$ matrix, and let $\lambda_1, \lambda_2, \ldots, \lambda_n$ be its *eigenvalues* (*counting multiplicities*). *Then*

1. trace$A = \lambda_1 + \lambda_2 + \ldots + \lambda_n$.

2. $\det A = \lambda_1\lambda_2 \ldots\lambda_n$.

Eigenvalues of a Triangular Matrix

Computing eigenvalues is equivalent to finding roots of a characteristic polynomial of a matrix (or using some numerical method), which can be quite time consuming. However, there is one particular case, when we can just read eigenvalues off the matrix. Namely

eigenvalues of a triangular matrix (counting multiplicities) are exactly the diagonal entries $a_{1,1}$, $a_{2,2}$, ..., a_n,n

By triangular here we mean either upper or lower triangular matrix. Since a diagonal matrix is a particular case of a triangular matrix (it is both upper and lower triangular the eigenvalues of a diagonal matrix are its diagonal entries

The proof of the statement about triangular matrices is trivial: we need to subtract λ from the diagonal entries of A, and use the fact that determinant of a triangular matrix is the product of its diagonal entries. We get the characteristic polynomial

$$\det(A - \lambda I) = (a_{1,1} - \lambda)(a_{2,2} - \lambda) \ldots(a_{n,n} - \lambda)$$

and its roots are exactly $a_{1,1}$, $a_{2,2}$, ..., $a_{n,n}$.

DIAGONALIZATION

Suppose an operator (matrix) a has a basis $b = v_1, v_2, \ldots$vn of eigenvectors, and let λ_1, λ_2, ..., n be the corresponding eigenvalues. Then the matrix of A in this basis is the diagonal matrix with 1, 2, ..., n on the diagonal

$$[A]_{BB} = \text{diag}\{\lambda_1, \lambda_2, \ldots, \lambda_n\} = \begin{pmatrix} \lambda_1 & & & 0 \\ & \lambda_2 & & \\ & & \ddots & \\ 0 & & & \lambda_n \end{pmatrix}.$$

Therefore, it is easy to find an Nth power of the operator A. Namely, its matrix in the basis B is

$$[A^N]_{BB} = \text{diag} = \{\lambda_1^N, \lambda_2^N, \ldots, \lambda_n^N\} = \begin{pmatrix} \lambda_1^N & & & 0 \\ & \lambda_2^N & & \\ & & \ddots & \\ 0 & & & \lambda_n^N \end{pmatrix}.$$

Moreover, functions of the operator are also very easy to compute: for example the operator (matrix) exponent e^{tA} is defined as $e^{tA} = I + tA + \frac{t_2 A_2}{2!} + \frac{t_3 A_3}{3!} \ldots = \sum_{k=0}^{\infty} \frac{t^k A^k}{k!}$, and its matrix in the basis B is

$$[A^N]_{BB} = \text{diag }\{e^{\lambda_1 t}, e^{\lambda_2 t}, \ldots, e^{\lambda_n t}\} = \begin{pmatrix} e^{\lambda_1 t} & & & 0 \\ & e^{\lambda_2 t} & & \\ & & \ddots & \\ 0 & & & e^{\lambda_n t} \end{pmatrix}.$$

To find the matrices in the standard basis S, we need to recall that the change of coordinate matrix $[I]_{SB}$ is the matrix with columns $v_1, v_2, \ldots, v_n$.

Let us call this matrix S, then

$$A = [A]_{ss} = s\begin{pmatrix} \lambda_1 & & & 0 \\ & \lambda_2 & & \\ & & \ddots & \\ 0 & & & \lambda_n \end{pmatrix} S^{-1} = SDS^{-1},$$

where we use D for the diagonal matrix in the middle.

Similarly

$$A^N = SD^N S^{-1} = S\begin{pmatrix} \lambda_1^N & & & 0 \\ & \lambda_2^N & & \\ & & \ddots & \\ 0 & & & \lambda_n^N \end{pmatrix} S^{-1}$$

and similarly for e^{tA}.

Another way of thinking about powers (or other functions) of diagonalizable operators is to see that if operator A can be represented as $A = SDS^{-1}$, then

$$A^N = \underbrace{(SDS^{-1})(SDS^{-1})\ldots}_{N \text{ Times}} = SD^N S^{-1}$$

and it is easy to compute the Nth power of a diagonal matrix. The following theorem is almost trivial.

Theorem. A matrix a admits a representation $A = SDS^{-1}$, where *D is a diagonal matrix if and only if there exists a basis of eigenvectors of A.*

Proof. We already discussed above that if there is a basis of eigenvectors, then the matrix admits the representation $A = SDS^{-1}$, where $S = [I]_{SB}$ is the change of coordinate matrix from coordinates in the basis B to the standard coordinates.

On the other hand if the matrix admits the representation a $= SDS^{-1}$ with a diagonal matrix D, then columns of S are eigenvectors of A (column number k corresponds to the kth diagonal entry of D). Since S is invertible, its columns form a basis.

Theorem. Let $\lambda_1, \lambda_2, \ldots, \lambda_r$ *be distinct eigenvalues of A, and let $v_1, v_2, \ldots, v_r$ be the corresponding eigenvectors. Then vectors $v_1, v_2, \ldots,$ vr are linearly independent.*

Proof. We will use induction on r. The case $r = 1$ is trivial, because by the definition an eigenvector is non–zero, and a system consisting of one non–zero vector is linearly independent.

Suppose that the statement of the theorem is true for $r - 1$. Suppose there exists a non–trivial linear combination

$$c_1v_1 + c_2v_2 + ... + c_rv_r = \sum_{k=1}^{r} c_k v_k = 0$$

Applying $A - \lambda_r I$ and using the fact that $(A - \lambda_r I)v_r = 0$ we get

$$\sum_{k=1}^{r-1} c_k(\lambda_k - \lambda_r)v_k = 0.$$

By the induction hypothesis vectors $v_1, v_2, ..., vr^{-1}$ are linearly independent, so

$$c_k(\lambda_k - r) = 0$$

for $k = 1, 2, ..., r - 1$. Since $\lambda_k \neq \lambda_r$ we can conclude that $c_k = 0$ for $k < r$. Then it follows from that $c_r = 0$, *i.e.*, we have the trivial linear combination.

Corollary. If an operator $A : V \to V$ has exactly $n = \dim V$ distinct then it is diagonalizable.

Proof. For each eigenvalue k let vk be a corresponding eigenvector (*just pick one eigenvector for each eigenvalue*). By Theorem the system $v_1, v_2, ..., v_n$ is linearly independent, and since it consists of exactly $n = \dim V$ vectors it is a basis.

Bases of Subspaces (*AKA* Direct Sums of Subspaces)

Let $V_1, V_2, ..., V_p$ be subspaces of a vector space V. We say that the system of subspaces is a basis in V if any vector $v \in V$ admits a unique representation as a sum

$$v = v_1 + v_2 + ... + vp = \sum_{k=1}^{p} v_k, v_k \in V_k.$$

We also say, that a system of subspaces $V_1, V_2, ..., V_p$ is linearly independent if the equation

$$v_1 + v_2 + ... + v_p = 0, \ vk \in V_k$$

has only trivial solution ($v_k = 0 \ \forall k = 1, 2, ..., p$).

Another way to phrase that is to say that a system of subspaces $V_1, V_2, ..., V_p$ is linearly independent if and only if any system of non–zero vectors vk, where $v_k \in V_k$, is linearly independent.

We say that the system of subspaces $V_1, V_2, ..., V_p$ is generating (or complete, or spanning) if any vector $v \in V$ admits representation.

Remark. From the above definition one can immediately see that Theorem states in fact that the system of eigenspaces E_k of an operator A

$$E_k := Ker(A - \lambda_k I), \lambda_k \in \sigma(A),$$

is linearly independent.

Remark. It is easy to see that similarly to the bases of vectors, a system of subspaces $V_1, V_2, ..., V_p$ is a basis if and only if it is generating and linearly independent.

There is a simple example of a basis of subspaces. Let V be a vector space with a basis $v_1, v_2, ..., v_n$. Split the set of indices 1, 2, ..., n into p subsets $\Lambda_1, \Lambda_2, ..., \Lambda_p$, and define subspaces $V_k := \text{span}\{v_j : j \in \Lambda_k\}$. Clearly the subspaces V_k form a basis of V.

The following theorem shows that in the finite–dimensional case it is essentially the only possible example of a basis of subspaces.

Theorem. Let $V_1, V_2, ..., V_p$ *be a basis of subspaces, and let us have in each subspace* V_k *a basis (of vectors)* B_k^2. Then the union $[k_B k$ of these bases is a basis in V. To prove the theorem we need the following lemma.

Lemma. Let $V_1, V_2, ..., V_p$ *be a linearly independent family of subspaces, and let us have in each subspace Vk a linearly independent system* B_k *of vectors* [3] *Then the union B* $:= U_k B_k$ *is a linearly independent system.*

Proof. The proof of the lemma is almost trivial, if one thinks a bit about it. The main diculty in writing the proof is a choice of a appropriate notation. Instead of using two indices (one for the number k and the other for the number of a vector in B_k, let us use "flat" notation.

Namely, let n be the number of vectors in $B := [U_k B_k$. Let us order the set B, for example as follows: first list all vectors from B_1, then all vectors in B_2, etc, listing all vectors from B_p last.

This way, we index all vectors in B by integers 1, 2, ..., n, and the set of indices {1, 2, ..., n} splits into the sets 1,2, ..., p such that the set B_k consists of vectors $b_j : j \in A_k$. Suppose we have a non–trivial linear combination

$$c_1 b_1 + c_2 b_2 + ... + c_n b_n = \sum_{j=1}^{n} c_j b_j = 0$$

Denote

$$v_k \sum_{j \in \Lambda_k} c_j b_j.$$

Then can be rewritten as

$$v_1 + v_2 + ... + v_p = 0.$$

Since $v_k \in v_k$ and the system of subspaces V_k is linearly independent, $v_k = 0 \ \forall k$. Than means that for every k

$$\sum_{j \in \Lambda_k} c_j b_j = 0,$$

and since the system of vectors $b_j : j \in \Lambda_k$ (*i.e.*, the system B_k) are linearly independent, we have $c_j = 0$ for all $j \in \Lambda_k$. Since it is true for all Λ_k, we can conclude that $c_j = 0$ for all j.

Proof. To prove the theorem we will use the same notation as in the proof of Lemma, *i.e.*, the system B_k consists of vectors b_j, $j \in \Lambda_k$.

Lemma asserts that the system of vectors $b_j, j = 12, ..., n$ is linearly independent, so it only remains to show that the system is complete.

Since the system of subspaces $V_1, V_2, ..., V_p$ is a basis, any vector $v \in V$ can be represented as

$$v = v_1 p_1 + v_2 p_2 + ... + v = p = \sum_{k=1}^{p} v_k, \; v_k \in V_k.$$

Since the vectors $b_j, j \in \Lambda_k$ form a basis in V_k, the vectors v_k can be represented as

$$v_k \sum_{j \in \Lambda_k} c_j b_j,$$

and therefore $v = \sum_{j=1}^{n} c_j b_j$.

Criterion of Diagonalizability. First of all let us mention a simple necessary condition. Since for a diagonal matrix $D = \text{diag}\{\lambda_1, \lambda_2, ..., \lambda_n\}$

$$\det(D - \lambda I) = (\lambda_1 - \lambda)(\lambda_2 - \lambda) ...(\lambda_n - \lambda),$$

we see that if an operator A is diagonalizable, its characteristic polynomial splits into the product of monomials. Note, that any polynomial can be decomposed into the product of monomials, if we allow complex coefficients (*i.e.*, complex eigenvalues).

In what follows we always assume that the characteristic polynomial splits into the product of monomials, either by working in a complex vector space, or simply assuming that a has exactly $n = \dim V$ eigenvalues(counting multiplicity).

Theorem. An operator $A : V \to V$ is diagonalizable if and only if for each eigenvalue λ the dimension of the eigenspace $Ker(A - \lambda I)$ (i.e., the geometric multiplicity) coincides with the algebraic multiplicity of λ.

Proof. First of all let us note, that for a diagonal matrix, the algebraic and geometric multiplicities of eigenvalues coincide, and therefore the same holds for the diagonalizable operators.

Let us now prove the other implication. Let $\lambda_1, \lambda_2, ..., \lambda_p$ be eigenvalues of A, and let $E_k := \text{Ker}(A - \lambda_k I)$ be the corresponding eigenspaces. The subspaces E_k, $k = 1, 2, ..., p$ are linearly independent.

Let B_k be a basis in E_k. By Lemma the system $B = [U_k B_k$ is a linearly independent system of vectors.

We know that each B_k consists of $\dim E_k$(= multiplicity of λ_k) vectors. So the number of vectors in b equal to the sum of multiplicities of eigenvectors k, which is exactly $n = \dim V$. So, we have a linearly independent system of $\dim V$ eigenvectors, which means it is a basis.

Some Examples

Real eigenvalues. Consider the matrix

$$A = \begin{pmatrix} 1 & 2 \\ 8 & 1 \end{pmatrix}.$$

Its characteristic polynomial is equal to

$$\begin{vmatrix} 1-\lambda & 2 \\ 8 & 1-\lambda \end{vmatrix} = (1-\lambda)^2 - 16$$

and its roots (eigenvalues) are $\lambda = 5$ and $\lambda = -3$. For the eigenvalue $\lambda = 5$

$$A - 5I = \begin{pmatrix} 1-5 & 2 \\ 8 & 1-5 \end{pmatrix} = \begin{pmatrix} -4 & 2 \\ 8 & -4 \end{pmatrix}$$

A basis in its nullspace consists of one vector $(1, 2)^T$, so this is the corresponding eigenvector. Similarly, for $\lambda = -3$

$$A - \lambda I = A + 3I = \begin{pmatrix} 4 & 2 \\ 8 & 4 \end{pmatrix}$$

and the eigenspace $\text{Ker}(A + 3I)$ is spanned by the vector $(1, -2)^T$. The matrix A can be diagonalized as

$$A = \begin{pmatrix} 1 & 2 \\ 8 & 1 \end{pmatrix} = \begin{pmatrix} 1 & 1 \\ 2 & -2 \end{pmatrix} \begin{pmatrix} 5 & 0 \\ 0 & -3 \end{pmatrix} \begin{pmatrix} 1 & 1 \\ 2 & -2 \end{pmatrix}^{-1}$$

Complex eigenvalues. Consider the matrix

$$A = \begin{pmatrix} 1 & 2 \\ -2 & 1 \end{pmatrix}.$$

Its characteristic polynomial is

$$\begin{vmatrix} 1-\lambda & 2 \\ -2 & 1-\lambda \end{vmatrix} = (1-\lambda)^2 + 2^2$$

and the eigenvalues (roots of the characteristic polynomial are $\lambda = 1 \pm 2i$. For

$$\lambda = 1 + 2i$$

$$A - \lambda I = \begin{pmatrix} -2i & 2 \\ -2 & -2i \end{pmatrix}$$

This matrix has rank 1, so the eigenspace $\text{Ker}(A - \lambda I)$ is spanned by one vector, for example by $(1, i)^T$.

Since the matrix A is real, we do not need to compute an eigenvector for $\lambda = 1-2i$: we can get it for free by taking the complex conjugate of the above eigenvector. So, for

$$\lambda = 1-2i$$

a corresponding eigenvector is $(1,-i)^T$, and so the matrix A can be diagonalized as

$$A = \begin{pmatrix} 1 & 1 \\ i & -i \end{pmatrix} \begin{pmatrix} 1+2i & 0 \\ 0 & 1-2i \end{pmatrix} \begin{pmatrix} 1 & 1 \\ i & -i \end{pmatrix}^{-1}.$$

A non–diagonalizable matrix. Consider the matrix

$$A = \begin{pmatrix} 1 & 1 \\ 0 & 1 \end{pmatrix}.$$

Its characteristic polynomial is

$$\begin{vmatrix} 1-\lambda & 1 \\ 0 & 1-\lambda \end{vmatrix} = (1-\lambda)^2,$$

so A has an eigenvalue 1 of multiplicity 2. But, it is easy to see that

$$\dim \operatorname{Ker}(A - I) = 1$$

(1 pivot, so $2 - 1 = 1$ free variable). Therefore, the geometric multiplicity of the eigenvalue 1 is different from its algebraic multiplicity, so a is not diagonalizable.

There is also an explanation which does not use Theorem. Namely, we got that the eigenspace $\operatorname{Ker}(A-1I)$ is one dimensional (spanned by the vector $(1, 0)^T$). If A were diagonalizable, it would have a diagonal form $\begin{pmatrix} 1 & 0 \\ 0 & 1 \end{pmatrix}$ in some basis, and so the dimension of the eigenspace wold be 2. Therefore A cannot be diagonalized.

Example. Consider a function $f : \mathbb{R}^2 \to \mathbb{R}^2$, dened for every $(x, y) \in \mathbb{R}^2$ by

$$f(x, y) = (s, t),$$

where

$$\begin{pmatrix} s \\ t \end{pmatrix} = \begin{pmatrix} 3 & 3 \\ 1 & 5 \end{pmatrix} \begin{pmatrix} x \\ y \end{pmatrix}.$$

Note that

$$\begin{pmatrix} 3 & 3 \\ 1 & 5 \end{pmatrix} \begin{pmatrix} 3 \\ -1 \end{pmatrix} = \begin{pmatrix} 6 \\ -2 \end{pmatrix} = 2\begin{pmatrix} 3 \\ -1 \end{pmatrix} \text{ and } \begin{pmatrix} 3 & 3 \\ 1 & 5 \end{pmatrix} \begin{pmatrix} 1 \\ 1 \end{pmatrix} = \begin{pmatrix} 6 \\ 6 \end{pmatrix} = 6\begin{pmatrix} 1 \\ 1 \end{pmatrix}.$$

On the other hand, note that

$$v_1 = \begin{pmatrix} 3 \\ -1 \end{pmatrix} \text{ and } v_2 = \begin{pmatrix} 1 \\ 1 \end{pmatrix}$$

form a basis for $\mathbb{R}^2$. It follows that every $u \in \mathbb{R}^2$ can be written uniquely in the form

$$u = c_1 v_1 + c_2 v_2,$$

where $c_1, c_2 \in \mathbb{R}$, so that

$$Au = A(c_1 v_1 + c_2 v_2) = c_1 A v_1 + c_2 A v_2 = 2c_1 v_1 + 6c_2 v_2.$$

Note that in this case, the function $f : \mathbb{R}^2 \to \mathbb{R}^2$ can be described easily in terms of the two special vectors v_1 and v_2 and the two special numbers 2 and 6. Let us now examine how these special vectors and numbers arise. We hope to find numbers $\lambda \in \mathbb{R}$ and non-zero vectors $v \in \mathbb{R}^2$ such that

$$\begin{pmatrix} 3 & 3 \\ 1 & 5 \end{pmatrix} v = \lambda v.$$

Since

$$\lambda v = \lambda \begin{pmatrix} 1 & 0 \\ 0 & 1 \end{pmatrix} v = \begin{pmatrix} \lambda & 0 \\ 0 & \lambda \end{pmatrix} v,$$

we must have

$$\left(\begin{pmatrix}3 & 3\\ 1 & 5\end{pmatrix}-\begin{pmatrix}\lambda & 0\\ 0 & \lambda\end{pmatrix}\right)v=0.$$

In other words, we must have

$$\begin{pmatrix}3-\lambda & 3\\ 1 & 5-\lambda\end{pmatrix}v=0.$$

In order to have non-zero $v \in \mathbb{R}^2$, we must therefore ensure that

$$\det\begin{pmatrix}3-\lambda & 3\\ 1 & 5-\lambda\end{pmatrix}=0.$$

Hence $(3-\lambda)(5-\lambda)-3=0$, with roots $\lambda_1 = 2$ and $\lambda_2 = 6$. Substituting $\lambda = 2$ into (1), we obtain

$$\begin{pmatrix}3 & 3\\ 1 & 3\end{pmatrix}v=0, \text{ with root } v_1=\begin{pmatrix}3\\ -1\end{pmatrix}.$$

Substituting $\lambda = 6$ into (1), we obtain

$$\begin{pmatrix}-3 & 3\\ 1 & -1\end{pmatrix}v=0, \text{ with root } v_2=\begin{pmatrix}1\\ 1\end{pmatrix}.$$

Definition. Suppose that

$$A=\begin{pmatrix}a_{11} & \cdots & a_{1n}\\ \vdots & & \vdots\\ a_{n1} & \cdots & a_{nn}\end{pmatrix}$$

is an $n \times n$ matrix with entries in $\mathbb{R}$. Suppose further that there exist a number $\in \mathbb{R}$ and a non-zero vector $v \in \mathbb{R}^n$ such that $Av = \lambda v$. Then we say that λ is an eigenvalue of the matrix A, and that v is an eigenvector corresponding to the eigenvalue λ.

Suppose that λ is an eigenvalue of the $n \times n$ matrix A, and that v is an eigenvector corresponding to the eigenvalue λ. Then $Av = \lambda v = \lambda Iv$, where I is the $n \times n$ identity matrix, so that $(A - \lambda I)v = 0$. Since $v \in \mathbb{R}^n$ is non-zero, it follows that we must have

$$\det(A-\lambda I)=0.$$

In other words, we must have

$$\det\begin{pmatrix}a_{11}-\lambda & a_{12} & \cdots & a_{1n}\\ a_{21} & a_{22}-\lambda & & a_{2n}\\ \vdots & & \ddots & \vdots\\ a_{n1} & a_{n2} & \cdots & a_{nn}-\lambda\end{pmatrix}=0.$$

that is a polynomial equation. Solving this equation gives the eigenvalues of the matrix A. On the other hand, for any eigenvalue λ of the matrix A, the set

$$\{v \in \mathbb{R}^n : (A-\lambda I)v=0\}$$

is the nullspace of the matrix $A - \lambda I$, a subspace of $\mathbb{R}^n$.

Definition. The polynomial is called the characteristic polynomial of the matrix A. For any root λ of equation, the space is called the eigenspace corresponding to the eigenvalue λ.

Example. The matrix

$$\begin{pmatrix} 3 & 3 \\ 1 & 5 \end{pmatrix}$$

has characteristic polynomial $(3 - \lambda)(5 - \lambda) - 3 = 0$, in other words, $\lambda^2 - 8\lambda + 12 = 0$. Hence the eigenvalues are $\lambda_1 = 2$ and $\lambda_2 = 6$, with corresponding eigenvectors

$$v_1 = \begin{pmatrix} 3 \\ -1 \end{pmatrix} \text{ and } v_2 = \begin{pmatrix} 1 \\ 1 \end{pmatrix}$$

respectively. The eigenspace corresponding to the eigenvalue 2 is

$$\left\{ v \in \mathbb{R}^2 : \begin{pmatrix} 1 & 3 \\ 1 & 3 \end{pmatrix} v = 0 \right\} = \left\{ c \begin{pmatrix} 3 \\ -1 \end{pmatrix} : c \in \mathbb{R} \right\}.$$

The eigenspace corresponding to the eigenvalue 6 is

$$\left\{ v \in \mathbb{R}^2 : \begin{pmatrix} -3 & 3 \\ 1 & -1 \end{pmatrix} v = 0 \right\} = \left\{ c \begin{pmatrix} 1 \\ 1 \end{pmatrix} : c \in \mathbb{R} \right\}.$$

Example. Consider the matrix

$$A = \begin{pmatrix} -1 & 6 & -12 \\ 0 & -13 & 30 \\ 0 & -9 & 20 \end{pmatrix}.$$

To find the eigenvalues of A, we need to nd the roots of

$$\det \begin{pmatrix} -1-\lambda & 6 & -12 \\ 0 & -13-\lambda & 30 \\ 0 & -9 & 20-\lambda \end{pmatrix} = 0;$$

in other words, $(\lambda + 1)(\lambda - 2)(\lambda - 5) = 0$. The eigenvalues are therefore

$$\lambda_1 = -1, \lambda_2 = 2$$

and

$$\lambda_3 = 5.$$

An eigenvector corresponding to the eigenvalue -1 is a solution of the system

$$(A + I)v = \begin{pmatrix} 0 & 6 & -12 \\ 0 & -12 & 30 \\ 0 & -9 & 21 \end{pmatrix} v = 0, \text{ with root } v_1 = \begin{pmatrix} 1 \\ 0 \\ 0 \end{pmatrix}.$$

An eigenvector corresponding to the eigenvalue 2 is a solution of the system

$$(A-2I)v=\begin{pmatrix}-3 & 6 & -12\\ 0 & -15 & 30\\ 0 & -9 & 18\end{pmatrix}v=0, \text{ with root } v_2=\begin{pmatrix}0\\2\\1\end{pmatrix}.$$

An eigenvector corresponding to the eigenvalue 5 is a solution of the system

$$(A-5I)v=\begin{pmatrix}-6 & 6 & -12\\ 0 & -18 & 30\\ 0 & -9 & 15\end{pmatrix}v=0, \text{ with root } v_3=\begin{pmatrix}1\\-5\\-3\end{pmatrix}.$$

Note that the three eigenspaces are all lines through the origin. Note also that the eigenvectors v_1, v_2 and v_3 are linearly independent, and so form a basis for $\mathbb{R}^3$.

Example. Consider the matrix

$$A=\begin{pmatrix}17 & -10 & -5\\ 45 & -28 & -15\\ -30 & 20 & 12\end{pmatrix}.$$

To find the eigenvalues of A, we need to nd the roots of

$$\det\begin{pmatrix}17-\lambda & -10 & -5\\ 45 & -28-\lambda & -15\\ -30 & 20 & 12-\lambda\end{pmatrix}=0;$$

in other words, $(\lambda+3)(\lambda-2)^2=0$. The eigenvalues are therefore $\lambda_1=-3$ and $\lambda_2=2$. An eigenvector corresponding to the eigenvalue -3 is a solution of the system

$$(A+3I)v=\begin{pmatrix}20 & -10 & -5\\ 45 & -25 & -15\\ -30 & 20 & 15\end{pmatrix}v=0, \text{ with root } v_1=\begin{pmatrix}1\\3\\-2\end{pmatrix}.$$

An eigenvector corresponding to the eigenvalue 2 is a solution of the system

$$(A-2I)v=\begin{pmatrix}15 & -10 & -5\\ 45 & -30 & -15\\ -30 & 20 & 10\end{pmatrix}v=0, \text{ with roots } v_2=\begin{pmatrix}1\\0\\3\end{pmatrix} \text{ and } v_3=\begin{pmatrix}2\\3\\0\end{pmatrix}.$$

Note that the eigenspace corresponding to the eigenvalue -3 is a line through the origin, while the eigenspace corresponding to the eigenvalue 2 is a plane through the origin. Note also that the eigenvectors v_1, v_2 and v_3 are linearly independent, and so form a basis for $\mathbb{R}^3$.

Example. Consider the matrix

$$A = \begin{pmatrix} 2 & -1 & 0 \\ 1 & 0 & 0 \\ 0 & 0 & 3 \end{pmatrix}.$$

To find the eigenvalues of A, we need to nd the roots of

$$\det \begin{pmatrix} 2-\lambda & -1 & 0 \\ 1 & 0-\lambda & 0 \\ 0 & 0 & 3-\lambda \end{pmatrix} = 0;$$

in other words, $(\lambda - 3)(\lambda - 1)^2 = 0$. The eigenvalues are therefore $\lambda_1 = 3$ and $\lambda_2 = 1$. An eigenvector corresponding to the eigenvalue 3 is a solution of the system

$$(A - 3I)v = \begin{pmatrix} -1 & -1 & 0 \\ 1 & -3 & 0 \\ 0 & 0 & 0 \end{pmatrix} v = 0, \text{ with root } v_1 = \begin{pmatrix} 0 \\ 0 \\ 1 \end{pmatrix}.$$

An eigenvector corresponding to the eigenvalue 1 is a solution of the system

$$(A - I)v = \begin{pmatrix} 1 & -1 & 0 \\ 1 & -1 & 0 \\ 0 & 0 & 2 \end{pmatrix} v = 0, \text{ with root } v_2 = \begin{pmatrix} 1 \\ 1 \\ 0 \end{pmatrix}.$$

Note that the eigenspace corresponding to the eigenvalue 3 is a line through the origin. On the other hand, the matrix

$$\begin{pmatrix} 1 & -1 & 0 \\ 1 & -1 & 0 \\ 0 & 0 & 2 \end{pmatrix}$$

has rank 2, and so the eigenspace corresponding to the eigenvalue 1 is of dimension 1 and so is also a line through the origin. We can therefore only nd two linearly independent eigenvectors, so that $\mathbb{R}^3$ does not have a basis consisting of linearly independent eigenvectors of the matrix A.

Example. Consider the matrix

$$A = \begin{pmatrix} 3 & -3 & 2 \\ 1 & -1 & 2 \\ 1 & -3 & 4 \end{pmatrix}.$$

To find the eigenvalues of A, we need to find the roots of

$$\det\begin{pmatrix}3-\lambda & -3 & 2\\ 1 & -1-\lambda & 2\\ 1 & -3 & 4-\lambda\end{pmatrix}=0;$$

in other words, $(\lambda - 2)^3 = 0$. The eigenvalue is therefore $\lambda = 2$. An eigenvector corresponding to the eigenvalue 2 is a solution of the system

$$(A-2I)v=\begin{pmatrix}1 & -3 & 2\\ 1 & -3 & 2\\ 1 & -3 & 2\end{pmatrix}v=0,\text{ with roots } v_1=\begin{pmatrix}2\\ 0\\ -1\end{pmatrix}\text{ and } v_2=\begin{pmatrix}3\\ 1\\ 0\end{pmatrix}.$$

Note now that the matrix

$$\begin{pmatrix}1 & -3 & 2\\ 1 & -3 & 2\\ 1 & -3 & 2\end{pmatrix}$$

has rank 1, and so the eigenspace corresponding to the eigenvalue 2 is of dimension 2 and so is a plane through the origin. We can therefore only nd two linearly independent eigenvectors, so that $\mathbb{R}^3$ does not have a basis consisting of linearly independent eigenvectors of the matrix A.

Example. Suppose that λ is an eigenvalue of a matrix A, with corresponding eigenvector v. Then

$$A^2v = A(Av) = A(\lambda v) = \lambda(Av) = \lambda(\lambda v) = \lambda^2 v.$$

Hence λ^2 is an eigenvalue of the matrix A^2, with corresponding eigenvector v. In fact, it can be proved by induction that for every natural number $k \in \mathbb{N}$, λ^k is an eigenvalue of the matrix A^k, with corresponding eigenvector v.

Example. Consider the matrix

$$\begin{pmatrix}1 & 5 & 4\\ 0 & 2 & 6\\ 0 & 0 & 3\end{pmatrix}.$$

To find the eigenvalues of A, we need to find the roots of

$$\det\begin{pmatrix}1-\lambda & 5 & 4\\ 0 & 2-\lambda & 6\\ 0 & 0 & 3-\lambda\end{pmatrix}=0;$$

in other words, $(\lambda - 1)(\lambda - 2)(\lambda - 3) = 0$. It follows that the eigenvalues of the matrix A are given by the entries on the diagonal. In fact, this is true for all triangular matrices.

The Diagonalization Problem

Example. Let us return to Examples are consider again the matrix

$$A = \begin{pmatrix} 3 & 3 \\ 1 & 5 \end{pmatrix}.$$

We have already shown that the matrix A has eigenvalues $\lambda_1 = 2$ and $\lambda_2 = 6$, with corresponding eigenvectors

$$v_1 = \begin{pmatrix} 3 \\ -1 \end{pmatrix} \text{ and } v_2 = \begin{pmatrix} 1 \\ 1 \end{pmatrix}$$

respectively. Since the eigenvectors form a basis for $\mathbb{R}^2$, every $u \in \mathbb{R}^2$ can be written uniquely in the form

$$u = c_1 v_1 + c_2 v_2, \text{ where } c_1, c_2 \in \mathbb{R},$$

and

$$Au = 2c_1 v_1 + 6c_2 v_2.$$

Write

$$c = \begin{pmatrix} c_1 \\ c_2 \end{pmatrix}, u = \begin{pmatrix} x \\ y \end{pmatrix}, Au = \begin{pmatrix} s \\ t \end{pmatrix}.$$

Then both can be rewritten as

$$\begin{pmatrix} x \\ y \end{pmatrix} = \begin{pmatrix} 3 & 1 \\ -1 & 1 \end{pmatrix} \begin{pmatrix} c_1 \\ c_2 \end{pmatrix}$$

and

$$\begin{pmatrix} s \\ t \end{pmatrix} = \begin{pmatrix} 3 & 1 \\ -1 & 1 \end{pmatrix} \begin{pmatrix} 2c_1 \\ 6c_2 \end{pmatrix} = \begin{pmatrix} 3 & 1 \\ -1 & 1 \end{pmatrix} \begin{pmatrix} 2 & 0 \\ 0 & 6 \end{pmatrix} \begin{pmatrix} c_1 \\ c_2 \end{pmatrix}$$

respectively. If we write

$$P = \begin{pmatrix} 3 & 1 \\ -1 & 1 \end{pmatrix} \text{ and } D = \begin{pmatrix} 2 & 0 \\ 0 & 6 \end{pmatrix},$$

then matrix become $u = Pc$ and $Au = PDc$ respectively, so that $APc = PDc$. Note that $c \in \mathbb{R}^2$ is arbitrary. This implies that $(AP - PD)c = 0$ for every $c \in \mathbb{R}^2$. Hence we must have $AP = PD$. Since P is invertible, we conclude that

$$P^{-1} AP = D.$$

Note here that

$$P = (v_1 \; v_2) \text{ and } D = \begin{pmatrix} \lambda_1 & 0 \\ 0 & \lambda_2 \end{pmatrix}.$$

Note also the crucial point that the eigenvectors of A form a basis for $\mathbb{R}^2$.

We now consider the problem in general.

Proposition. Suppose that A is an nn matrix, with entries in $\mathbb{R}$. Suppose further that

A has eigenvalues $\lambda_1, \ldots, \lambda_n \in \mathbb{R}$, not necessarily distinct, with corresponding eigenvectors $v_1, \ldots, v_n \in \mathbb{R}^n$, and that $v_1, \ldots, v_n$ are linearly independent. Then

$$P^{-1}AP = D,$$

where

$$P = (v_1 \ \ldots \ v_n) \text{ and } D = \begin{pmatrix} \lambda_2 & & \\ & \ddots & \\ & & \lambda_n \end{pmatrix}$$

Proof. Since $v_1, \ldots, v_n$ are linearly independent, they form a basis for $\mathbb{R}^n$, so that every $u \in \mathbb{R}^n$ can be written uniquely in the form

$$u = c_1v_1 + \ldots + c_nv_n, \text{ where } c_1, \ldots, c_n \in \mathbb{R}^n,$$

and

$$Au = A(c_1v_1 + \ldots + c_nv_n) = c_1Av_1 + \ldots + c_nAv_n = \lambda_1c_1v_1 + \ldots + \lambda_nc_nv_n.$$

Writing

$$c = \begin{pmatrix} c_1 \\ \vdots \\ c_n \end{pmatrix},$$

we see that both equations can be rewritten as

$$u = Pc \text{ and } Au = P = \begin{pmatrix} \lambda_1c_1 \\ \vdots \\ \lambda_nc_n \end{pmatrix} = PDc$$

respectively, so that

$$APc = PDc.$$

Note that $c \in \mathbb{R}^n$ is arbitrary. This implies that $(AP - PD)c = 0$ for every $c \in \mathbb{R}^n$. Hence we must have $AP = PD$. Since the columns of P are linearly independent, it follows that P is invertible. Hence $P^{-1}AP = D$ as required.

Example. Consider the matrix

$$A = \begin{pmatrix} -1 & 6 & -12 \\ 0 & -13 & 30 \\ 0 & -9 & 20 \end{pmatrix},$$

as in Example. We have $P^{-1}AP = D$, where

$$P = \begin{pmatrix} 1 & 0 & 1 \\ 0 & 2 & -5 \\ 0 & 1 & -3 \end{pmatrix} \text{ and } D = \begin{pmatrix} -1 & 0 & 0 \\ 0 & 2 & 0 \\ 0 & 0 & 5 \end{pmatrix}.$$

Example. Consider the matrix

$$A = \begin{pmatrix} 17 & -10 & -5 \\ 45 & -28 & -15 \\ -30 & 20 & 12 \end{pmatrix},$$

as in Example. We have $P^{-1}AP = D$, where

$$P = \begin{pmatrix} 1 & 1 & 2 \\ 3 & 0 & 3 \\ -2 & 3 & 0 \end{pmatrix} \text{ and } D = \begin{pmatrix} -3 & 0 & 0 \\ 0 & 2 & 0 \\ 0 & 0 & 2 \end{pmatrix}.$$

Definition. Suppose that A is an $n \times n$ matrix, with entries in $\mathbb{R}$. We say that A is diagonalizable if there exists an invertible matrix P, with entries in $\mathbb{R}$, such that $P^{-1}AP$ is a diagonal matrix, with entries in $\mathbb{R}$. It follows from Proposition that an $n \times n$ matrix A with entries in $\mathbb{R}$ is diagonalizable if its eigenvectors form a basis for $\mathbb{R}^n$. In the opposite direction, we establish the following result.

Proposition. Suppose that A is an nn matrix, with entries in $\mathbb{R}$. Suppose further that A is diagonalizable. Then A has n linearly independent eigenvectors in $\mathbb{R}^n$.

Proof. Suppose that A is diagonalizable. Then there exists an invertible matrix P, with entries in $\mathbb{R}$, such that $D = P^{-1}AP$ is a diagonal matrix, with entries in $\mathbb{R}$.

Denote by $v_1, ..., v_n$ the columns of P; in other words, write

$$P = (v_1 \ldots v_n).$$

Also write

$$D = \begin{pmatrix} \lambda_1 & & \\ & \ddots & \\ & & \lambda_n \end{pmatrix}.$$

Clearly we have $AP = PD$. It follows that

$$(Av_1 \ldots Av_n) = A(v_1 \ldots v_n) = (v_1 \ldots v_n)\begin{pmatrix} \lambda_1 & & \\ & \ddots & \\ & & \lambda_n \end{pmatrix}$$

$$= (\lambda_1 v_1 \ldots \lambda l_n v_n).$$

Equating columns, we obtain

$$Av_1 = \lambda_1 v_1, ..., Av_n = \lambda_n v_n.$$

It follows that A has eigenvalues $\lambda_1, ..., \lambda_n \in \mathbb{R}$, with corresponding eigenvectors $v_1, ..., v_n \in \mathbb{R}^n$. Since P is invertible and $v_1, ..., v_n$ are the columns of P, it follows that the eigenvectors $v_1, ..., v_n$ are linearly independent.

In view of Propositions, the question of diagonalizing a matrix A with entries in $\mathbb{R}$ is reduced to one of linear independence of its eigenvectors.

Proposition. Suppose that A is an $n \times n$ matrix, with entries in $\mathbb{R}$. Suppose further that A has distinct eigenvalues $\lambda_1, ..., \lambda_n \in \mathbb{R}$, with corresponding eigenvectors $v_1, ..., v_n \in \mathbb{R}^n$. Then $v_1, ..., v_n$ are linearly independent.

Proof. Suppose that $v_1, ..., v_n$ are linearly dependent. Then there exist $c_1, ..., c_n \in \mathbb{R}$, not all zero, such that

$$c_1v_1 + ... + c_nv_n = 0.$$

Then

$$A(c_1v_1 + ... + c_nv_n) = c_1Av_1 + ... + c_nAv_n = \lambda_1c_1v_1 + ... + \lambda_nc_nv_n = 0.$$

Since $v_1, ..., v_n$ are all eigenvectors and hence non-zero, it follows that at least two numbers among $c_1, ..., c_n$ are non-zero, so that $c_1, ..., c_{n-1}$ are not all zero. Multiplying by λ_n and subtracting, we obtain

$$(\lambda_1 - n)c_1v_1 + ... + (\lambda_{n-1} - \lambda_n)c_{n-1}v_{n-1} = 0.$$

Note that since $\lambda_1, ..., \lambda_n$ are distinct, the numbers $\lambda_1 - \lambda_n, ..., \lambda_{n-1} - \lambda_n$ are all non-zero. It follows that $v_1, ..., v_{n-1}$ are linearly dependent. To summarize, we can eliminate one eigenvector and the remaining ones are still linearly dependent. Repeating this argument a finite number of times, we arrive at a linearly dependent set of one eigenvector, clearly an absurdity.

We now summarize our discussion in this section.

Diagonalization Process. Suppose that A is an $n \times n$ matrix with entries in $\mathbb{R}$.

(1) Determine whether the n roots of the characteristic polynomial $\det(A - \lambda I)$ are real.

(2) If not, then A is not diagonalizable. If so, then nd the eigenvectors corresponding to these eigenvalues. Determine whether we can find n linearly independent eigenvectors.

(3) If not, then A is not diagonalizable. If so, then write

$$P = (v_1 \ldots v_n) \text{ and } D = \begin{pmatrix} \lambda_1 & & \\ & \ddots & \\ & & \lambda_n \end{pmatrix},$$

where $\lambda_1, ..., \lambda_n \in \mathbb{R}$ are the eigenvalues of A and where $v_1, ..., v_n \in \mathbb{R}^n$ are respectively their corresponding eigenvectors. Then $P^{-1}AP = D$.

Some Remarks

In all the examples we have discussed, we have chosen matrices A such that the characteristic polynomial $\det(A - \lambda I)$ has only real roots. However, there are matrices A where the characteristic polynomial has non-real roots. If we permit $\lambda_1, ..., \lambda_n$ to take values in C and permit "eigenvectors" to have entries in $\mathbb{C}$, then we may be able to "diagonalize" the matrix A, using matrices P and D with entries in $\mathbb{C}$. The details are similar.

Example. Consider the matrix

$$A = \begin{pmatrix} 1 & -5 \\ 1 & -1 \end{pmatrix}.$$

To find the eigenvalues of A, we need to find the roots of

$$\det = \begin{pmatrix} 1-\lambda & -5 \\ 1 & -1-\lambda \end{pmatrix} = 0;$$

in other words, $\lambda^2 + 4 = 0$. Clearly there are no real roots, so the matrix A has no eigenvalues in $\mathbb{R}$. Try to show, however, that the matrix A can be "diagonalized" to the matrix

$$D = \begin{pmatrix} 2i & 0 \\ 0 & -2i \end{pmatrix}.$$

We also state without proof the following useful result which will guarantee many examples where the characteristic polynomial has only real roots.

Proposition. Suppose that A is an $n \times n$ matrix, with entries in $\mathbb{R}$. Suppose further that A is symmetric. Then the characteristic polynomial $\det(A - \lambda I)$ has only real roots. We conclude this section by discussing an application of diagonalization. We illustrate this by an example.

Example. Consider the matrix

$$A = \begin{pmatrix} 17 & -10 & -5 \\ 45 & -28 & -15 \\ -30 & 20 & 12 \end{pmatrix},$$

as in Example. Suppose that we wish to calculate A^{98}. Note that $P^{-1}AP = D$, where

$$P = \begin{pmatrix} 1 & 1 & 2 \\ 3 & 0 & 3 \\ -2 & 3 & 0 \end{pmatrix} \text{ and } D = \begin{pmatrix} -3 & 0 & 0 \\ 0 & 2 & 0 \\ 0 & 0 & 2 \end{pmatrix}.$$

It follows that $A = PDP^{-1}$, so that

$$A^{98} = \underbrace{(PDP^{-1})\ldots(PDP^{-1})}_{98} = PD^{98}P^{-1} = P\begin{pmatrix} 3^{98} & 0 & 0 \\ 0 & 2^{98} & 0 \\ 0 & 0 & 2^{98} \end{pmatrix}P^{-1}.$$

This is much simpler than calculating A^{98} directly.

An Application to Genetics

In this section, we discuss very briey the problem of autosomal inheritance. Here we consider a set of two genes designated by G and g. Each member of the population inherits one from each parent, resulting in possible genotypes GG, Gg and gg. Furthermore, the gene G dominates the gene g, so that in the case of human eye colours, for example, people with genotype GG or Gg have brown eyes while people with genotype gg have blue eyes. It is also believed that each member of the population has equal probability of inheriting one or the other gene from each parent. The table below gives these peobabilities in detail. Here the genotypes of the parents are listed on top, and the genotypes of the ospring are listed on the left.

	$GG-GG$	$GG-Gg$	$GG-gg$	$Gg-Gg$	$Gg-gg$	$gg-gg$
GG	1	$\frac{1}{2}$	0	$\frac{1}{4}$	0	0
Gg	0	$\frac{1}{2}$	1	$\frac{1}{2}$	$\frac{1}{2}$	0
gg	0	0	0	$\frac{1}{4}$	$\frac{1}{2}$	1

Example. Suppose that a plant breeder has a large population consisting of all three genotypes. At regular intervals, each plant he owns is fertilized with a plant known to have genotype GG, and is then disposed of and replaced by one of its osprings. We would like to study the distribution of the three genotypes after n rounds of fertilization and replacements, where n is an arbitrary positive integer. Suppose that $GG(n)$, $Gg(n)$ and $gg(n)$ denote the proportion of each genotype after n rounds of fertilization and replacements, and that $GG(0)$, $Gg(0)$ and gg(0) denote the initial proportions. Then clearly we have

$$GG(n) + Gg(n) + gg(n) = 1 \text{ for every } n = 0, 1, 2, \ldots$$

On the other hand, the left hand half of the table above shows that for every $n = 1, 2, 3, \ldots$, we have

$$GG(n) = GG(n-1) + \frac{1}{2}Gg(n-1),$$

$$Gg(n) = \frac{1}{2}Gg(n-1) + gg(n-1),$$

and

$$gg(n) = 0,$$

so that

$$\begin{pmatrix} GG(n) \\ Gg(n) \\ gg(n) \end{pmatrix} = \begin{pmatrix} 1 & 1/2 & 0 \\ 0 & 1/2 & 1 \\ 0 & 0 & 0 \end{pmatrix} \begin{pmatrix} GG(n-1) \\ Gg(n-1) \\ gg(n-1) \end{pmatrix}.$$

It follows that

$$\begin{pmatrix} GG(n) \\ Gg(n) \\ gg(n) \end{pmatrix} = A^n \begin{pmatrix} GG(0) \\ Gg(0) \\ gg(0) \end{pmatrix} \text{ for every } n = 1, 2, 3, \ldots,$$

where the matrix

$$A = \begin{pmatrix} 1 & 1/2 & 0 \\ 0 & 1/2 & 1 \\ 0 & 0 & 0 \end{pmatrix}$$

has eigenvalues $\lambda_1 = 1, \lambda_2 = 0, \lambda_3 = 1/2$, with respective eigenvectors

$$P = \begin{pmatrix} 1 & 1 & 1 \\ 0 & -2 & -1 \\ 0 & 1 & 0 \end{pmatrix}$$

We therefore write

$$P = \begin{pmatrix} 1 & 1 & 1 \\ 0 & -2 & -1 \\ 0 & 1 & 0 \end{pmatrix} \text{ and } D = \begin{pmatrix} 1 & 0 & 0 \\ 0 & 0 & 0 \\ 0 & 0 & 1/2 \end{pmatrix},$$

$$\text{with } P^{-1} = \begin{pmatrix} 1 & 1 & 1 \\ 0 & 0 & 1 \\ 0 & -1 & -2 \end{pmatrix}.$$

Then $P^{-1}AP = D$, so that $A = PDP^{-1}$, and so

$$A^n = PD^nP^{-1} = \begin{pmatrix} 1 & 1 & 1 \\ 0 & -2 & -1 \\ 0 & 1 & 0 \end{pmatrix} \begin{pmatrix} 1 & 0 & 0 \\ 0 & 0 & 0 \\ 0 & 0 & 1/2^n \end{pmatrix} \begin{pmatrix} 1 & 1 & 1 \\ 0 & 0 & 1 \\ 0 & -1 & -2 \end{pmatrix}$$

$$= \begin{pmatrix} 1 & 1-1/2^n & 1-1/2^{n-1} \\ 0 & 1/2^n & 1/2^{n-1} \\ 0 & 0 & 0 \end{pmatrix}$$

It follows that

$$\begin{pmatrix} GG(n) \\ Gg(n) \\ gg(n) \end{pmatrix} = \begin{pmatrix} 1 & 1-1/2^n & 1-1/2^{n-1} \\ 0 & 1/2^n & 1/2^{n-1} \\ 0 & 0 & 0 \end{pmatrix} \begin{pmatrix} GG(0) \\ Gg(0) \\ gg(0) \end{pmatrix}$$

$$= \begin{pmatrix} GG(0)+Gg(0)+gg(0)-Gg(0)/2^n - gg(0)/2^{n-1} \\ Gg(0)/2^n + gg(0)/2^{n-1} \\ 0 \end{pmatrix}$$

$$= \begin{pmatrix} 1-Gg(0)/2^n - gg(0)/2^{n-1} \\ Gg(0)/2^n + gg(0)/2^{n-1} \\ 0 \end{pmatrix} \to \begin{pmatrix} 1 \\ 0 \\ 0 \end{pmatrix} \text{ as } n \to \infty$$

This means that nearly the whole crop will have genotype *GG*.

Chapter 6

Inner Product Spaces

Inner Product in $\mathbb{R}^n$ and $\mathbb{C}^n$

Inner product and norm in $\mathbb{R}^n$. In dimensions 2 and 3, we defined the length of a vector x (*i.e.*, the distance from its endpoint to the origin) by the Pythagorean rule, for example in $\mathbb{R}^3$ the length of the vector is defined as

$$\| x \| = \sqrt{x_1^2 + x_2^2 + x_3^2}.$$

It is natural to generalize this formula for all n, to define the norm of the vector $x \in \mathbb{R}^n$ as

$$\| x \| = \sqrt{x_1^2 + x_2^2 + \ldots + x_n^2}.$$

The word *norm* is used as a fancy replacement to the word length.

The *dot product* in $\mathbb{R}^3$ was defined as $x \cdot y = x_1y_2 + x_2y_2 + x_3y_3$, where

$$x = (x_1, x_2, x_3)^T \text{ and } y = (y_1, y_2, y_3)^T.$$

Similarly, in $\mathbb{R}^n$ one can define the *inner product* (x, y) of two vectors

$$x = (x_1, x_2, \ldots, x_n)^T, y = (y_1, y_2, \ldots, y_n)^T \text{ by}$$

$$(x, y) := x_1y_1 + x_2y_2 + \ldots + x_ny_n = y^T x,$$

so $\| x \| = \sqrt{(x,x)}$.

Note, that $y^T x = x^T y$, and we use the notation $y^T x$ only to be consistent.

Inner product and norm in $\mathbb{C}^n$. Let us now define norm and inner product for $\mathbb{C}^n$. The complex space $\mathbb{C}^n$ is the most natural space from the point of view of spectral theory: even if one starts from a matrix with real coefficients (or operator on a real vectors space), the eigenvalues can be complex, and one needs to work in a complex space.

For a complex number $z = x + iy$, we have $| z |^2 = x^2 + y^2 = z\bar{z}$. If $z \in \mathbb{C}^n$ is given by

$$z = \begin{pmatrix} z_1 \\ z_2 \\ \vdots \\ z_n \end{pmatrix} = \begin{pmatrix} x_1 + iy_1 \\ x_2 + iy_2 \\ \vdots \\ x_n + iy_n \end{pmatrix},$$

it is natural to define its norm $\| z \|$ by

$$\| z \|^2 = \sum_{k=1}^{n} (x_k^2 + y_k^2) = \sum_{k=1}^{n} | z_k |^2.$$

Let us try to define an inner product on $\mathbb{C}^n$ such that $\| z \|^2 = (z, z)$. One of the choices is to define (z, w) by

$$(z,w) = z_1 \overline{w}_1 + z_2 \overline{w}_2 + ... + z_n \overline{w}_n = \sum_{k=1}^{n} z_k w_k,$$

and that will be our definition of the inner product in $\mathbb{C}^n$.

To simplify the notation, let us introduce a new notion. For a matrix A let us define its *Hermitian adjoint*, or simply *adjoint* A^* by $A^* = A^{-T}$, meaning that we take the transpose of the matrix, and then take the complex conjugate of each entry. Note, that for a real matrix A, $A^* = AT$.

Using the notion of A^*, one can write the inner product in $\mathbb{C}^n$ as

$$(z, w) = w^*z.$$

Remark. It is easy to see that one can define a different inner product in $\mathbb{C}^n$ such that $\| z \|_2 = (z, z)$, namely the inner product given by

$$(z, w)_1 = \overline{z}_1 w_1 + \overline{z}_2 w_2 + ... + \overline{z}_n w_n = z^*w.$$

We did not specify what properties we want the inner product to satisfy, but z^*w and w^*z are the only reasonable choices giving $\| z \|^2 = (z, z)$.

Note, that the above two choices of the inner product are essentially equivalent: the only difference between them is notational, because $(z, w)_1 = (w, z)$.

While the second choice of the inner product looks more natural, the first one, $(z, w) = w^*z$ is more widely used, so we will use it as well.

Inner Product Spaces. The inner product we defined for $\mathbb{R}^n$ and $\mathbb{C}^n$ satisfies the following properties:

1. (Conjugate) symmetry: $(x, y) = \overline{(x, y)}$; note, that for a real space, this property is just symmetry, $(x, y) = (y, x)$;
2. Linearity: $(\alpha x + \alpha y, z) = (x, z) + (y, z)$ for all vector x, y, z and all scalars α, β;
3. Non–negativity: $(x, x) \geq 0 \ \forall x$;
4. Non–degeneracy: $(x, x) = 0$ if and only if $x = 0$.

Let V be a (complex or real) vector space. An *inner product* on V is a function, that assign to each pair of vectors x, y a scalar, denoted by (x, y) such that the properties 1–4 from the previous section are satisfied.

Note that for a real space V we assume that (x, y) is always real, and for a complex space the inner product (x, y) can be complex.

A space V together with an inner product on it is called an *inner product space*. Given an inner product space, one defines the norm on it by

$$\| x \| = \sqrt{(x,x)}.$$

Example. Let V be $\mathbb{R}^n$ or $\mathbb{C}^n$. We already have an inner product

$(x, y) = y^*x = y^*x = \sum_{k=1}^{n} x_k y_k$

defined above.

This inner product is called the standard inner product in $\mathbb{R}^n$ or $\mathbb{C}^n$ We will use symbol F to denote both $\mathbb{C}$ and $\mathbb{R}$. When we have some statement about the space F_n, it means the statement is true for both $\mathbb{R}^n$ and $\mathbb{C}^n$.

Example. Let V be the space Pn of polynomials of degree at most n. Define the inner product by

$$(f,g) = \int_{-1}^{1} f(t)\overline{g(t)}dt.$$

It is easy to check, that the above properties 1–4 are satisfied.

This definition works both for complex and real cases. In the real case we only allow polynomials with real coefficients, and we do not need the complex conjugate here.

Let us recall, that for a square matrix A, its trace is defined as the sum of the diagonal entries,

$$\text{trace } A = \sum_{k=1}^{n} a_{k,k}.$$

Example. For the space $M_m \times n$ of $m \times n$ matrices let us define the so–called Frobenius inner product by

$$(A, B) = \text{trace } (B^*A).$$

Again, it is easy to check that the properties, *i.e.*, that we indeed defined an inner product.

Note, that

$$\text{trace } (B^* A) = \sum_{j,k} A_{j,k} \overline{B}_{j,k},$$

so this inner product coincides with the standard inner product in $\mathbb{C}^{mn}$.

Properties of Inner Product. The statements we get in this section are true for any inner product space, not only for F_n. To prove them we use only properties 1–4 of the inner product.

First of all let us notice, that properties 1 and 2 imply that

$$2'.\ (x, \alpha y + \beta z) = \bar{\alpha}(x, y) + \bar{\beta}(x, z).$$

Indeed,

$$(x, \alpha y + \beta z) = \overline{(\alpha y + \beta z, x)} = \overline{\alpha(y,x) + \beta(z,x)}$$
$$= \overline{\alpha(y,x)} + \overline{\beta(z,x)} = \bar{\alpha}(x,y) + \bar{\beta}(x,z)$$

Note also that property 2 implies that for all vectors x $(0, x) = (x, 0) = 0$.

Lemma. Let x be a vector in an inner product space V. Then $x = 0$ if and only if

$$(x, y) = 0 \ \forall y \in V.$$

Proof. Since $(0, y) = 0$ we only need to show that implies $x = 0$. Putting $y = x$ in) we get $(x, x) = 0$, so $x = 0$.

Applying the above lemma to the difference $x - y$ we get the following

Corollary. Let x, y be vectors in an inner product space V. The equality $x = y$ holds if and only if

$$(x, z) = (y, z) \ \forall z \in V.$$

The following corollary is very simple, but will be used a lot

Corollary. Suppose two operators $A, B : x \to Y$ satisfy

$$(Ax, y) = (Bx, y) \quad \forall x \in X, \ \forall y \in V.$$

Then $A = B$

Proof. By the previous corollary (fix x and take all possible y's) we get $Ax = B_x$. Since this is true for all $x \in X$, the transformations A and B coincide.

The following property relates the norm and the inner product.

Theorem. (Cauchy–Schwarz inequality).

$$|(x, y)| \leq \|x\| \cdot \|y\|.$$

Proof. The proof we are going to present, is not the shortest one, but it gives a lot for the understanding.

Let us consider the real case first. If $y = 0$, the statement is trivial, so we can assume that $y \neq 0$. By the properties of an inner product, for all scalar t

$$0 \leq \|x - ty\|^2 = (x - ty, x - ty) = \|x\|^2 - 2t(x, y) + t^2\|y\|^2.$$

In particular, this inequality should hold for $t = \dfrac{(x,y)}{\|y\|^2}1$, and for this point the inequality becomes

$$0 \leq \|x\|^2 - 2\frac{(x,y)^2}{\|y\|^2} + \frac{(x,y)^2}{\|y\|^2} = \|x\|^2 - \frac{(x,y)^2}{\|y\|^2},$$

which is exactly the inequality we need to prove.

There are several possible ways to treat the complex case. One is to replace x by αx, where α is a complex constant, $|\alpha| = 1$ such that $(\alpha x, y)$ is real, and then repeat the proof for the real case.

The other possibility is again to consider

$$0 \leq \| x - ty \|^2 = (x - ty, x - ty) = (x, x - ty) - t(y, x - ty)$$
$$= \| x \|^2 - t(y, x) - \bar{t}(x, y) + |t|^2 \| y \|^2.$$

Substituting $t = \dfrac{(x,y)}{\| y \|^2} + \dfrac{\overline{(x,y)}}{\| y \|^2}$ into this inequality, we get

$$0 \leq \| x \|^2 - \frac{|(x,y)|^2}{\| y \|^2}$$

which is the inequality we need.

Note, that the above paragraph is in fact a complete formal proof of the theorem. The reasoning before that was only to explain why do we need to pick this particular value of t.

An immediate Corollary of the Cauchy–Schwarz Inequality is the following lemma.

Lemma. (Triangle inequality). *For any vectors x, y in an inner product space*

$$\| x + y \| \leq \| x \| + \| y \|.$$

Proof. $\| x + y \|^2 = (x + y, x + y) = \| x \|^2 + \| y \|^2 + (x, y) + (y, x)$

$$\leq \| x \|^2 + \| y \|^2 + 2\| x \| \cdot \| y \| = (\| x \| + \| y \|)^2.$$

The following polarization *identities* allow one to reconstruct the inner product from the norm:

Lemma (Polarization identities). For $x, y \in V$

$$(x,y) = \frac{1}{4}(\| x + y \|^2 - \| x - y \|^2)$$

if V is a real inner product space, and

$$(x,y) = \frac{1}{4} \sum_{a=\pm 1, +i} \alpha \| x + \alpha y \|^2$$

if V is a *complex space*.

The lemma is proved by direct computation.

Another important property of the norm in an inner product space can be also checked by direct calculation.

Lemma. (Parallelogram Identity). *For any vectors u, v*

$$\| u + v \|^2 + \| u - v \|^2 = 2(\| u \|^2 + \| v \|^2).$$

In 2–dimensional space this lemma relates sides of a parallelogram with its diagonals, which explains the name. It is a well–known fact from planar geometry.

NORMS

Normed spaces

We have proved before that the norm $\| v \|$ satisfies the following properties:

1. Homogeneity: $\| v \| = \| \cdot \| \, \| v \|$ for all vectors v and all scalars .
2. Triangle inequality: $\| u + v \| \leq \| u \| + \| v \|$.
3. Non–negativity: $\| v \| \geq 0$ for all vectors v.
4. Non–degeneracy: $\| v \| = 0$ if and only if $v = 0$.

Suppose in a vector space V we assigned to each vector v a number $\| v \|$ such that above properties 1–4 are satisfied. Then we say that the function $v \mapsto \| v \|$ is a norm. *A* vector space V equipped with a norm is called a *normed space.*

Any inner product space is a normed space, because the norm $\| v \| = \sqrt{(v, v)}$ satisfies the above properties 1–4. However, there are many other normed spaces. For example, given p, $1 < p < \infty$ one can define the norm $\| \cdot \|_p$ on $\mathbb{R}^n$ or $\mathbb{C}^n$ by

$$\| x \|_p = (| x_1 |^p + | x_2 |^p + ... + \| x_n \|^p)^{1/p} = \left(\sum_{k=1}^{n} | x_k |^p \right)^{1/p}.$$

One can also define the norm $\| . \|_\infty$ $(p = \infty)$ by
$\| x \|_\infty = \max\{| x_k |: k = 1, 2, ..., n\}$.

The norm $\| \cdot \|_p$ for $p = 2$ coincides with the regular norm obtained from the inner product.

To check that $\| \cdot \|_p$ is indeed a norm one has to check that it satisfies all the above properties 1–4. Properties 1, 3 and 4 are very easy to check. The triangle inequality (property 2) is easy to check for $p = 1$ and $p = 1$ (and we proved it for $p = 2$).

For all other p the triangle inequality is true, but the proof is not so simple, and we will not present it here. The triangle inequality for $k \cdot$ kp even has special name: its called *Minkowski inequality,* after the German mathematician H. Minkowski.

Note, that the norm $\| \cdot \|_p$ for $p \neq 2$ cannot be obtained from an inner product. It is easy to see that this norm is not obtained from the standard inner product in R_n (C_n). But we claim more! We claim that it is impossible to introduce an inner product which gives rise to the norm $\| . \|_p, p \neq 2$.

This statement is actually quite easy to prove. It is easy to see that the Parallelogram Identity fails for the norm $\| . \|_p, p \neq 2$. and one can easily find a counter example in $\mathbb{R}^2$, which then gives rise to a counter example in all other spaces.

In fact, the Parallelogram Identity, as the theorem below asserts completely characterizes norms obtained from an inner product.

Theorem. A norm in a normed space is obtained from some inner product if and only if it satisfies the Parallelogram Identity

$$\| u+v \|^2 + \| u-v \|^2 = 2(\| u \|^2 + \| v \|^2) \ \forall u,v \in V.$$

The inverse implication is more complicated. If we are given a norm, and this norm came from an inner product, then we do not have any choice; this inner product must be given by the polarization identities.

But, we need to show that (x, y) we got from the polarization identities is indeed an inner product, *i.e.*, that it satisfies all the properties. It is indeed possible to check if the norm satisfies the parallelogram identity, but the proof is a bit too involved, so we do not present it here.

ORTHOGONALITY

Orthogonal and Orthonormal Bases

Definition. Two vectors u and v are called *orthogonal* (also *perpendicular*) if $(u, v) = 0$. We will write $u \perp v$ to say that the vectors are orthogonal. Note, that for orthogonal vectors u and v we have the following Pythagorean identity:

$$\| u+v \|^2 = \| u \|^2 + \| v \|^2 \text{ if } u \perp v.$$

The proof is straightforward computation,

$$\| u+v \|^2 = (u+v, u+v) = (u, u) + (v, v) + (u, v) + (v, u) = \| u \|^2 + \| v \|^2$$

($(u, v) = (v, u) = 0$ because of orthogonality).

Definition. We say that a vector v is orthogonal to a subspace E if v is orthogonal to all vectors w in E.

We say that subspaces E and F are orthogonal if all vectors in E are orthogonal to F, *i.e.*, all vectors in E are orthogonal to all vectors in F

The following lemma shows how to check that a vector is orthogonal to a subspace.

Lemma. Let E be spanned by vectors $v_1, v_2, \ldots, v_r$. Then v ? E if and only if

$$v \perp v_k, \ \forall k = 1, 2, \ldots, r.$$

Proof. By the definition, if $v \perp E$ then v is orthogonal to all vectors in E. In particular, $v \perp v_k$, $k = 1, 2, \ldots, r$.

On the other hand, let $v \perp v_k$, $k = 1, 2, \ldots, r$. Since the vectors v_k span E, any vector $w \in E$ can be represented as a linear combination $\sum_{k=1}^{r} a_k V_k$. Then so $v \perp w$.

Definition. A system of vectors $v_1, v_2, \ldots, v_n$ is called *orthogonal* if any two vectors are orthogonal to each other (*i.e.*, if $(v_j , v_k) = 0$ for $j \neq k$).

If, in addition $\|\, v_k \| = 1$ for all k, we call the system *orthonormal*.

Lemma. (Generalized Pythagorean identity). Let $v_1, v_2, \ldots, v_n$ *be an orthogonal system. Then*

$$\left\| \sum_{k=1}^{n} \alpha_k v_k \right\|^2 = \sum_{k=1}^{n} |\alpha_k|^2 \| v_k \|^2$$

This formula looks particularly simple for orthonormal systems, where $\| v_k \| = 1$.

Proof of the Lemma.

$$\left\| \sum_{k=1}^{n} \alpha_k v_k \right\|^2 = \left(\sum_{k=1}^{n} \alpha_k v_k, \sum_{j=1}^{n} \alpha_j v_j \right) = \sum_{k=1}^{n} \sum_{j=1}^{n} a_k \bar{a}_j (v_k, v_j).$$

Because of orthogonality $v_k, v_j) = 0$ if $j \neq$ k. Therefore we only need to sum the terms with $j = k$, which gives exactly

$$\sum_{k=1}^{n} |\alpha_k|^2 (v_k, v_k) = \sum_{k=1}^{n} |a_k|^2 \| v_k \|^2 .$$

Corollary. Any orthogonal system $v_1, v_2, ..., v_n$ of non–zero vectors is linearly independent.

Proof. Suppose for some $\alpha_1, ..., \beta_2, ..., \alpha_n$ we have $\sum_{k=1}^{n} \alpha_k v_k = 0$. Then by the Generalized Pythagorean identity

$$0 = \| 0 \|^2 = \sum_{k=1}^{n} |\alpha_k|^2 \| v_k \|^2 .$$

Since $\| v_k \| \neq 0$ $(v_k \neq 0)$ we conclude that

$$\alpha_k = 0 \ \forall k,$$

so only the trivial linear combination gives 0.

Remark. In what follows we will usually mean by an orthogonal system an orthogonal system of non–zero vectors. Since the zero vector 0 is orthogonal to everything, it always can be added to any orthogonal system, but it is really not interesting to consider orthogonal systems with zero vectors.

Orthogonal and Orthonormal Bases

Definition. An orthogonal (orthonormal) system $v_1, v_2, ..., v_n$ which is also a basis is called an orthogonal (orthonormal) basis.

It is clear that in dim $V = n$ then any orthogonal system of n non–zero vectors is an orthogonal basis. As we studied before, to find coordinates of a vector in a basis one needs to solve a linear system. However, for an orthogonal basis finding coordinates of a vector is much easier. Namely, suppose $v_1, v_2, ..., v_n$ is an orthogonal basis, and let

$$x = a_1 v_1 + 2 v_2 + ... + a_n v_n = \sum_{j=1}^{n} \alpha_j v_j .$$

Taking inner product of both sides of the equation with v_1 we get

$$(x, v_1) = \sum_{j=1}^{n} \alpha_j (v_j, v_1) = \alpha_1(v_1, v_1) = \alpha_1 \| v_1 \|^2$$

(all inner products $(v_j, v_1) = 0$ if $j \neq 1$), so

$$\alpha_1 = \frac{(x, v_1)}{\| v_1 \|^2}.$$

Similarly, multiplying both sides by v_k we get

$$(x, v_k) = \sum_{j=1}^{n} \alpha_j (v_j, v_k) = \alpha_k (v_k, v_k) = a_k \| v_k \|^2$$

so

$$\alpha_k = \frac{(x, v_k)}{\| v_k \|^2}.$$

Therefore,

to find coordinates of a vector in an orthogonal basis one does not need to solve a linear system, the coordinates are determined by the formula.

This formula is especially simple for orthonormal bases, when $\| v_k \| = 1$.

Orthogonal Projection and Gram–Schmidt Orthogonalization

Recalling the definition of orthogonal projection from the classical planar (2–dimensional) geometry, one can introduce the following definition. Let E be a subspace of an inner product space V.

Definition. For a vector v its orthogonal projection $P_E v$ onto the subspace E is a vector w such that

1. $w \in E$;
2. $v - w \perp E$.

We will use notation $w = P_E v$ for the orthogonal projection.

After introducing an object, it is natural to ask:

1. Does the object exist?
2. Is the object unique?
3. How does one find it?

We will show first that the projection is unique. Then we present a method of finding the projection, proving its existence.

The following theorem shows why the orthogonal projection is important and also proves that it is unique.

Theorem. The orthogonal projection $w = P_E v$ minimizes the distance from v to E, i.e., for all $x \in E$

$$\| v - w \| \leq \| v - x \|.$$

Moreover, if for some $x \in E$

$$\| v - w \| = \| v - x \|,$$

then $x = v$.

Proof. Let $y = w - x$. Then

$$v - x = v - w + w - x = v - w + y.$$

Since $v - w \perp E$ we have $y \perp v - w$ and so by Pythagorean Theorem

$$\| v - x \|^2 = \| v - w \|^2 + \| y \|^2 \geq \| v - w \|^2.$$

Note that equality happens only if $y = 0$ *i.e.*, if $x = w$.

The following proposition shows how to find an orthogonal projection if we know an orthogonal basis in E.

Proposition. Let $v_1, v_2, \ldots, v_r$ be an orthogonal basis in E. Then the orthogonal projection $P_E v$ of a vector v is given by the formula

$$P_E v = \sum_{k=1}^{r} \alpha_k v_k, \text{ where } \alpha_k = \frac{(v, v_k)}{\| v_k \|^2}.$$

In other words

$$P_E v = \sum_{k=1}^{r} \frac{(v, v_k)}{\| v_k \|^2} v_k.$$

Note that the formula for k coincides with, *i.e.*, this formula for an orthogonal system (not a basis) gives us a projection onto its span.

Remark. It is easy to see now from formula that the orthogonal projection P_E is a linear transformation.

One can also see linearity of P_E directly, from the definition and uniqueness of the orthogonal projection. Indeed, it is easy to check that for any x and y the vector

$$\alpha x + \beta y - (\alpha P_E x - \beta P_E y)$$

is orthogonal to any vector in E, so by the definition

$$P_E(\alpha x + \beta y) = \alpha P_E x + \beta P_E y.$$

Remark. Recalling the definition of inner product in $\mathbb{C}^n$ and $\mathbb{R}^n$ one can get from the above formula the matrix of the orthogonal projection P_E onto E in $\mathbb{C}^n$ $(\mathbb{R}^n)$ is given by

$$P_E = \sum_{k=1}^{r} \frac{1}{\| V_k \|^2} v_k v_k^*$$

where columns $v_1, v_2, \ldots, v_r$ form an orthogonal basis in E.

Proof of Proposition. Let

$$w:\sum_{k=1}^{r}\alpha_k v_k, \text{ where } \alpha_k=\frac{(v,v_k)}{\|v_k\|^2}.$$

We want to show that $v - w \perp E$. By Lemma it is sucient to show that $v - w \perp v_k$, $k = 1, 2, ..., n$. Computing the inner product we get for $k = 1, 2, ..., r$

$$(v - w, v_k) = (v, v_k) - (w, v_k) = (v, v_k) - \sum_{j=1}^{r}\alpha_j(v_j, v_k)$$

$$=(v,v_k)-a_k(v_k,v_k)=\frac{(v,v_k)}{\|v_k\|^2}\|v_k\|^2=0.$$

So, if we know an orthogonal basis in E we can find the orthogonal projection onto E. In particular, since any system consisting of one vector is an orthogonal system, we know how to perform orthogonal projection onto one–dimensional spaces.

But how do we find an orthogonal projection if we are only given a basis in E? Fortunately, there exists a simple algorithm allowing one to get an orthogonal basis from a basis.

Gram–Schmidt Orthogonalization Algorithm. Suppose we have a linearly independent system $x_1, x_2, ..., x_n$. The Gram–Schmidt method constructs from this system an orthogonal system $v_1, v_2, ..., v_n$ such that span$\{x_1, x_2, ..., x_n\}$ = span $\{v_1, v_2, ..., v_n\}$.

Moreover, for all $r \leq n$ we get

span $\{x_1, x_2, ..., x_r\}$ = span $\{v_1, v_2, ..., v_r\}$

Now let us describe the algorithm.

Step 1. Put $v_1 := x_1$. Denote by $E_1 :=$ span$\{x_1\}$ = span$\{v_1\}$.

Step 2. Define v_2 by

$$v_2 = x_2 - P_{E_1}x_2 = x_2 - \frac{(x_2,v_1)}{\|v_1\|^2}v_1.$$

Define E_2 = span $\{v_1, v_2\}$. Note that span $\{x_1, x_2\} = E_2$.

Step 3. Define v_3 by

$$v_3 := x_3 - P_{E_2}x_3 = x_3 - \frac{(x_3,v_1)}{\|v_1\|^2}v_1 - \frac{(x_3,v_2)}{\|v_2\|^2}v_2$$

Put $E_3 :=$ span $\{v_1, v_2, v_3\}$. Note that span $\{x_1, x_2, x_3\} = E_3$. Note also that $x_3 \notin E_2$ so $v_3 \neq 0$.

Step $r + 1$. Suppose that we already made r steps of the process, constructing an orthogonal system (consisting of non–zero vectors) $v_1, v_2, ..., v_r$ such that $E_r :=$ span$\{v_1, v_2, ..., v_r\}$ = span$\{x_1, x_2, ..., x_r\}$. Define

$$v_{r+1} := x_{r+1} - P_{E_r}x_{r+1} = x_{r+1} - \sum_{k=1}^{r}\frac{(x_{r+1}.v_k)}{\|v_k\|^2}v_k$$

Note, that $x_{r+1} \notin E_r$ so $v_{r+1} \neq 0$.
Continuing this algorithm we get an orthogonal system $v_1, v_2, ..., v_n$.
An example. Suppose we are given vectors

$$x_1 = (1, 1, 1)^T, x_2 = (0, 1, 2)^T, x_3 = (1, 0, 2)^T,$$

and we want to orthogonalize it by Gram–Schmidt. On the first step define

$$v_1 = x_1 = (1, 1, 1)^T.$$

On the second step we get

$$v_2 = x_2 - P_{E_1}x_2 = x_2 - \frac{(x_2, v_1)}{\| v_1 \|^2} v_1.$$

Computing

$$(x_2, v_1) = \left(\begin{pmatrix} 0 \\ 1 \\ 2 \end{pmatrix}, \begin{pmatrix} 1 \\ 1 \\ 1 \end{pmatrix} \right) = 3, \| v_1 \|^2 = 3,$$

we get

$$v_2 = \begin{pmatrix} 0 \\ 1 \\ 2 \end{pmatrix} - \frac{3}{3} \begin{pmatrix} 1 \\ 1 \\ 1 \end{pmatrix} = \begin{pmatrix} -1 \\ 0 \\ 0 \end{pmatrix}.$$

Finally, define

$$v_3 = x_3 - P_{E_2}x_3 = x_3 - \frac{(x_3, v_1)}{\| v_1 \|^2} v_1 - \frac{(x_3, v_2)}{\| v_2 \|^2} v_2.$$

Computing

$$\left(\begin{pmatrix} 1 \\ 0 \\ 2 \end{pmatrix}, \begin{pmatrix} 1 \\ 1 \\ 1 \end{pmatrix} \right) = 3, \left(\begin{pmatrix} 1 \\ 0 \\ 2 \end{pmatrix}, \begin{pmatrix} -1 \\ 0 \\ 1 \end{pmatrix} \right) = 1, \| v_1 \|^2 = 3, \ \| v_2 \|^2 = 2$$

($\| v_1 \|^2$ was already computed before) we get

$$v_3 = \begin{pmatrix} 1 \\ 0 \\ 2 \end{pmatrix} - \frac{3}{3} \begin{pmatrix} 1 \\ 1 \\ 1 \end{pmatrix} - \frac{1}{2} \begin{pmatrix} -1 \\ 0 \\ 1 \end{pmatrix} = \begin{pmatrix} \frac{1}{2} \\ -1 \\ \frac{1}{2} \end{pmatrix}$$

Remark. Since the multiplication by a scalar does not change the orthogonality, one can multiply vectors v_k obtained by Gram–Schmidt by any non–zero numbers.

In particular, in many theoretical constructions one normalizes vectors v_k by dividing them by their respective norms $\| v_k \|$. Then the resulting system will be orthonormal, and the formulas will look simpler.

On the other hand, when performing the computations one may want to avoid fractional entries by multiplying a vector by the least common denominator of its entries. Thus one may want to replace the vector v_3 from the above example by $(1, -2, 1)^T$.

Orthogonal Complement. Decomposition $V = E \oplus E^{\perp}$.

Definition. For a subspace E its orthogonal complement E? is the set of all vectors orthogonal to E,

$$E^{\perp} := \{x : x \perp E\}.$$

If $x, y \perp E$ then for any linear combination $\alpha x + \beta y \perp E$. Therefore $E^{\perp}$ is a subspace.

By the definition of orthogonal projection any vector in an inner product space V admits a unique representation

$$v = v_1 + v_2,\ v_1 \in E,\ v_2 \perp E \text{ (eqv. } v_2 \in E^{\perp})$$

(where clearly $v_1 = P_E v$).

This statement is often symbolically written as $V = E \oplus E^{\perp}$, which mean exactly that any vector admits the unique decomposition above.

The following proposition gives an important property of the orthogonal complement.

Proposition. For a subspace E

$$(E^{\perp})^{\perp} = E.$$

Least Square Solution

The equation

$$Ax = b$$

has a solution if and only if $b \in R$ an A. But what do we do to solve an equation that does not have a solution?

This seems to be a silly question, because if there is no solution, then there is no solution. But, situations when we want to solve an equation that does not have a solution can appear naturally, for example, if we obtained the equation from an experiment. If we do not have any errors, the right side b belongs to the column space Ran A, and equation is consistent. But, in real life it is impossible to avoid errors in measurements, so it is possible that an equation that in theory should be consistent, does not have a solution. So, what one can do in this situation?

Least square solution. The simplest idea is to write down the error

$$\| Ax - b \|$$

and try to find x minimizing it. If we can find x such that the error is 0, the system is consistent and we have exact solution. Otherwise, we get the so–called *least square* solution.

The term *least square* arises from the fact that minimizing $\| Ax - b \|$ is equivalent to minimizing

$$\| Ax - b \|^2 = \sum_{k=1}^{m} |(Ax)_k - b_k|^2 = \sum_{k=1}^{m} \left| \sum_{j=1}^{n} A_{k,j} x_j - b_k \right|^2$$

i.e., to minimizing the sum of squares of linear functions.

There are several ways to find the least square solution. If we are in $\mathbb{R}^n$, and everything is real, we can forget about absolute values. Then we can just take partial derivatives with respect to x_j and find the where all of them are 0, which gives us minimum.

Geometric approach. However, there is a simpler way of finding the minimum. Namely, if we take all possible vectors x, then Ax gives us all possible vectors in Ran A, so minimum of $\| Ax - b \|$ is exactly the distance from b to Ran A. Therefore the value of $\| Ax - b \|$ is minimal if and only if $Ax = P_{Ran\,A}b$, where P_{RanA} stands for the orthogonal projection onto the column space Ran A.

So, to find the least square solution we simply need to solve the equation

$$Ax = P_{RanA}b.$$

If we know an orthogonal basis $v_1, v_2, \ldots,$ vn in Ran A, we can find vector $P_{Ran}Ab$ by the formula

$$P_{\text{Ran}\,A}\, b = \sum_{k=1}^{n} \frac{(b, v_k)}{\| v_k \|^2} v_k.$$

If we only know a basis in Ran A, we need to use the Gram–Schmidt orthogonalization to obtain an orthogonal basis from it.

So, theoretically, the problem is solved, but the solution is not very simple: it involves Gram–Schmidt orthogonalization, which can be computationally intensive. Fortunately, there exists a simpler solution.

Normal equation. Namely, Ax is the orthogonal projection $P_{Ran}Ab$ if and only if $b - Ax \perp$ Ran A ($Ax \in$ Ran A for all x).

If $a_1, a_2, \ldots, a_n$ are columns of A, then the condition $Ax \perp$ Ran A can be rewritten as

$$b - Ax \perp a_k, \ \forall k = 1, 2, \ldots, n.$$

That means

$$0 = (b - Ax, a_k) = a_k^*(b - Ax) \quad \forall k = 1, 2, \ldots, n.$$

Joining rows a_k^* together we get that these equations are equivalent to

$$A^*(b - Ax) = 0,$$

which in turn is equivalent to the so–called normal equation

$$A^*Ax = A^*b.$$

A solution of this equation gives us the least square solution of $Ax = b$.

Note, that the least square solution is unique if and only if A^*A is invertible.

Formula for the orthogonal projection. As we already discussed above, if x is a solution of the *normal equation* $A^*Ax = A^*b$ (*i.e.*, a least square solution of $Ax = b$), then $Ax = P_{\text{Ran}\,A}b$. So, to find the orthogonal projection of b onto the column space Ran A we need to solve the normal equation $A^*Ax = A^*b$, and then multiply the solution by A.

If the operator A^*A is invertible, the solution of the normal equation $A^*Ax = A^*b$ is given by $x = (A^*A)^{-1}A^*b$, so the orthogonal projection $P_{\text{Ran}}Ab$ can be computed as

$$P_{\text{Ran}}Ab = A(AA)^{-1}A^*b.$$

Since this is true for all b,

$$P_{Ran}A = A(A^*A)^{-1}A^*$$

is the formula for the matrix of the orthogonal projection onto Ran A.

The following theorem implies that for an $m \times n$ matrix A the matrix A^*A is invertible if and only if rank $A = n$.

Theorem. For *an* $m \times n$ *matrix* A

$$\text{Ker}A = \text{Ker}(A^*A).$$

Indeed, according to the rank theorem Ker$A = \{0\}$ if and only rank A is n. Therefore Ker$(A^*A) = \{0\}$ if and only if rank $A = n$. Since the matrix A^*A is square, it is invertible if and only if rank $A = n$.

To prove the equality Ker A = Ker (A^*A) one needs to prove two inclusions Ker(A^*A) Ker$A$ and Ker$A$ Ker(A^*A). One of the inclusion is trivial, for the other one use the fact that

$$\| Ax \|^2 = (Ax, Ax) = (A^*Ax, x).$$

Example. line fitting. Let us introduce a few examples where the least square solution appears naturally. Suppose that we know that two quantities x and y are related by the law $y = a + bx$. The coefficients a and b are unknown, and we would like to find them from experimental data.

Suppose we run the experiment n times, and we get n pairs (x_k, y_k), $k = 1, 2, ..., n$. Ideally, all the points (x_k, y_k) should be on a straight line, but because of errors in measurements, it usually does not happen: the point are usually close to some line, but not exactly on it. That is where the least square solution helps!

Ideally, the coefficients a and b should satisfy the equations

$$a + bx_k = y_k,\ k = 1, 2, ..., n$$

(note that here, x_k and y_k are some fixed numbers, and the unknowns are a and b). If it is possible to find such a and b we are lucky. If not, the standard thing to do, is to minimize the total quadratic error

$$\sum_{k=1}^{n} | a + bx_k - y_k |^2 .$$

But, minimizing this error is exactly finding the least square solution of the system

$$\begin{pmatrix} 1 & x_1 \\ 1 & x_2 \\ \vdots & \vdots \\ 1 & x_n \end{pmatrix} \begin{bmatrix} a \\ b \end{bmatrix} = \begin{pmatrix} y_1 \\ y_2 \\ \vdots \\ y_n \end{pmatrix}$$

(recall, that $x_k\, y_k$ are some given numbers, and the unknowns are a and b).

Example. Suppose our data (x_k, y_k) consist of pairs

$$(-2, 4), (-1, 2), (0, 1), (2, 1), (3, 1).$$

Then we need to find the least square solution of

$$\begin{pmatrix} 1 & - \\ 1 & -1 \\ 1 & 0 \\ 1 & 2 \\ 1 & 3 \end{pmatrix} \begin{bmatrix} a \\ b \end{bmatrix} = \begin{pmatrix} 4 \\ 2 \\ 1 \\ 1 \\ 1 \end{pmatrix}$$

Then

$$A^*A = \begin{pmatrix} 1 & 1 & 1 & 1 & 1 \\ -2 & -1 & 0 & 2 & 3 \end{pmatrix} \begin{pmatrix} 1 & -2 \\ 1 & -1 \\ 1 & 0 \\ 1 & 2 \\ 1 & 3 \end{pmatrix} = \begin{pmatrix} 5 & 2 \\ 5 & 18 \end{pmatrix}$$

and

$$A^*b = \begin{pmatrix} 1 & 1 & 1 & 1 & 1 \\ -2 & -1 & 0 & 2 & 3 \end{pmatrix} \begin{pmatrix} 4 \\ 2 \\ 1 \\ 1 \\ 1 \end{pmatrix} = \begin{pmatrix} 9 \\ -5 \end{pmatrix}$$

so the normal equation $A^*Ax = A^*b$ is rewritten as

$$\begin{pmatrix} 5 & 2 \\ 2 & 18 \end{pmatrix} \begin{pmatrix} a \\ b \end{pmatrix} = \begin{pmatrix} 9 \\ -5 \end{pmatrix}.$$

The solution of this equation is

$$a = 2,\ b = -1/2,$$

so the best fitting straight line is

$$y = 2 - 1/2x.$$

Examples. **Curves and Planes.** The least square method is not limited to the line fitting. It can also be applied to more general curves, as well as to surfaces in higher dimensions.

The only constraint here is that the parameters we want to find be involved linearly. The general algorithm is as follows:

1. Find the equations that your data should satisfy if there is exact fit;
2. Write these equations as a linear system, where unknowns are the parameters you want to find. Note, that the system need not to be consistent (and usually is not);
3. Find the least square solution of the system.

An example: curve fitting. For example, suppose we know that the relation between x and y is given by the quadratic law $y = a + bx + cx_2$, so we want to fit a parabola $y = a + bx + cx^2$ to the data. Then our unknowns a, b, c should satisfy the equations

$$a + bx_k + cx_k^2 = yk,\ \ldots,\ k = 1, 2, \ldots, n$$

or, in matrix form

$$\begin{pmatrix} 1 & x_1 & x_1^2 \\ 1 & x_2 & x_2^2 \\ \vdots & \vdots & \vdots \\ 1 & x_n & x_n^2 \end{pmatrix} \begin{pmatrix} a \\ b \\ c \end{pmatrix} = \begin{pmatrix} y_1 \\ y_2 \\ \vdots \\ y_n \end{pmatrix}$$

For example, for the data from the previous example we need to find the least square solution of

$$\begin{pmatrix} 1 & -2 & 4 \\ 1 & -1 & 1 \\ 1 & 0 & 0 \\ 1 & 2 & 4 \\ 1 & 3 & 9 \end{pmatrix} \begin{pmatrix} a \\ b \\ c \end{pmatrix} = \begin{pmatrix} 4 \\ 2 \\ 1 \\ 1 \\ 1 \end{pmatrix}.$$

Then

$$A^*A = \begin{pmatrix} 1 & 1 & 1 & 1 & 1 \\ -2 & -1 & 0 & 2 & 3 \\ 4 & 1 & 0 & 4 & 9 \end{pmatrix} \begin{pmatrix} 1 & -2 & 4 \\ 1 & -1 & 1 \\ 1 & 0 & 0 \\ 1 & 2 & 4 \\ 1 & 3 & 9 \end{pmatrix} = \begin{pmatrix} 5 & 2 & 18 \\ 2 & 18 & 26 \\ 18 & 26 & 114 \end{pmatrix}$$

and

$$A^*b = \begin{pmatrix} 1 & 1 & 1 & 1 & 1 \\ -2 & -1 & 0 & 2 & 3 \\ 4 & 1 & 0 & 4 & 9 \end{pmatrix} = \begin{pmatrix} 4 \\ 2 \\ 1 \\ 1 \\ 1 \end{pmatrix} = \begin{pmatrix} 9 \\ -5 \\ 31 \end{pmatrix}.$$

Therefore the normal equation $A^*Ax = A^*b$ is

$$\begin{pmatrix} 5 & 2 & 18 \\ 2 & 18 & 26 \\ 18 & 26 & 114 \end{pmatrix} \begin{pmatrix} a \\ b \\ c \end{pmatrix} = \begin{pmatrix} 9 \\ -5 \\ 31 \end{pmatrix}$$

which has the unique solution

$$a = 86/77,\ b = -62/77,\ c = 43/154.$$

Therefore,

$$y = 86/77 - 62\,x/77 + 43\,x^2/154$$

is the best fitting parabola.

Plane fitting. As another example, let us fit a plane $z = a + b_x + c_y$ to the data

$$(x_k, y_k, z_k) \in \in \mathbb{R}^3,\ k = 1, 2, \ldots n.$$

The equations we should have in the case of exact fit are

$$a + bx_k + cy_k = z_k,\ k = 1, 2, \ldots, n,$$

or, in the matrix form

$$\begin{pmatrix} 1 & x_1 & y_1 \\ 1 & x_2 & y_2 \\ \vdots & \vdots & \vdots \\ 1 & x_n & y_n \end{pmatrix} \begin{pmatrix} a \\ b \\ c \end{pmatrix} = \begin{pmatrix} x_1 \\ x_2 \\ \vdots \\ z_n \end{pmatrix}.$$

So, to find the best fitting plane, we need to find the best square solution of this system (the unknowns are a, b, c).

Fundamental Subspaces Revisited

Adjoint matrices and adjoint operators. Let as recall that for an $m \times n$ matrix A its *Hermitian adjoint* (or simply adjoint) A^* is defined by $A^* := \overline{A^T}$. In other words, the matrix A^* is obtained from the transposed matrix A^T by taking complex conjugate of each entry. The following identity is the main property of adjoint matrix:

$$\boxed{(Ax, y) = (x, A^* y) \forall x, y \in V.}$$

Before proving this identity, let us introduce some useful formulas. Let us recall that for transposed matrices we have the identity $(AB)^T = B^T A^T$. Since for complex numbers z and w we have $\overline{zw} = \overline{z}\overline{w}$, the identity

$$(AB)^* = B^* A^*$$

holds for the adjoint.

Also, since $(A^T)^T = A$ and $z = \overline{\overline{z}}, z$,

$$(A^*) = A^* = A.$$

Now, we are ready to prove the main identity:

$$(Ax, y) = y^* Ax = (A^* y)^* x = (x, A^* y);$$

the first and the last equalities here follow from the definition of inner product in F_n, and the middle one follows from the fact that

$$(Ax) = x(A) = xA.$$

Uniqueness of the adjoint. The above main identity $(Ax, y) = (x, A^*y)$ is often used as the definition of the adjoint operator. Let us first notice that the adjoint operator is unique: if a matrix B satisfies

$$(Ax, y) = (x, By) \ \forall x, y,$$

then $B = A^*$. Indeed, by the definition of A^*

$$(x, A^*y) = (x, By) \ \forall x$$

and therefore by Corollary $A^*y = By$. Since it is true for all y, the linear transformations, and therefore the matrices A and B coincide.

Adjoint transformation in abstract setting. The above main identity $(Ax, y) = (x, A^*y)$ can be used to define the adjoint operator in abstract setting, where $A : V \rightarrow W$ is an operator acting from one inner product space to another. Namely, we define $A^* : W \rightarrow V$ to be the operator satisfying

$$(Ax, y) = (x, A^*y) \ \forall x \in V, \forall y \in W.$$

Why does such an operator exists? We can simply construct it: consider orthonormal bases $A = v_1, v_2, \ldots, v_n$ in V and $B = w_1, w_2, \ldots, w_m$ in W. If $[A]_{BA}$ is the matrix of A with respect to these bases, we define the operator A^* by defining its matrix $[A^*]_{AB}$ as

$$[A^*]_{AB} = ([A]_{BA})^*.$$

Useful formulas. Below we present the properties of the adjoint operators (matrices) we will use a lot.

1. $(A + B) = A^* + B^*$;
2. $(\alpha A)^* = \overline{\alpha} A^*$;
3. $(AB) = B^*A^*$;
4. $(A^*)^* = A$;
5. $(y, Ax) = (A^*y, x)$.

Relation Between Fundamental Subspaces

Theorem. Let $A : V \to W$ *be an operator acting from one inner product space to another. Then*

1. $\text{Ker } A = (\text{Ran } A)^\perp$;
2. $\text{Ker } A = (\text{Ran } A)^\perp$;
3. $\text{Ran } A = (\text{Ker } A)^\perp$;
4. $\text{Ran } A = (\text{Ker } A)^\perp$.

Proof. First of all, let us notice, that since for a subspace E we have $(E^\perp)^\perp = E$, the statements 1 and 3 are equivalent. Similarly, for the same reason, the statements 2 and 4 are equivalent as well. Finally, statement 2 is exactly statement 1 applied to the operator A^* (here we use the fact that

$$(A^*)^* = A)$$

So, to prove the theorem we only need to prove statement 1. We will present 2 proofs of this statement: a "matrix" proof, and an "invariant", or "coordinate–free" one.

In the "matrix" proof, we assume that A is an $m \times n$ matrix, *i.e.*, that $A : F^n \to F^m$. The general case can be always reduced to this one by picking orthonormal bases in V and W, and considering the matrix of A in this bases.

Let $a_1, a_2, ..., a_n$ be the columns of A. Note, that $x \in (\text{Ran } A)^\perp$ if and only if $x \perp a_k$ (*i.e.*, $(x, a_k) = 0$) $\forall k = 1, 2, ..., n$. By the definition of the inner product in F_n, that means

$$0 = (x, a_k) = A_k^* = 1, 2, ..., n.$$

Since A_k^* is the row number k of A^*, the above n equalities are equivalent to the equation

$$A^*x = 0.$$

So, we proved that $x \in (\text{Ran } A)^\perp$ if and only if $A^*x = 0$, and that is exactly the statement 1.

Now, let us present the "coordinate–free" proof. The inclusion $x \in (\text{Ran } A)^{\perp}$ means that x is orthogonal to all vectors of the form A_y, *i.e.*, that

$$(x, Ay) = 0 \ \forall_y .$$

Since $(x, A_y) = (A^*x, y)$, the last identity is equivalent to

$$(A^*x, y) = 0 \ \forall y,$$

and by Lemma this happens if and only if $Ax = 0$. So we proved that $x \in (\text{Ran } A)^{\perp}$ if and only if $A^*x = 0$, which is exactly the statement 1 of the theorem.

The above theorem makes the structure of the operator A and the geometry of fundamental subspaces much more transparent. It follows from this theorem that the operator A can be represented as a composition of orthogonal projection onto Ran A^* and an isomorphism from Ran A^* to Ran A.

Isometrices and Unitary Operators. Unitary and Orthogonal Matrices

Main definitions

Definition. An operator $U : x \to Y$ is called an isometry, if it preserves the norm,

$$\| Ux \| = \| x \| \ \forall x \in X, \ldots$$

The following theorem shows that an isometry preserves the inner product.

Theorem. An operator $U : x \to Y$ is an isometry if and only if it preserves the inner product, i.e., if and only if

$$(x, y) = (Ux, Uy) \ \forall x, y \in X.$$

Proof. The proof uses the polarization identities. For example, if X is a complex space

$$(U_x, U_y) = \frac{1}{4} \sum_{a=\pm 1, \pm i} \alpha \| Ux + \alpha Uy \|^2$$

$$= \frac{1}{4} \sum_{a=\pm 1, \pm i} \alpha \| U(x + \alpha Uy) \|^2$$

$$= \frac{1}{4} \sum_{a=\pm 1, \pm i} \alpha \| x + \alpha y \|^2 = (x, y).$$

Similarly, for a real space x

$$(U_x, U_y) = \frac{1}{4}(\| Ux + Uy \|^2 - \| Ux - Uy \|^2)$$

$$= \frac{1}{4}(\| U(x + y) \|^2 - \| Ux - Uy \|^2)$$

$$= \frac{1}{4}(\| x + y \|^2 - \| x - y \|^2) = (x, y).$$

Lemma. An operator $U : X \to Y$ *is an isometry if and only if* $U^*U = I$.

Proof. By the definitions of the isometry and of the adjoint operator

$$(x, y) = (Ux, Uy) = (U^*Ux, y) \ \forall x, y \in X.$$

Therefore, if $U^*U = I$, we have $(x, y) = (U_x, U_y)$ and therefore U is an isometry. On the other hand, if U is an isometry, then for all $x \in X$

$$(U^*Ux, y) = (x, y) \ \forall y \in X,$$

and therefore by Corollary $U^*Ux = x$. Since it is true for all $x \in X$, we have $U^*U = I$.

The above lemma implies that an isometry is always left invertible (U^* being a left inverse).

Definition. An isometry $U : X \to Y$ is called a unitary operator if it is invertible.

Proposition. An isometry $U : X \to Y$ is a unitary operator if and only if $\dim X = \dim Y$.

Proof. Since U is an isometry, it is left invertible, and since $\dim X = \dim Y$, it is invertible (a left invertible square matrix is invertible). On the other hand, if $U : x \to Y$ is invertible, $\dim X = \dim Y$ (only quare matrices are invertible, isomorphic spaces have equal dimensions).

A square matrix U is called *unitary* if $U^*U = I$, *i.e.*, a unitary matrix is a matrix of a unitary operator acting in F_n. A unitary matrix with real entries is called an orthogonal matrix. In other words, an orthogonal matrix is a matrix of a unitary operator acting in the real space $\mathbb{R}^n$.

Few properties of unitary operators:

1. For a unitary transformation U, $U^{-1} = U^*$;
2. If U is unitary, $U^* = U^{-1}$ is also unitary;
3. If U is a isometry, and $v_1, v_2, \ldots, v_n$ is an orthonormal basis, then $Uv_1, Uv_2, \ldots, U_{vn}$ is an orthonormal system. Moreover, if U is unitary, $Uv_1, Uv_2, \ldots, Uv_n$ is an orthonormal basis.
4. A product of unitary operators is a unitary operator as well.

Examples. First of all, let us notice, that

a matrix U is an isometry if and only if its columns form an orthonormal system.

This statement can be checked directly by computing the product U^*U. It is easy to check that the columns of the rotation matrix

$$\begin{pmatrix} \cos\alpha & -\sin\alpha \\ \sin\alpha & \cos\alpha \end{pmatrix}$$

are orthogonal to each other, and that each column has norm 1. Therefore, the rotation matrix is an isometry, and since it is square, it is unitary. Since all entries of the rotation matrix are real, it is an orthogonal matrix.

The next example is more abstract. Let x and Y be inner products paces, $\dim X = \dim$

$Y = n$, and let $x_1, x_2, ..., x_n$ and $y_1, y_2, ..., y_n$ be orthonormal bases in X and Y respectively. Define an operator $U : X \to Y$ by

$$U_{xk} = y_k, \; k = 1, 2, ..., n.$$

Since for a vector $x = c_1 x_1 + c_2 x_2 + ... + c_n x_n$

$$\| x \|^2 = | c_1 |^2 + | c_2 |^2 + ... + | c_n |^2$$

and

$$\| Ux \|^2 = U\left(\sum_{k=1}^{n} c_k x_k\right)\|^2 = \| \sum_{k=1}^{n} c_k x_k \|^2 = \sum_{k=1}^{n} | c_k |^2,$$

one can conclude that $\| Ux \| = \| x \|$ for all $x \in X$, so U is a unitary operator.

Properties of Unitary Operators

Proposition. Let U be a unitary matrix. Then

1. $| \det U | = 1$. *In particular, for an orthogonal matrix* $\det U = \pm 1$;
2. If λ *is an eigenvalue of* U, *then* $| \lambda | = 1$

Remark. Note, that for an orthogonal matrix, an eigenvalue (unlike the determinant) does not have to be real. Our old friend, the rotation matrix gives an example.

Proof. Let $\det U = z$. Since $\det (U) = \overline{\det(U)}$, we have

$$| z |^2 = \bar{z} z = \det (U^*U) = \det I = 1,$$

so $| \det U | = | z | = 1$. Statement 1 is proved.

To prove statement 2 let us notice that if $Ux = \lambda x$ then $\| Ux \| = \| \lambda x \| = | \lambda | \cdot \| x \|$, so $| \lambda | = 1$.

Unitary Equivalent Operators

Definition. Operators (matrices) a and b are called unitarily equivalent if there exists a unitary operator U such that $A = UBU)^*$. Since for a unitary U we have $U^{-1} = U^*$, any two unitary equivalent matrices are similar as well.

The converse is not true, it is easy to construct a pair of similar matrices, which are not unitarily equivalent.

The following proposition gives a way to construct a counterexample.

Proposition. A matrix a is unitarily equivalent to a diagonal one if and only if it has an orthogonal (orthonormal) basis of eigenvectors.

Proof. Let $A = UBU^*$ and let $Ax = \lambda x$. Then $BUx = UAU^*Ux = UAx = U(\lambda x) = \lambda Ux$, *i.e.*, Ux is an eigenvector of B.

So, let A be unitarily equivalent to a diagonal matrix D, *i.e.*, let $D = UAU^*$. The vectors e_k of the standard basis are eigenvectors of D, so the vectors Ue_k are eigenvectors of A. Since U is unitary, the system $Ue_1, Ue_2, ..., Ue_n$ is an orthonormal basis.

Let now a has an orthogonal basis $u_1, u_2, ..., u_n$ of eigenvectors. Dividing each vector u_k by its norm if necessary, we can always assume that the system $u_1, u_2, ..., u_n$ is an *orthonormal* basis. Let D be the matrix of A in the basis $B = u_1, u_2, ..., u_n$. Clearly, D is a diagonal matrix.

Denote by U the matrix with columns $u_1, u_2, ..., u_n$. Since the columns form an orthonormal basis, U is unitary. The standard change of coordinate formula implies

$$A = [A]_{SS} = [I]_{SB} [A]_{BB} [I]_{BS} = UDU^{-1}$$

and since U is unitary, $A = UDU^*$.

COMPLEX INNER PRODUCTS

Our task in this section is to dene a suitable complex inner product. We begin by giving a reminder of the basics of complex vector spaces or vector spaces over $\mathbb{C}$.

Definition. A complex vector space V is a set of objects, known as vectors, together with vector addition + and multiplication of vectors by elements of $\mathbb{C}$, and satisfying the following properties:

($VA1$) For every $u, v \in V$, we have $u + v \in V$.

(VA2) For every $u, v, w \in V$, we have $u + (v + w) = (u + v) + w$.

($VA3$) There exists an element $0 \in V$ such that for every $u \in V$, we have $u + 0 = 0 + u = u$.

($VA4$) For every $u \in V$, there exists $u \in V$ such that $u + (u) = 0$.

($VA5$) For every $u, V \in V$, we have $u + V = V + u$.

($SM1$) For every $c \in \mathbb{C}$ and $u \in V$, we have $cu \in V$.

($SM2$) For every $c \in \mathbb{C}$ and $u, V \in V$, we have $c(u + v) = cu + cv$.

($SM3$) For every $a, b \in C$ and $u \in V$, we have $(a + b)u = au + bu$.

($SM4$) For every $a, b \in C$ and $u \in V$, we have $(ab)u = a(bu)$.

($SM5$) For every $u \in V$, we have $1u = u$.

Remark. Subspaces of complex vector spaces can be dened in a similar way as for real vector spaces. An example of a complex vector space is the euclidean space C_n consisting of all vectors of the form $u = (u_1, ..., u_n)$, where $u_1, ... , u_n \in \mathbb{C}$. We shall first generalize the concept of dot product, norm and distance, rst developed for $\mathbb{R}^n$ in Chapter 9.

Definition. Suppose that $u = (u_1, ... , u_n)$ and $v = (v_1, ..., v_n)$ are vectors in $\mathbb{C}$. The complex euclidean inner product of u and v is dened by

$$u \,.\, v = u_1 \overline{v_1} + ... + u_n \overline{v_n} \,;$$

the complex euclidean norm of u is dened by

$$\| u \| = (u\ u)^{1/2} = (| u_1 |^2 + ... + | u_n |^2)^{1/2};$$

and the complex euclidean distance between u and v is dened by

$$d(u, v) = \| u - v \| = (| u_1 - v_1 |^2 + ... + | u_n - v_n |^2)^{1/2}.$$

Corresponding to Proposition, we have the following result.

Proposition. Suppose that $u, v, w \in \mathbb{C}^n$ and $c \in C$. *Then*

(a) $u \,.\, v = \overline{v.u}$;

(b) $u \ (v + w) = (u \ \ v) + (u \ \ w)$;

(c) $c(u \ \ v) = (c_u) \ \ v$, and

(d) $u \,.\, u \geq 0$, and $u \,.\, u = 0$ *if and only if* $u = 0$.

The following definition is motivated by Proposition.

Definition. Suppose that V is a complex vector space. By a complex inner product on V, we mean a function $\langle , \rangle : V \times V \rightarrow \mathbb{C}$ which satises the following conditions:

(*IP*1) For every u, $V \in V$, we have $hu, vi = hv, ui$.

(*IP*2) For every $u, v, w \in V$, we have $\langle u,v+w \rangle = \langle u,v \rangle + \langle u,w \rangle$.

(*IP*3) For every $u, v \in V$ and $c \in \mathbb{C}$, we have chu, $c\langle u,v \rangle = \langle cu.v \rangle$.

(*IP*4) For every $u \in V$, we have $\langle u,u \rangle \geq 0$, and $\langle u,u \rangle = 0$ if and only if $u = 0$.

Definition. A complex vector space with an inner product is called a complex inner product space or a unitary space.

Definition. Suppose that u and v are vectors in a complex inner product space V. Then the norm of u is defined by

$$\| u \| = \langle u,u \rangle^{1/2},$$

and the distance between u and v is defined by

$$d(u, v) = \| u - v \|.$$

Using this inner product, we can discuss orthogonality, orthogonal and orthonormal bases, the Gram Schmidt orthogonalization process, as well as orthogonal projections, in a similar way as for real inner product spaces. In particular, the results in Sections can be generalized to the case of complex inner product spaces.

Unitary Matrices

For matrices with real entries, orthogonal matrices and symmetric matrices play an important role in the orthogonal diagonalization problem. For matrices with complex entries, the analogous roles are played by unitary matrices and hermitian matrices respectively.

Definition. Suppose that A is a matrix with complex entries. Suppose further that the matrix A is obtained from the matrix A by replacing each entry of A by its complex conjugate. Then the matrix

$$A = \overline{A}^{-t}$$

is called the conjugate transpose of the matrix A.

Proposition. Suppose that A and B are matrices with complex entries, and that $c \in \mathbb{C}$. *Then*

(a) $(A^*)^* = A$;

(b) $(A + B)^* = A^* + B^*$;

(c) $(cA^*) = \bar{c}A^*$; *and*

(d) $(AB)^* = B^*A^*$.

Definition. A square matrix A with complex entries and satisfying the condition $A^{-1} = A^*$ is said to be a unitary matrix.

Corresponding to Proposition, we have the following result.

Proposition. Suppose that A is an $n \times n$ matrix with complex entries. Then

(*a*) A is unitary if and only if the row vectors of A form an orthonormal basis of $\mathbb{C}^n$ under the complex euclidean inner product.

(*b*) A is unitary if and only if the column vectors of A form an orthonormal basis of $\mathbb{C}^n$ under the complex euclidean inner product.

Unitary Diagonalization

Corresponding to the orthogonal disgonalization problem in Section 10.3, we now discuss the following unitary diagonalization problem.

Definition. A square matrix A with complex entries is said to be unitarily diagonalizable if there exists a unitary matrix P with complex entries such that $P^{-1}AP = P^*AP$ is a diagonal matrix with complex entries.

First of all, we would like to determine which matrices are unitarily diagonalizable. For those that are, we then need to discuss how we may nd a unitary matrix P to carry out the diagonalization. As before, we study the question of eigenvalues and eigenvectors of a given matrix; these are dened as for the real case without any change. We have indicated that a square matrix with real entries is orthogonally diagonalizable if and only if it is symmetric. The most natural extension to the complex case is the following.

Definition. A square matrix A with complex entries is said to be hermitian if $A = A$.

Unfortunately, it is not true that a square matrix with complex entries is unitarily diagonalizable if and only if it is hermitian. While it is true that every hermitian matrix is unitarily diagonalizable, there are unitarily diagonalizable matrices that are not hermitian. The explanation is provided by the following.

Definition. A square matrix A with complex entries is said to be normal if $AA^* = A^*A$.

Remark. Note that every hermitian matrix is normal and every unitary matrix is normal.

Corresponding to Propositions, we have the following results.

Proposition. Suppose that A is an n n matrix with complex entries. Then it is unitarily diagonalizable if and only if it is normal.

Proposition. Suppose that u_1 *and* u_2 *are eigenvectors of a normal matrix A with complex entries, corresponding to distinct eigenvalues* λ_1 *and* λ_2 *respectively. Then* $u_1 \; u_2 = 0$*. In other words, eigenvectors of a normal matrix corresponding to distinct eigenvalues are orthogonal.*

We can now follow the procedure below.

Unitary Diagonalization Process. Suppose that A is a normal n n matrix with complex entries.

(1) *Determine the n complex roots $l_1, \ldots, l_n$ of the characteristic polynomial det(A – lI), and n linearly independent eigenvectors $u_1, \ldots, u_n$ of A corresponding to these eigenvalues as in the Diagonalization process.*

(2) *Apply the Gram-Schmidt orthogonalization process to the eigenvectors $u_1, \ldots, u_n$ to obtain orthogonal eigenvectors $v_1, \ldots, v_n$ of A, noting that eigenvectors corresponding to distinct eigenvalues are already orthogonal.*

(3) *Normalize the orthogonal eigenvectors $v_1, \ldots, v_n$ to obtain orthonormal eigenvectors $w_1, \ldots, w_n$ of A. These form an orthonormal basis of $\mathbb{C}^n$. Furthermore, write*

$$P = (w_1 \ldots w_n) \text{ and } D = \begin{pmatrix} \lambda_1 & & \\ & \ddots & \\ & & \lambda_n \end{pmatrix},$$

where $\lambda_1, \ldots, \lambda_n \in \mathbb{C}$ are the eigenvalues of A and where $w_1, \ldots, w_n \in \mathbb{C}^n$ are respectively their orthogonalized and normalized eigenvectors. Then $P^*AP = D$.

We conclude this chapter by discussing the following important result which implies Proposition, that all the eigenvalues of a symmetric real matrix are real.

Proposition. Suppose that A is a hermitian matrix. Then all the eigenvalues of A are real.

Proof. Suppose that A is a hermitian matrix. Suppose further that is an eigenvalue of A, with corresponding eigenvector v. Then

$$Av = \lambda v.$$

Multiplying on the left by the conjugate transpose v^* of v, we obtain

$$v^*Av = v^*\lambda v = \lambda v^* v.$$

To show that λ is real, it suces to show that the 1×1 matrices v^*Av and v^*v both have real entries. Now

$$(v^*Av)^* = v^*A^*(v^*)^* = v^*Av$$

and

$$(v^*v)^* = v^*(v^*)^* = v^*v.$$

It follows that both v^*Av and v^*v are hermitian. It is easy to prove that hermitian matrices must have real entries on the main diagonal. Since v^*Av and v^*v are 1×1, it follows that they are real.

APPLICATIONS OF REAL INNER PRODUCT SPACES

Least Squares Approximation

Given a continuous function $f : [a; b] \to \mathbb{R}$, we wish to approximate f by a polynomial

$$g : [a; b] \to \mathbb{R}$$

of degree at most k, such that the error

$$\int_a^b | f(x) - g(x)^2 dx$$

is minimized. The purpose of this section is to study this problem using the theory of real inner product spaces. Our argument is underpinned by the following simple result in the theory.

Proposition. Suppose that V is a real inner product space, and that W is a finite-dimensional subspace of V. Given any $u \in V$, the inequality

$$\| u \ \text{proj}_W u \| \leq \| u \ w \|$$

holds for every $w \in W$.

In other words, the distance from u to any $w \in W$ is minimized by the choice $w = \text{proj}_W u$, the orthogonal projection of u on the subspace W. Alternatively, $\text{proj}_W u$ can be thought of as the vector in W closest to u.

Proof. Note that

$$u \ \text{proj}_W u \in W^{\perp} \text{ and proj } W_u \ \ w \in W.$$

It follows from Pythagoras's theorem that

$$\| u - w \|^2 = \|(u - proj_W u - w) + (proj_W u - w)\|^2$$
$$= \| u - \text{proj}_W u \|^2 + \| \text{proj}_W u - w \|^2; \text{ so that}$$
$$\| u - w \|^2 \ \ \| u - \text{proj}_W u \|^2 = \| \text{proj}_W u - w \|^2 \geq 0:$$

The result follows immediately.

Let V denote the vector space $C[a, b]$ of all continuous real valued functions on the closed interval $[a, b]$, with inner product

$$(f, g) = \int_a^b f(x)g(x)dx.$$

Then

$$\int_a^b | f(x)g(x)^2 dx = (f - g, f - g) = \| f - g \|^2 .$$

It follows that the least squares approximation problem is reduced to one of nding a suitable polynomial g to minimize the norm $\| f - g \|$.

Now let $W = P_k [a, b]$ be the collection of all polynomials $g : [a, b] \to \mathbb{R}$ with real coecients and of degree at most k. Note that W is essentially P_k, although the variable is restricted to the closed interval $[a, b]$. It is easy to show that W is a subspace of V. In view of Proposition 11A, we conclude that

$$g = \text{proj}_W f$$

gives the best least squares approximation among polynomials in $W = P_k [a, b]$. This subspace is of dimension $k + 1$. Suppose that $\{v_0, v_1, \ldots, v_k\}$ is an orthogonal basis of $W = P_k [a, b]$. Then by Proposition, we have

$$g = \frac{\langle f, v_0 \rangle}{\| v_0 \|^2} v_0 + \frac{\langle f, v_1 \rangle}{\| v_1 \|^2} v_1 + \ldots + \frac{\langle f, v_k \rangle}{\| v_k \|^2} v_k.$$

Example. Consider the function $f(x) = x^2$ in the interval [0, 2]. Suppose that we wish to nd a least squares approximation by a polynomial of degree at most 1. In this case, we can take $V = C[0, 2]$, with inner product

$$\langle f, g \rangle = \int_0^2 f(x)g(x)dx,$$

and $W = P_1\,[0, 2]$, with basis f_1, xg. We now apply the Gram-Schmidt orthogonalization process to this basis to obtain an orthogonal basis $\{1, x - 1\}$ of W, and take

$$g = \frac{\langle x^2, 1 \rangle}{\| 1 \|^2} 1 + \frac{\langle x^2, x-1 \rangle}{\| x-1 \|^2} (x-1).$$

$$\langle x^2, 1 \rangle = \int_0^2 x^2 dx = \frac{8}{3} \text{ and } \| 1 \|^2 = \int_0^2 dx = 2,$$

while

$$\langle x^2, x-1 \rangle = \int_0^2 x^2 (x-1) dx = \frac{4}{3} \text{ and } \| x-1 \|^2 = \langle x-1, x-1 \rangle$$

$$= \int_0^2 (x-1)^2 dx = \frac{2}{3}.$$

It follows that

$$g = \frac{4}{3} + 2(x-1) = 2x - \frac{2}{3}.$$

Example. Consider the function $f(x) = e^x$ in the interval [0, 1]. Suppose that we wish to find a least squares approximation by a polynomial of degree at most 1. In this case, we can take $V = C[0, 1]$, with inner product

$$\langle f, g \rangle = \int_0^1 f(x)g(x)d\,|\,x,$$

and $W = P_1[0, 1]$, with basis $\{1, x\}$. We now apply the Gram-Schmidt orthogonalization process to this basis to obtain an orthogonal basis $\{1, x - 1/2\}$ of W, and take

$$g = \frac{\langle e^x, 1 \rangle}{\| 1 \|^2} 1 + \frac{\langle e^x, x-1/2 \rangle}{\| x-1/2 \|^2} \left(x - \frac{1}{2} \right).$$

Now

$$\langle e^x, 1 \rangle = \int_0^1 e^x dx = e - 1 \text{ and } \langle e^x, x \rangle = \int_0^1 e^x dx = 1,$$

so that

$$\left\langle e^x, x - \frac{1}{2} \right\rangle = \langle e^x, x \rangle - \frac{1}{2} \langle e^x, 1 \rangle = \frac{3}{2} - \frac{e}{2}.$$

Also

$$\| 1 \|^2 = \langle 1,1 \rangle = \int_0^1 dx = 1 \text{ and } \left\| x\frac{1}{2} \right\|^2 = \left\langle x - \frac{1}{1}, x - \frac{1}{1} \right\rangle = \int_0^1 \left(x - \frac{1}{2} \right)^2 dx = \frac{1}{12}.$$

It follows that

$$g = (e-1) + (18-6e)\left(x - \frac{1}{2} \right) = (18-6e)x + (4e-10).$$

Remark. From the proof of Proposition, it is clear that $\| u - w \|$ is minimized by the unique choice $w = \text{proj}_W u$. It follows that the least squares approximation problem posed here has a unique solution.

Quadratic Form

A real quadratic form in n variables $x_1, \ldots, x_n$ is an expression of the form

$$\sum_{i=1}^{n} \sum_{\substack{j=1 \\ i \le j.}}^{n} C_{ij} x_i x_j,$$

where $c_{ij} \in \mathbb{R}$ for every $i, j = 1; \ldots ; n$ satisfying $i \le j$.

Example. The expression $5x_1^2 + 6x_1x_2 + 7x_2^2$ is a quadratic form in two variables x_1 and x_2. It can be written in the form

$$5x_1^2 + 6x_1x_2 + 7x_2^2 (x_1x_2)\begin{pmatrix} 5 & 3 \\ 3 & 7 \end{pmatrix}\begin{pmatrix} x_1 \\ x_2 \end{pmatrix}.$$

Example. The expression $4x_1^2 + 5x_2^2 + 3x_3^2 + 2x_1x_2 + 4x_1x_3 + 6x_2x_3$ is a quadratic form in three variables x_1, x_2 and x_3. It can be written in the form

$$4x_1^2 + 5x_2^2 + 3x_3^2 + 2x_1x_2 + 4x_1x_3 + 6x_2x_3 = (x_1x_2x_3)\begin{pmatrix} 4 & 1 & 2 \\ 1 & 5 & 3 \\ 2 & 3 & 3 \end{pmatrix}\begin{pmatrix} x_1 \\ x_2 \\ x_3 \end{pmatrix}.$$

Note that in both examples, the quadratic form can be described in terms of a real symmetric matrix. In fact, this is always possible. To see this, note that given any quadratic form (1), we can write, for every $i, j = 1, \ldots, n$,

$$a_{ij} = \begin{cases} c_{ij} & \text{if } i = j, \\ \frac{1}{2} c_{ij} & \text{if } i > j, \\ \frac{1}{2} c_{ji} & \text{if } i > j. \end{cases}$$

Then

$$\sum_{i=1}^{n}\sum_{j=1}^{n} c_{ij}x_i x_j = \sum_{i=1}^{n}\sum_{j=1}^{n} a_{ij}x_i x_j = (x_1 \dots x_n)\begin{pmatrix} a_{11} & \dots & a_{1n} \\ \vdots & & \vdots \\ a_{n1} & \dots & a_{nm} \end{pmatrix}\begin{pmatrix} x_1 \\ \vdots \\ x_n \end{pmatrix}.$$

The matrix

$$A = \begin{pmatrix} a_{11} & \dots & a_{1n} \\ \vdots & & \vdots \\ a_{n1} & \dots & a_{nm} \end{pmatrix}$$

We are interested in the case when $x_1, \dots, x_n$ take real values. In this case, we can write

$$x = \begin{pmatrix} x_1 \\ \vdots \\ x_n \end{pmatrix}.$$

It follows that a quadratic form can be written as

$$x^t Ax,$$

where A is an $n \times n$ real symmetric matrix and x takes values in $\mathbb{R}^n$.

Many problems in mathematics can be studied using quadratic forms. Here we shall restrict our attention to two fundamental problems which are in fact related. The rst is the question of what conditions the matrix A must satisfy in order that the inequality

$$x^t Ax > 0$$

holds for every non-zero $x \in \mathbb{R}^n$. The second is the question of whether it is possible to have a change of variables of the type $x = P_y$, where P is an invertible matrix, such that the quadratic form $x_t Ax$ can be represented in the alternative form $y^t Dy$, where D is a diagonal matrix with real entries.

Definition. A quadratic form $x^t Ax$ is said to be positive denite if $x^t Ax > 0$ for every non-zero $x \in \mathbb{R}^n$. In this case, we say that the symmetric matrix A is a positive denite matrix. To answer our rst question, we shall prove the following result.

Proposition. A quadratic form $x^t Ax$ *is positive denite if and only if all the eigenvalues of the symmetric matrix A are positive.*

Our strategy here is to prove Proposition by first studying our second question. Since the matrix A is real and symmetric, it follows from Proposition 10E that it is orthogonally diagonalizable. In other words, there exists an orthogonal matrix P and a diagonal matrix D such that $P^t AP = D$, and so $A = PDP^t$. It follows that

$$x^t Ax = x^t PDP^t x,$$

and so, writing

$$y = P^t x;$$

we have

$$x^tAx = y^tDy.$$

Also, since P is an orthogonal matrix, we also have $x = Py$. This answers our second question. Furthermore, in view of the Orthogonal diagonalization process, the diagonal entries in the matrix D can be taken to be the eigenvalues of A, so that

$$D = \begin{pmatrix} \lambda_1 & & \\ & \ddots & \\ & & \lambda_n \end{pmatrix},$$

where $\lambda_1, \ldots, n \in \mathbb{R}$ are the eigenvalues of A. Writing

$$y = \begin{pmatrix} y_1 \\ \vdots \\ t_n \end{pmatrix},$$

we have

$$x^tAx = y^tDy = \lambda_1 y_1^2 + \ldots + \lambda_n y_n^2.$$

Note now that $x = 0$ if and only if $y = 0$, since P is an invertible matrix.

Example. Consider the quadratic form $2x_1^2 + 5x_2^2 + 2x_3^2 + 4x_1x_2 + 2x_1x_3 + 4x_2x_3$. This can bewritten in the form x^tAx, where

$$A = \begin{pmatrix} 2 & 2 & 1 \\ 2 & 5 & 2 \\ 1 & 2 & 2 \end{pmatrix} \text{ and } x = \begin{pmatrix} x_1 \\ x_2 \\ x_3 \end{pmatrix}.$$

The matrix A has eigenvalues 1 = 7 and (double root) 2 = 3 = 1, see Example. Furthermore, we have $P^tAP = D$, where

$$P = \begin{pmatrix} 1/\sqrt{6} & 1/\sqrt{2} & 1/\sqrt{3} \\ 2/\sqrt{6} & 0 & -1/\sqrt{3} \\ 1/\sqrt{6} & -1/\sqrt{2} & 1/\sqrt{3} \end{pmatrix} \text{ and } D = \begin{pmatrix} 7 & 0 & 0 \\ 0 & 1 & 0 \\ 0 & 0 & 1 \end{pmatrix}.$$

Writing $y = P^tx$, the quadratic form becomes $7x_1^2 + y_2^2 + y_3^2$ which is clearly positive defnite.

Example. Consider the quadratic from $5x_1^2 + 6x_2^2 + y_3^2 - 4x_1x_2 + 4x_2x_3$. This cn be written in the form x^tAx, where

$$A = \begin{pmatrix} 5 & -2 & 0 \\ -2 & 6 & 2 \\ 0 & 2 & 7 \end{pmatrix} \text{ and } x = \begin{pmatrix} x_1 \\ x_2 \\ x_3 \end{pmatrix}.$$

The matrix A has eigenvalues $\lambda_1 = \lambda_3$, $\lambda_2 = 6$ and $\lambda_3 = 9$, See Example, furthermore, we have $P^tAP = D$, where

$$P = \begin{pmatrix} 2/3 & 2/3 & -1/3 \\ 2/3 & -1/3 & 2/3 \\ -1/3 & 2/3 & 2/3 \end{pmatrix} \text{ and } D = \begin{pmatrix} 3 & 0 & 0 \\ 0 & 6 & 0 \\ 0 & 0 & 9 \end{pmatrix}.$$

Writing $y = P^tx$, the quadratic form becomes $3y_1^2 + 6y_3^2 + 9y_3^2$ which is clearly positive definite.

Example. Consider the quadratic form $x_1^2 + x_2^2 + 2x_1x_2$. Clearly this is equal to $(x_1 + x_2)^2$ and is therefore not positive defnite. The quadratic form can be written in the form x^tAx, where

$$A = \begin{pmatrix} 1 & 1 \\ 1 & 1 \end{pmatrix} \text{ and } x = \begin{pmatrix} x_1 \\ x_2 \end{pmatrix}.$$

It follows from Proposition that the eigenvalues of A are not all positive. Indeed, the matrix A has eigenvalues $\lambda_1 = 2$ and $\lambda_2 = 0$, with corresponding eigenvectors

$$\begin{pmatrix} 1 \\ 1 \end{pmatrix} \text{ and } \begin{pmatrix} 1 \\ -1 \end{pmatrix}.$$

Hence we may take

$$P = \begin{pmatrix} 1/\sqrt{2} & 1/\sqrt{2} \\ 1/\sqrt{2} & -1/\sqrt{2} \end{pmatrix} \text{ and } D = \begin{pmatrix} 2 & 0 \\ 0 & 0 \end{pmatrix}.$$

Writing $y = P^tx$, the quadratic form becomes $2y_1^2$ which is not positive denite.

Real Fourier Series

Let E denote the collection of all functions $f : [-\pi, \pi] \to \mathbb{R}$ which are piecewise continuous on the interval $[-\pi, \pi]$. This means that any $f \in E$ has at most a finite number of points of discontinuity, ateach of which f need not be dened but must have one sided limits which are finite. We further adopt the convention that any two functions $f, g \in E$ are considered equal, denoted by $f = g$, if $f(x) = g(x)$ for every $x \in [-\pi, \pi]$ with at most a finite number of exceptions.

It is easy to check that E forms a real vector space. More precisely, let $\in E$ denote the function $\lambda : [-\pi, \pi] \in \mathbb{R}$, where $(x) = 0$ for every $x \in [-\pi, \pi]$. Then the following conditions hold:

For every $f, g \in E$, we have $f, g \in E$.
For every $f, g, h \in E$, we have $f + (g + h) = (f + g) + h$.
For every $f \in E$, we have $f + \lambda = \lambda + f = f$.
For every $f \in E$, we have $f + (-f) = \lambda$.
For every $f, g \in E$, we have $f + g = g + f$.

For every $c \in \mathbb{R}$ and $f \in E$, we have $cf \in E$.

For every $c \in \mathbb{R}$ and $f, g \in E$, we have $c(f + g) = cf + cg$.

For every $a, c \in \mathbb{R}$ and $f \in E$, we have $(a + b)f = af + bf$.

For every $a, c \in \mathbb{R}$ and $f \in E$, we have $(a + b)f = a(bf)$.

For every $f \in E$, we have $1f = f$.

We now give this vector space E more structure by introducing an inner product. For every $f, g \in E$,

$$\langle f, g\rangle = \frac{1}{\pi}\int_{-\pi}^{\pi} f(x)g(x)\,dx.$$

The integral exists since the function $f(x)g(x)$ is clearly piecewise continuous on $[-\pi', \pi]$. It is easy to check that the following conditions hold:

For every $f, g \in E$, we have $\langle f, g\rangle = \langle g, f\rangle$.

For every $f, g, h \in E$, we have $\langle f, g+h\rangle = \langle f, g\rangle + \langle f, h\rangle$.

For every $f, g \in E$ and $c \in \mathbb{R}$, we have $c\langle f, g\rangle = \langle cf, g\rangle$.

For every $f \in E$, we have $\langle f, f\rangle \geq 0$, and ,$\langle f, f\rangle = 0$ if and only if $f = \lambda$. Hence E is a real inner product space.

The diculty here is that the inner product space E is not finite-dimensional. It is not straightforward to show that the set

$$\left\{\frac{1}{\sqrt{2}} \sin x, \cos x, \sin 2x, \cos 2x, \sin 3x, \cos 3x, \ldots\right\}$$

in E forms an orthonormal "basis" for E. The diculty is to show that the set spans E.

Remark. It is easy to check that the elements in (4) form an orthonormal "system". For every $k, m \in \mathbb{N}$,

we have

$$\left\langle \frac{1}{\sqrt{2}}, \frac{1}{\sqrt{2}} \right\rangle = \frac{1}{\pi}\int_{-\pi}^{\pi} \frac{1}{2}\,dx = 1;$$

$$\left\langle \frac{1}{\sqrt{2}}, \sin kx \right\rangle = \frac{1}{\pi}\int_{-\pi}^{\pi} \frac{1}{\sqrt{2}} \sin kx = 0;$$

$$\left\langle \frac{1}{\sqrt{2}}, \cos kx \right\rangle = \frac{1}{\pi}\int_{-\pi}^{\pi} \frac{1}{\sqrt{2}} \cos kx = 0;$$

as well as

$$\langle \sin kx, \sin mx\rangle = \frac{1}{\pi}\int_{-\pi}^{\pi} \sin kx \sin mx\,dx = \frac{1}{\pi}$$

$$\int_{-\pi}^{\pi} \frac{1}{2}(\cos(k-m)x - \cos\,(k+m)x)\,dx = \begin{cases} 1 & if\ k = m \\ 0 & if\ k \neq m \end{cases}$$

$$\langle \cos kx, \cos mx \rangle = \frac{1}{\pi}\int_{-\pi}^{\pi} \cos kx \cos mx dx = \frac{1}{\pi}$$

$$\int_{-\pi}^{\pi} \frac{1}{2}(\cos(k-m)x - \cos(k+m)x)dx = \begin{cases} 1 & if\ k = m \\ 0 & if\ k \neq m \end{cases}$$

$$\langle \sin kx, \cos mx \rangle = \frac{1}{\pi}\int_{-\pi}^{\pi} \sin kx \cos mx dx = \frac{1}{\pi}\int_{-\pi}^{\pi} \frac{1}{2}(\sin(k-m)x - \sin(k+m)x)dx$$

Let us assume that we have established that the set (4) forms an orthonormal basis for E. Then a natural extension of Proposition 9*H* gives rise to the following: Every function $f \in E$ can be written uniquely in the form

$$\frac{a_0}{2} + \sum_{n=1}^{\infty}(a_n \cos nx + b_n \sin nx)$$

known usually as the (trigonometric) Fourier series of the function f, with Fourier coecients

$$\frac{a_0}{\sqrt{2}}\left\langle f, \frac{1}{\sqrt{2}} \right\rangle = \frac{1}{\pi}\int_{-\pi}^{\pi} f(x)dx,$$

and, for every $n \in N$,

$$a_n = \langle f, \cos nx \rangle = \frac{1}{x}\int_{-\pi}^{\pi} f(x)\cos nx dx \text{ and}$$

$$b_n = \langle f, \sin nx \rangle = \frac{1}{x}\int_{-\pi}^{\pi} f(x)\sin nx dx$$

Note that the constant term in the Fourier series (5) is given by

$$\left\langle f, \frac{1}{\sqrt{2}} \right\rangle = \frac{1}{\sqrt{2}} = \frac{a_0}{2}.$$

Example. Consider the function $f : [-\pi, \pi] \to \mathbb{R}$, given by $f(x) = x$ for every $x \in [-\pi, \pi]$. For every $n \in \mathbb{N} \cup \{0\}$, we have

$$a_n = \frac{1}{\pi}\int_{-\pi}^{\pi} x \cos nx dx = 0$$

since the integrand is an odd function. On the other hand, for every $n \in N$, we have

$$b_n = \frac{1}{\pi}\int_{-\pi}^{\pi} x \sin nx dx = \frac{1}{\pi}\int_{0}^{\pi} x \sin nx dx,$$

since the integrand is an even function. On integrating by parts, we have

$$b_n = \frac{2}{\pi}\left(-\left[\frac{x \cos nx}{n}\right]_0^{\pi} + \int_0^{\pi} \frac{\cos nx}{n} dx \right) = \frac{2}{\pi}\left(-\left[\frac{x \cos nx}{n}\right]_0^{\pi} + \left[\frac{\sin nx}{n^2}\right]_0^{\pi} \right) = \frac{2(-1)^{n+1}}{n}.$$

We therefore have the (trigonometric) Fourier series

$$\sum_{n=1}^{\infty} \frac{2(-1)^{n+1}}{n} \sin nx.$$

Note that the function f is odd, and this plays a crucial role in eschewing the Fourier coefficients a_n corresponding to the even part of the Fourier series.

Example. Consider the function $f: [-\pi, \pi] \to \mathbb{R}$, given by $f(x) = |x|$ for every $x \in [-\pi, \pi]$. For every $n \in N \cup \{0\}$, we have

$$a_n = \frac{1}{\pi}\int_{-\pi}^{\pi} |x| \cos nx dx = \frac{2}{\pi}\int_0^{\pi} x \operatorname{cox} nx\, dx,$$

since the integrand is an even function. Clearly

$$a_0 = \frac{2}{\pi}\int_0^{\pi} x dx = \pi.$$

Furthermore, for every $n \in \mathbb{N}$, on integrating by parts, we have

$$a_n = \frac{2}{\pi}\left(\left[\frac{x \sin nx}{n}\right]_0^{\pi} - \int_0^{\pi} \frac{\sin nx}{n} dx\right)$$

$$= \frac{2}{\pi}\left(\left[\frac{x \sin nx}{n}\right]_0^{\pi} + \left[\frac{\cos nx}{n^2}\right]_0^{\pi}\right) = \begin{cases} 0 & \text{if } n \text{ is even,} \\ -\dfrac{4}{\pi n^2} & \text{if } n \text{ is odd.} \end{cases}$$

On the other hand, for every $n \in \mathbb{N}$, we have

$$b_n = \frac{1}{\pi}\int_{-\pi}^{\pi} |x| \sin nx dx = 0$$

since the integrand is an odd function. We therefore have the (trigonometric) Fourier series

$$\frac{\pi}{2} - \sum_{\substack{n-1 \\ n \text{ odd}}}^{\pi} \frac{4}{\pi n^2} \cos nx = \frac{\pi}{2} - \sum_{k=1}^{\infty} \frac{4}{\pi(2k-1)^2} \cos(2k-1)x.$$

Note that the function f is even, and this plays a crucial role in eschewing the Fourier coefficients b_n corresponding to the odd part of the Fourier series.

Example. Consider the function $f: [-\pi, \pi] \to \mathbb{R}$, given for every $x \in [-\pi, \pi]$ by

$$f(x) = \operatorname{sgn}(x) = \begin{cases} +1 & \text{if } 0 < x \le \pi, \\ 0 & \text{if } x = 0 \\ -1 & \text{if } -\pi \le x < 0, \end{cases}$$

For every $n \in \mathbb{N} \cup \{0\}$ we have

$$a_n = \frac{1}{\pi}\int_{-\pi}^{\pi} \operatorname{sgn}(x) \cos nx dx = 0,$$

since the integrand is an odd function. On the other hand, for every $n \in \mathbb{N}$, we have

$$a_n = \frac{1}{\pi}\int_{-\pi}^{\pi} \operatorname{sgn}(x) \sin nx dx \frac{2}{\pi}\int_0^{\pi} \sin n, x dx,$$

since the integrand is an even function. It is easy to see that

$$b_n = -\frac{2}{\pi}\left[\frac{\cos nx}{m}\right]_0^{\pi} = \begin{cases} 0 & \text{if } n \text{ is even} \\ \dfrac{4}{\pi n} & \text{if } n \text{ is odd.} \end{cases}$$

We therefore have the (trigonometric) Fourier series

$$\sum_{\substack{n=1 \\ n\,\text{odd}}}^{\infty} \frac{4}{\pi n}\sin nx = \sum_{k=1}^{\infty} \frac{4}{\pi(2k-1)}\sin(2k-1)x.$$

Example. Consider the function $f: [-\pi, \pi] \to \mathbb{R}$, given by $f(x) = x^2$ for every $x \in [-\pi, \pi]$ For every $n \in \mathbb{N} \cup \{0\}$, we have

$$a_n = \frac{1}{\pi}\int_{-\pi}^{\pi} x^2 \cos nx dx = \frac{2}{\pi}\int_0^{\pi} x^2 \cos nx\, dx,$$

since the integrand is an even function. Clearly

$$a_0 = \frac{2}{\pi}\int_0^{\pi} x^2 dx = \frac{3\pi^2}{3}.$$

Furthermore, for every $n \in \mathbb{N}$, on integrating by parts, we have

$$a_n = \frac{2}{\pi}\left(\left[\frac{x^2 \sin nx}{n}\right]_0^{\pi} - \int_0^{\pi} \frac{2x \sin nx}{n} dx\right)$$

$$= \frac{2}{\pi}\left(\left[\frac{x^2 \sin nx}{n}\right]_0^{\pi} + \left[\frac{2x \cos nx}{n^2}\right]_0^{\pi} - \int_0^{\pi} \frac{2\cos nx}{n^2} dx\right)$$

$$= \frac{2}{\pi}\left(\left[\frac{x^2 \sin nx}{n}\right]_0^{\pi} - \left[\frac{2x \sin nx}{n^2}\right]_0^{\pi} - \left[\frac{2x \sin nx}{n^3}\right]_0^{\pi}\right) = \frac{4(-1)^n}{n^2}.$$

On the other hand, for every $n \in \mathbb{N}$, we have

$$b_n = \frac{1}{\pi}\int_{-\pi}^{\pi} x^2 \sin nx dx = 0$$

since the integrand is an odd function. We therefore have the (trigonometric) Fourier series

$$\frac{\pi^2}{3} + \sum_{n=1}^{\infty} \frac{4(-1)^n}{n^2}\cos nx.$$

Chapter 7

Structure of Operators in Inner Product Spaces

In this chapter we again assuming that all spaces are finite–dimensional.

1. Upper triangular (Schur) representation of an operator.

Theorem. Let $A : X \to X$ be an operator acting in a complex inner product space. There exists an orthonormal basis $u_1, u_2, ..., u_n$ in X such that the matrix of A in this basis is upper triangular.

In other words, any $n \times n$ matrix A can be represented as $T = UTU^*$, where U is a unitary, and T is an upper triangular matrix.

Proof. We prove the theorem using the induction in dim$X$. If dim$X = 1$ the theorem is trivial, since any 1×1 matrix is upper triangular.

Suppose we proved that the theorem is true if dim$X = n - 1$, and we want to prove it for dim $X = n$. Let λ_1 be an eigenvalue of A, and let u_1, $\| u_1 \| = 1$ be a corresponding eigenvector, $Au_1 = \lambda_1 u_1$. Denote $E = u_1^{\perp}$, and let $v_2, ..., v_n$ be some orthonormal basis in E (clearly, dim$E$ = dim $X_{-1} = n^{-1}$), so $u_1, v_2, ..., v_n$ is an orthonormal basis in X. In this basis the matrix of A has the form

$$\left(\begin{array}{c|c} \lambda_1 & * \\ \hline 0 & \\ \vdots & A_1 \\ 0 & \end{array}\right);$$

here all entries below λ_1 are zeroes, and * means that we do not care what entries are in the first row right of λ_1.

We do care enough about the lower right $(n - 1) \times (n - 1)$ block, to give it name: we denote it as A_1.

Note, that A_1 defines a linear transformation in E, and since dim$E = n - 1$, the induction hypothesis implies that there exists an orthonormal basis (let us denote is as $u_2, ..., u_n$) in which the matrix of A_1 is upper triangular.

So, matrix of A in the orthonormal basis $u_1, u_2, ..., u_n$ has the form), where matrix A_1 is upper triangular. Therefore, the matrix of a in this basis is upper triangular as well.

Remark. Note, that the subspace $E = u_1^\perp$ introduced in the proof is not invariant under A, i.e. the inclusion $AE \subset E$ does not necessarily holds. That means that A_1 is not a part of A, it is some operator constructed from A.

Note also, that $AE \subset E$ if and only if all entries denoted by * (i.e. all entries in the first row, except λ_1) are zero.

Remark. Note, that even if we start from a real matrix A, the matrices U and T can have complex entries. The rotation matrix

$$\begin{pmatrix} \cos\alpha & -\sin\alpha \\ \sin\alpha & \cos\alpha \end{pmatrix}, a \neq k\pi, k \in \mathbb{Z}$$

is not unitarily equivalent (not even similar) to a real upper triangular matrix. Indeed, eigenvalues of this matrix are complex, and the eigenvalues of an upper triangular matrix are its diagonal entries.

Remark. An analogue of Theorem can be stated and proved for an arbitrary vector space, without requiring it to have an inner product. In this case the theorem claims that any operator have an upper triangular form in some basis. a proof can be modeled after the proof of Theorem. An alternative way is to equip V with an inner product by fixing a basis in V and declaring it to be an orthonormal one.

Note, that the version for inner product spaces is stronger than the one for the vector spaces, because it says that we always can find an orthonormal basis, not just a basis.

The following theorem is a real–valued version of Theorem

Theorem. Let $A : X \to X$ be an operator acting in a real inner product space. Suppose that all eigenvalues of A are real. Then there exists an orthonormal basis $u_1, u_2, ..., u_n$ in X such that the matrix of A in this basis is upper triangular.

In other words, any real $n \times n$ matrix A can be represented as $T = UTU^* = UTU^T$, where U is an orthogonal, and T is a real upper triangular matrices.

Proof. To prove the theorem we just need to analyse the proof of Theorem. Let us assume (we can always do that without loss of generality, that the operator (matrix) A acts in $\mathbb{R}^n$.

Suppose, the theorem is true for $(n-1) \times (n-1)$ matrices. As in the proof of Theorem let 1 be a real eigenvalue of A, $u_1 \in \mathbb{R}^n$, $\| u_1 \| = 1$ be a corresponding eigenvector, and let $v_2, ..., v_n$ be on orthonormal system (in $\mathbb{R}^n$) such that $u_1, v_2, ..., v_n$ is an orthonormal basis in $\mathbb{R}^n$. The matrix of a in this basis has form equation, where A_1 is some real matrix.

If we can prove that matrix A_1 has only real eigenvalues, then we are done. Indeed, then by the induction hypothesis there exists an orthonormal basis $u_2, ..., u_n$ in $E = u_1^\perp$

such that the matrix of A_1 in this basis is upper triangular, so the matrix of a in the basis u_1, u_2, ..., u_n is also upper triangular.

To show that A_1 has only real eigenvalues, let us notice that

$$\det(A - \lambda I) = (\lambda_1 - \lambda) \det(A_1 - \lambda)$$

(take the cofactor expansion in the first row, for example), and so any eigenvalue of A_1 is also an eigenvalue of A. But a has only real eigenvalues!

Spectral Theorem for self–adjoint and normal operators.

In this section we deal with matrices (operators) which are unitarily equivalent to diagonal matrices.

Let us recall that an operator is called self–adjoint if $A = A^*$. A matrix about of a self–adjoint operator (in some orthonormal basis), i.e. a matrix satisfying $A = a$ is called a Hermitian matrix. Since we usually do not distinguish between operators and their matrices, we will use both terms.

Theorem. Let $A = A^*$ be a self–adjoint operator in an inner product space X (the space can be complex or real). Then all eigenvalues of A are real, and there exists and orthonormal basis of eigenvectors of A in X.

This theorem can be restated in matrix form as follows

Theorem. Let $A = A$ be a self–adjoint (and therefore square) matrix. Then A can be represented as

$$A = UDU,$$

where U is a unitary matrix and D is a diagonal matrix with real entries.

Moreover, if the matrix A is real, matrix U can be chosen to be real (i.e. orthogonal).

Proof. To prove Theorems let us first apply Theorem if X is a real space) to find an orthonormal basis in X such that the matrix of a in this basis is upper triangular. Now let us ask ourself a question: What upper triangular matrices are self–adjoint?

The answer is immediate: an upper triangular matrix is self–adjoint if and only if it is a diagonal matrix with real entries. Theorem is proved.

Let us give an independent proof to the fact that eigenvalues of a selfadjoint operators are real. Let $A = A^*$ and $Ax = \lambda x$, $x \neq 0$. Then

$$(Ax, x) = (x, x) = (x, x) = \| x \|^2.$$

On the other hand,

$$(Ax, x) = (x, A^*x) = (x, Ax) = (x, \lambda x) = \overline{\lambda}\,(x, x) = \overline{\lambda}\,\| x \|^2,$$

so $\lambda\| x \|^2 = \overline{\lambda}\,\| x \|^2$. Since $\|x\|$ 6= 0 ($x \neq 0$),

we can conclude $\lambda = \overline{\lambda}$, so is real.

It also follows from Theorem that eigenspaces of a self–adjoint operator are orthogonal. Let us give an alternative proof of this result.

Proposition. Let $A = A^*$ be a self–adjoint operator, and let u, v be its eigenvectors, $Au = \lambda u$, $Av = \lambda v$. Then, if $\lambda \neq \mu$, the eigenvectors u and v are orthogonal.

Proof. This proposition follows from the spectral theorem, but here we are giving a direct proof. Namely,

$$(Au, v) = (lu, v) = l(u, v).$$

On the other hand

$$(Au, v) = (u, A^*v) = (u, Av) = (u, \mu v) = \overline{\mu}\,(u, v) = \mu(u, v)$$

(the last equality holds because eigenvalues of a self–adjoint operator are real), so $\lambda(u, v) = \mu(u, v)$. If $\lambda \neq \mu$ it is possible only if $(u, v) = 0$.

Now let us try to find what matrices are unitarily equivalent to a diagonal one. It is easy to check that for a diagonal matrix D

$$D^*D = DD^*.$$

Therefore $AA = AA$ if the matrix of a in some orthonormal basis is diagonal.

Definition. An operator (matrix) N is called *normal* if $N^*N = NN$.

Clearly, any self–adjoint operator ($AA^* = AA^*$) is normal. Also, any unitary operator $U : X \to X$ is normal since $U^*U = UU^* = I$.

Note, that a normal operator is an operator acting in one space, not from one space to another. So, if U is a unitary operator acting from one space to another, we cannot say that U is normal.

Theorem. Any normal operator N in a complex vector space has an orthonormal basis of eigenvectors.

In other words, any matrix N satisfying $N^*N = NN^*$ can be represented as

$$N = UDU^*,$$

where U is a unitary matrix, and D is a diagonal one.

Remark. Note, that in the above theorem even if N is a real matrix, we did not claim that matrices U and D are real. Moreover, it can be easily shown, that if D is real, N must be self–adjoint.

Proof. To prove Theorem we apply Theorem to get an orthonormal basis, such that the matrix of N in this basis is upper triangular. To complete the proof of the theorem we only need to show that an upper triangular normal matrix must be diagonal.

We will prove this using induction in the dimension of matrix. The case of 1×1 matrix is trivial, since any 1×1 matrix is diagonal.

Suppose we have proved that any $(n - 1) \times (n - 1)$ upper triangular normal matrix is diagonal, and we want to prove it for $n \times n$ matrices. Let N be $n \times n$ upper triangular normal matrix. We can write it as

$$N = \left(\begin{array}{c|ccc} a_{1,1} & a_{1,2} & \cdots & a_{1,n} \\ \hline 0 & & & \\ \vdots & & N_1 & \\ 0 & & & \end{array}\right)$$

where N_1 is an upper triangular $(n - 1) \times (n - 1)$ matrix.

Let us compare upper left entries (first row first column) of N^*N and NN^*. Direct computation shows that that

$$(NN)_{1,1} = \bar{a}_{1,1} a_{1,1} = |a_{1,1}|^2$$

and

$$(NN)_{1,1} = |a_{1,1}|^2 + |a_{1,2}|^2 + \ldots + |a_{1,n}|^2.$$

So, $(N^*N)_{1,1} = (NN)_{1,1}$ if and only if $a_{1,2} = \ldots = a_{1,n} = 0$. Therefore, the matrix N has the form

$$N = \left(\begin{array}{c|ccc} a_{1,1} & 0 & \ldots & 0 \\ \hline 0 & & & \\ \vdots & & N_1 & \\ 0 & & & \end{array}\right)$$

It follows from the above representation that

$$N^*N = \left(\begin{array}{c|ccc} |a_{1,1}|^2 & 0 & \ldots & 0 \\ \hline 0 & & & \\ \vdots & & N_1^*N_1 & \\ 0 & & & \end{array}\right), \quad NN^* = \left(\begin{array}{c|ccc} |a_{1,1}|^2 & 0 & \ldots & 0 \\ \hline 0 & & & \\ \vdots & & N_1N_1^* & \\ 0 & & & \end{array}\right)$$

so $N_1^*N_1 = N_1N_1^*$. That means the matrix N_1 is also normal, and by the induction hypothesis it is diagonal. So the matrix N is also diagonal.

The following proposition gives a very useful characterization of normal operators.

Proposition. An operator $N : X \to X$ is normal if and only if

$$\| Nx \| = \| N^*x \| \quad \forall x \in X.$$

Proof. Let N be normal, $N^*N = NN^*$. Then

$\| Nx \|^2 = (Nx,Nx) = (N^*Nx, x) = (NN^*x, x) = (N^*x,N^*x) = \| Nx \|^2$

so $\| Nx \| = \| N^*x \|$.

Now let

$$\| Nx \| = \| N^*x \| \quad \forall x \in X.$$

The Polarization Identities imply that for all $x, y \in x$

$$(N^*N x, y) = (Nx, Ny) = \sum_{a=\pm 1,\pm i} \alpha \| Nx + \alpha Ny \|^2$$

$$= \sum_{a=\pm 1,\pm i} \alpha \| N(x+\alpha y) \|^2 = \sum_{\alpha=\pm 1,\pm i} \alpha \| N^*(x+\alpha y) \|^2$$

$$= \sum_{\alpha=\pm 1,\pm i} \alpha \| N^*(x+\alpha N^* y) \|^2 = (N^*x, N^*y) = (NN^* x, y)$$

and therefore $N^*N = NN^*$.

Polar and Singular Value Decompositions

Definition. a self adjoint operator $A : X \to X$ is called *positive definite* if

$$(Ax, x) > 0 \quad \forall x \neq 0,,$$

and it is called *positive semidefinite* if

$$(Ax, x) \geq 0 \quad \forall x \in X.$$

We will use the notation $A > 0$ for positive definite operators, and $A \geq 0$ for positive semi–definite.

The following theorem describes positive definite and semidefinite operators.

Theorem. Let $a = A^*$. Then

1. *$A > 0$ if and only if all eigenvalues of A are positive.*
2. *A A 0 if and only if all eigenvalues of A are non–negative.*

Proof. Pick an orthonormal basis such that matrix of a in this basis is diagonal. To finish the proof it remains to notice that a diagonal matrix is positive definite (positive semidefinite) if and only if all its diagonal entries are positive (non–negative).

Corollary. Let $A = A^* \geq 0$ be a positive semidefinite operator. There exists a unique positive semidefinite operator B such that $B^2 = A$

Such B is called (positive) square root of A and is denoted as $\sqrt{A}$ or $A^{1/2}$.

Proof. Let us prove that $\sqrt{A}$ exists. Let $v_1, v_2, \ldots, v_n$ be an orthonormal basis of eigenvectors of A, and let $\lambda_1, \lambda_2, \ldots, \lambda_n$ be the corresponding eigenvalues. Note, that since $A \geq 0$, all $\lambda_k \geq 0$.

In the basis $v_1, v_2, \ldots, v_n$ the matrix of a is a diagonal matrix diag$\{1, 2, \ldots, n\}$ with entries $\lambda_1, \lambda_2, \ldots, \lambda_n$ on the diagonal. Define the matrix of B in the same basis as diag $\{\sqrt{\lambda_1}, \sqrt{\lambda_2} \ldots \sqrt{\lambda_n}\}$. Clearly, $B = B$ and $B^2 = A$.

To prove that such B is unique, let us suppose that there exists an operator $C = C^* \geq 0$ such that $C^2 = A$. Let $u_1, u_2, \ldots, u_n$ be an orthonormal basis of eigenvalues of C, and let $\mu_1, \mu_2, \ldots, \mu_n$ be the corresponding eigenvalues (note that $u_k \geq 0 \forall_k$). The matrix of C in the basis $u_1, u_2, \ldots, u_n$ is a diagonal one diag $\{\mu_1, \mu_2, \ldots, \mu_n\}$ and therefore the matrix of $A = C^2$ in the same basis is diag$\{\mu_1^2, \mu_2^2 \ldots \mu_n^2\}$ This implies that any eigenvalue λ of A is of form μ_k^2, and, moreover, if $Ax = \lambda x$, then $Cx = \sqrt{\lambda} x$.

Therefore in the basis $v_1, v_2, \ldots, v_n$ above, the matrix of C has the diagonal form diag$\{\sqrt{\lambda_1}, \sqrt{\lambda_1}, \ldots, \sqrt{\lambda_n}\}$, i.e. $B = C$.

Modulus of an operator. Singular values. Consider an operator $A : X \to Y$. Its *Hermitian square* A^*A is a positive semidefinite operator acting in X. Indeed,

$$(A^*A^*) = A^*A^{**} = A^*A$$

and

$$(A^*Ax, x) = (Ax, Ax) = \| Ax \|^2 \geq 0 \ \forall x \in X.$$

Therefore, there exists a (unique) positive–semidefinite square root $R = \sqrt{A^*A}$. This operator R is called the *modulus* of the operator A, and is often denoted as $| A |$.

The modulus of A shows how "big" the operator A is: Proposition. For a linear operator $A : X \to Y$

$$\| |A| x \| = \| Ax \| \ \forall x \in X.$$

Proof. For any $x \in X$

$$\| |A| x \|^2 = (|A| x, |A| x) = (|A|^* |A| x, x) = (|A|^2 x, x)$$
$$= (A^*Ax, x) = (Ax, Ax) = \| Ax \|^2$$

Corollary.

$$\text{Ker}\, A = \text{Ker}\, |A| = (\text{Ran}\, |A|)^{\perp}.$$

Proof. The first equality follows immediately from Proposition, the second one follows from the identity Ker T = (Ran T^*)$_\perp$ ($|A|$ is self–adjoint).

Theorem. (Polar decomposition of an operator). Let $A : X \to X$ *be an operator* (*square matrix*). *Then A can be represented as*

$$A = U |A|,$$

where U is a unitary operator.

Remark. The unitary operator U is generally not unique. As one will see from the proof of the theorem, U is unique if and only if a is invertible.

Remark. The polar decomposition $A = U|A|$ also holds for operators $A : X \to Y$ acting from one space to another. But in this case we can only guarantee that U is an *isometry* from Ran $|A| = (\text{Ker}A)^{\perp}$ to Y.

If dim $X \leq$ dim Y this isometry can be extended to the isometry from the whole X to Y (if dim X = dim Y this will be a unitary operator).

Proof. Consider a vector $x \in$ Ran $|A|$. Then vector x can be represented as $x = |A|v$ for some vector $v \in X$.

Define $U_0 x := A_v$. By Proposition

$$\| U_0 x \| = \| A v \| = \| |A| v \| = \| x \|$$

so it looks like U is an isometry from Ran $|A|$ to X.

But first we need to prove that U_0 is well defined. Let v_1 be another vector such that

$$x = |A| v_1.$$

But $x = |A| v = |A| v_1$ means that

$$v - v_1 \in \text{Ker}\, |A| = \text{Ker}\, A$$

so

$$Av = Av_1,$$

meaning that $U_0 x$ is well defined.

By the construction

$$A = U_0 |A|.$$

To extend U_0 to a unitary operator U, let us find some unitary transformation

$$U_1 : \text{Ker } A \to (\text{Ran } A)^{\perp} = \text{Ker } A^*.$$

It is always possible to do this, since for square matrices dim Ker$A$ = dim Ker A^* (the Rank Theorem). It is easy to check that $U = U_0 + U_1$ is a unitary operator, and that

$$A = U|A|.$$

Singular Values

Eigenvalues of $|A|$ are called the singular values of A. In other words, if $\lambda_1, \lambda_2, \ldots, \lambda_n$ are eigenvalues of A^*A, then

$$\sqrt{\lambda_1}, \sqrt{\lambda_2}, \ldots, \sqrt{\lambda_n}.$$

are singular values of A.

Consider an operator $A : X \to Y$, and let $\sigma_1, \sigma_2, \ldots, \sigma_n$ be the singular values of a counting multiplicities. Assume also that $\sigma_1, \sigma_2, \ldots, \sigma_r$ are the non–zero singular values of A, counting multiplicities. That means $\sigma_k = 0$ for $k > r$.

By the definition of singular values $\sigma_2^2, \sigma_2^2, \ldots, \sigma_n^2$ are eigenvalues of A^*A, and let $v_1, v_2, \ldots, v_n$ be an orthonormal basisof eigenvectors of

$$A^*A,\ A^*Av_k = \sigma_k^2 v_k$$

Proposition. The system

$$w_k := \frac{1}{\sigma_k} Av_k,\ k = 1, 2 \ldots, r$$

is an orthonormal system.

Proof. $(Av_j, Av_k) = (A^*Av_j, v_k) = \sigma_j^2 (v_j, v_k) = \begin{cases} 0, & j \neq k \\ \sigma_j^2, & j = k \end{cases}$

since $v_1, v_2, \ldots, v_r$ is an orthonormal system.

In the notation of the above proposition, the operator a can be represented as

$$A = \sum_{k=1}^{r} \sigma_k w_k v_k^*,$$

or, equivalently

$$Ax = \sum_{k=1}^{r} \sigma_k (x, v_k) w_k.$$

Indeed, we know that $v_1, v_2, \ldots, v_n$ is an orthonormal basis in X. Then

$$\sum_{k=1}^{r} \sigma_k w_k v_k^* v_j = \sigma_j w_j v_J^* v_j = \sigma_j w_j = Av_j \text{ if } j = 1, 2, \ldots, r,$$

and

$$\sum_{k=1}^{r} \sigma_k w_k v_k^* v_j = 0 = Av_j \quad \text{for } j > r.$$

So the operators in the left and right sides of equation coincide on the basis $v_1, v_2, ..., v_n$, so they are equal.

Definition

Remark. Singular value decomposition is not unique. Why?

Lemma. Let a can be represented as

$$A = \sum_{k=1}^{r} \sigma_k w_k v_k^*$$

where $\sigma_k > 0$ and $v_1, v_2, ..., v_r, w_1, w_2, ..., w_r$ *are some orthonormal systems. Then this representation gives a singular value decomposition of A.*

Proof. We only need to show that v_k are eigenvalues of

$$A^* A, A^* A_{vk} = \sigma_k^2 v_k.$$

Since $v_1, v_2, ..., v_r$ is an orthonormal system,

$$w_k^* w_j = (w_j, w_k) = \delta_{kj} := \begin{cases} 0, & j \neq k \\ 1, & j = k, \end{cases}$$

and therefore

$$A^* A = \sum_{k=1}^{r} \sigma_k^2 v_k v_k^*.$$

Since $v_1, v_2, ..., v_r$ is an orthonormal system

$$A^* Av_j = \sum_{k=1}^{r} \sigma_k^2 v_k v_k^* v_j = \sigma_j^2 v_j$$

thus v_k are eigenvectors of A^*A.

Corollary. Let

$$A = \sum_{k=1}^{r} \sigma_k w_k v_k^*$$

be a singular value decomposition of A. Then

$$A = \sum_{k=1}^{r} \sigma_k v_k w_k^*$$

*is a singular value decomposition of A**

Matrix representation of the singular value decomposition. The singular value decomposition can be written in a nice matrix form. It is especially easy to do if the operator

A is invertible. In this case dim X = dim $Y = n$, and the operator A has n non–zero singular values (counting multiplicities), so the singular value decomposition has the form

$$A = \sum_{k=1}^{n} \sigma_k w_k v_k^*$$

where $v_1, v_2, ..., v_n$ and $w_1, w_2, ..., w_n$ are orthonormal bases in x and Y respectively. It can be rewritten as

$$A = W \sum V^*,$$

where $\sum$ = diag$\{\sigma_1, \sigma_2, ..., \sigma_n\}$ and V and W are unitary matrices with columns $v_1, v_2, ..., v_n$ and $w_1, w_2, ..., w_n$ respectively.

Such representation can be written even if A is not invertible. Let us first consider the case dim X = dim $Y = n$, and let $\sigma_1, \sigma_2, ..., \sigma_r$, $r < n$ be non–zero singular values of A. Let

$$A = \sum_{k=1}^{n} \sigma_k w_k v_k^*$$

be a singular value decomposition of A. To represent A as $W \sum V$ let us complete the systems $\{v_k\}_{k=1}^{r}, \{w_k\}_{k=1}^{r}$ to orthonormal bases. Namely, let $v_r + 1, ..., v_n$ and $w_{r+1}, ..., w_n$ be an orthonormal bases in Ker A = Ker $|A|$ and (Ran $A)^{\perp}$ respectively. Then $v_1, v_2, ..., v_n$ and $w_1, w_2, ..., w_n$ are orthonormal bases in X and Y respectively and A can be represented as

$$A = W \sum V^*,$$

where $\sum$ is $n \times n$ diagonal matrix diag $\{\sigma_1, ..., \sigma_r, 0, ..., 0\}$, and V, W are $n \times n$ unitary matrices with columns $v_1, v_2, ..., v_n$ and $w_1, w_2, ..., w_n$ respectively.

Remark. Another way to interpret the singular value decomposition $A = W \sum V^*$ is to say that $\sum$ is the matrix of A in the (orthonormal) bases $A = v_1, v_2, ..., v_n$ and $B := w_1, w_2, ..., w_n$, *i.e*, that $= [A]_{B,A}$. We will use this interpretation later.

From singular value decomposition to the polar decomposition. Note, that if we know the singular value decomposition $A = W \sum V^*$ of a square matrix A, we can write a polar decomposition of A:

$$A = W \sum V^* = (WV)(V \sum V^*) = U|A|$$

so $|A| = V \sum V^*$ and $U = WV$.

General matrix form of the singular value decomposition. In the general case when dim $X = n$, dim $Y = m$ (i.e. A is an $m \times n$ matrix), the above representation $A = V \sum V^*$ is also possible. Namely, if

$$A = \sum_{k=1}^{r} \sigma_k w_k v_k^*$$

is a singular value decomposition of A, we need to complete the systems $v_1, v_2, \ldots, v_r$ and $w_1, w_2, \ldots, w_r$ to orthonormal bases in X and Y respectively. Then a can be represented as

$$A = W \sum V^*,$$

where $V \in M_{n \times n}$ and $W \in M_{m \times m}$ are unitary matrices with columns $v_1, v_2, \ldots, v_n$ and $w_1, w_2, \ldots, w_m$ respectively, and $\sum$ is a "diagonal" $m \times n$ matrix

$$\sum_{j,k} = \begin{cases} \sigma_k & j = k \le r: \\ 0 & \text{otherwise.} \end{cases}$$

In other words, to get the matrix one has to take the diagonal matrix diag$\{\sigma_1, \sigma_2, \ldots, r\}$ and make it to an $m \times n$ matrix by adding extra zeroes "south and east".

SINGULAR VALUES

As we discussed above, the singular value decomposition is simply diagonalization with respect to two dierent orthonormal bases. Since we have two dierent bases here, we cannot say much about spectral properties of an operator from its singular value decomposition. For example, the diagonal entries of $\sum$ in the singular value decomposition are not the eigenvalues of A. Note, that for a $= W \sum V^*$ as in we generally have $A^n \neq W \sum^n V^*$, so this diagonalization does not help us in computing functions of a matrix.

However, as the examples below show, singular values tell us a lot about so–called *metric properties* of a linear transformation.

Final Remark: performing singular value decomposition requires finding eigenvalues and eigenvectors of the Hermitian (self–adjoint) matrix A^*A. To find eigenvalues we usually computed characteristic polynomial, found its roots, and so on... This looks like quite a complicated process, especially if one takes into account that there is no formula for finding roots of polynomials of degree 5 and higher.

However, there are very eective numerical methods of find eigenvalues and eigenvectors of a hermitian matrix up to any given precision. These methods do not involve computing the characteristic polynomial and finding its roots. They compute approximate eigenvalues and eigenvectors directly by an iterative procedure. Because a Hermitian matrix has an orthogonal basis of eigenvectors, these methods work extremely well.

We will not discuss these methods here, it goes beyond the scope of this book. However, you should believe me that there are very eective numerical methods for computing eigenvalues and eigenvectors of a Hermitian matrix and for finding the singular value decomposition. These methods are extremely eective, and just a little more computationally intensive than solving a linear system.

Image of the unit ball. Consider for example the following problem: let $A : \mathbb{R}^n \to \mathbb{R}^m$ be a linear transformation, and let $B = \{x \in \mathbb{R}^n : \| x \| \leq 1\}$ be the closed unit ball in $\mathbb{R}^n$. We want to describe $A(B)$, i.e. we want to find out how the unit ball is transformed under the linear transformation.

Let us first consider the simplest case when A is a diagonal matrix $A =$ diag$\{\sigma_1, \sigma 2, ..., \sigma_n\}, \sigma_k > 0, k = 1, 2, ..., n$. Then for $v = (x_1, x_2, ..., x_n)^T$ and $(y_1, y_2, ..., y_n)^T = y = Ax$ we have $yk = \sigma_k x_k$ (equivalently, $x_k = yk/\sigma_k$) for $k = 1, 2, ..., n$, so

$$y = (y_1, y_2, ..., y_n)^T = Ax \text{ for } \| x \| \leq 1,$$

if and only if the coordinates $y_1, y_2, ..., y_n$ satisfy the inequality

$$\frac{y_1^2}{\sigma_1^2} + \frac{y_2^2}{\sigma_2^2} + \ldots + \frac{y_n^2}{\sigma_n^2} = \sum_{k=1}^{n} \frac{y_k^2}{\sigma_k^2} \leq 1$$

(this is simply the inequality $\| x \|^2 = \sum_k | xk |^2 \leq 1$).

The set of points in $\mathbb{R}^n$ satisfying the above inequalities is called an ellipsoid. If $n = 2$ it is an ellipse with half–axes σ_1 and σ_2, for $\sigma_n = 3$ it is an ellipsoid with half–axes σ_1, σ_2 and σ_2. In $\mathbb{R}^n$ the geometry of this set is also easy to visualize, and we call that set an ellipsoid with half axes $\sigma_1, \sigma_2, ..., \sigma_n$. The vectors $e_1, e_2, ..., e_n$ or, more precisely the corresponding lines are called the *principal axes* of the ellipsoid.

The singular value decomposition essentially says that any operator in an inner product space is diagonal with respect to a pair of orthonormal bases, see Remark 3.9. Namely, consider the orthogonal bases $A = v_1, v_2, ..., v_n$ and $B = w_1, w_2, ..., w_n$ from the singular value decomposition. Then the matrix of A in these bases is diagonal

$$[A]_{B,A} = \text{diag}\{s_n : n = 1, 2, ..., n\}.$$

Assuming that all $\sigma_k > 0$ and essentially repeating the above reasoning, it is easy to show that any point $y = Ax \in A(B)$ if and only if it satisfies the inequality

$$\frac{y_1^2}{\sigma_1^2} + \frac{y_2^2}{\sigma_2^2} + \ldots + \frac{y_n^2}{\sigma_n^2} = \sum_{k=1}^{n} \frac{y_k^2}{\sigma_k^2} \leq 1.$$

where $y_1, y_2, ..., y_n$ are coordinates of y in the orthonormal basis $B = w_1, w_2, ..., w_n$, not in the standard onc. Similarly, $(x_1, x_2, ..., x_n)^T = [x]_A$.

But that is essentially the same ellipsoid as before, only "rotated" (with dierent but still orthogonal principal axes)!

There is also an alternative explanation which is presented below.

Consider the general case, when the matrix A is not necessarily square, and (or) not all singular values are non–zero. Consider first the case of a "diagonal" matrix Σ of form. It is easy to see that the image ΣB of the unit ball B is the ellipsoid (not in the whole space but in the Ran Σ) with half axes $\sigma_1, \sigma_2, ..., \sigma_r$.

Consider now the general case, $A = W\Sigma V^*$, where V, W are unitary operators. Unitary transformations do not change the unit ball (because they preserve norm), so $V^*(B) = B$. We know that $\Sigma(B)$ is an ellipsoid in Ran Σ with half–axes $\sigma_1, \sigma_2, ..., \sigma_r$. Unitary transformations do not change geometry of objects, so $W(\Sigma(B))$ is also an ellipsoid with the same half–axes. It is not hard to see from the decomposition $A = W\Sigma V^*$ (using the fact that both W and V are invertible) that W transforms RanΣ to Ran A, so we can conclude:

the image $A(B)$ of the closed unit ball B is an ellipsoid in Ran A with half axes $\sigma_1, \sigma_2, ..., \sigma_r$. Here r is the number of non-zero singular values, i.e. the rank of A.

Operator norm of a linear transformation. Given a linear transformation A : $x \to Y$ let us consider the following optimization problem: find the maximum of kAxk on the closed unit ball $B = \{x \in X : \| x \leq 1\}$.

Again, singular value decomposition allows us to solve the problem. For a diagonal matrix A with non–negative entries the maximum is exactly maximal diagonal entry. Indeed, let $s_1, s_2, ..., s_r$ be non–zero diagonal entries of A and let s_1 be the maximal one. Since

$$Ax = \sum_{k=1}^{r} x_k e_k,$$

we can conclude that

$$\| Ax \| \leq \sum_{k=1}^{r} s_k^2 \mid x_k \mid^2 \leq s_1^2 \sum_{k=1}^{r} \mid x_k \mid^2 = s_1^1 . \| x \|^2,$$

so $\| Ax \| \leq s_1 \| x \|$. On the other hand, $\| Ae_1 \| = \| s_1 e_1 \| = s_1 \| e_1 |$, so indeed s_1 is the maximum of $\| Ax \|$ on the closed unit ball B. Note, that in the above reasoning we did not assume that the matrix A is square; we only assumed that all entries outside the "main diagonal" are 0, so formula holds.

To treat the general case let us consider the singular value decomposition, $A = W\Sigma V$, where W, V are unitary operators, and Σ is the diagonal matrix with non–negative entries. Since unitary transformations do not change the norm, one can conclude that the maximum of $\| Ax \|$ on the unit ball B is the maximal diagonal entry of Σ i.e. that

the maximum of $\|Ax\|$ on the unit ball B is the maximal singular value of A.

Definition. The quantity max $\{\| Ax \| : x \in X, \| x \| \leq 1\}$ is called the operator norm of a and denoted $\| A \|$.

It is an easy exercise to see that $\| A \|$ satisfies all properties of the norm:

1. $\| \alpha A \| = | \alpha | \cdot \| A \|$.
2. $\| A + B \| \leq \| A \| + \| B \|$.
3. $\| A \| \geq 0$ for all A.
4. $\| A \| = 0$ if and only if $A = 0$.

So it is indeed a norm on a space of linear transformations from from x to Y. One of the main properties of the operator norm is the inequality

$$\| Ax \| \geq \| A \| \cdot \| x \|,$$

which follows easily from the homogeneity of the norm $\| x \|$.

In fact, it can be shown that the operator norm $\| A \|$ is the best (smallest) number $C \geq 0$ such that

$$\| Ax \| \leq C \| x \| \ \forall x \in X.$$

This is often used as a definition of the operator norm.

On the space of linear transformations we already have one norm, the Frobenius, or Hilbert–Schmidt norm $\| A \|_2$,

$$\| A \|_2^2 = trace(A^* A).$$

So, let us investigate how these two norms compare.

Let s_1, s_2, ..., s_r be non–zero singular values of A (counting multiplicities), and let s_1 be the largest eigenvalues. Then $s_1^2, s_2^2, \ldots, s_r^2$ are non-zero eigenvalues of A^*A (again counting multiplicities). Recalling that the trace equals the sum of the eigenvalues we conclude that

$$\| A \|_2 = \text{trace}(A^* A) = \sum_{k=1}^{r} s_k^2.$$

On the other hand we know that the operator norm of a equals its largest singular value, i.e. $\| A \| = s_1$. So we can conclude that $\| A \| \leq \| A \|_2$, i.e. that the operator norm of a matrix cannot be more than its.

This statement also admits a direct proof using the Cauchy–Schwarz inequality, and such a proof is presented in some textbooks. The beauty of the proof we presented here is that it does not require any computations and illuminates the reasons behind the inequality.

Condition number of a matrix. Suppose we have an invertible matrix A and we want to solve the equation $Ax = b$. The solution, of course, is given by $x = A^{-1}b$, but we want to investigate what happens if we know the data only approximately.

That happens in the real life, when the data is obtained, for example by some experiments. But even if we have exact data, round–o errors during computations by a computer may have the same effect of distorting the data.

Let us consider the simplest model, suppose there is a small error in the right side of the equation. That means, instead of the equation $Ax = b$ we are solving

$$Ax = b + \Delta b,$$

where Δb is a small perturbation of the right side b.

So, instead of the exact solution x of $Ax = b$ we get the approximate solution $x + \Delta x$ of $A(x + \Delta x) = b + \Delta b$. We are assuming that A is invertible, so $x = A^{-1}\Delta b$. We want to know how big is the relative error in the solution $\| \Delta x \| / \| x \|$ in comparison with the relative error in the right side $\| \Delta b \| / \| b \|$. It is easy to see that

$$\frac{\| \Delta x \|}{\| x \|} = \frac{\| A^{-1} \Delta b \|}{\| x \|} = \frac{\| A^{-1} \Delta b \|}{\| b \|} \frac{\| b \|}{\| x \|} = \frac{\| A^{-1} \Delta b \|}{\| b \|} \frac{\| Ax \|}{\| x \|}.$$

Since $\| A^{-1}\Delta b \| \leq \| A^{-1} \| . \| \Delta b \|$ and $\| A x \| \leq \| A \| \cdot \| x \|$ we can conclude that

$$\frac{\| \Delta x \|}{\| x \|} \leq \| A^{-1} \| . \| A \| . \frac{\| \Delta b \|}{\| b \|}.$$

The quantity $\| A \| \cdot \| A^{-1} \|$ is called the *condition number* of the matrix A. It estimates how the relative error in the solution x depends on the relative error in the right side b.

Let us see how this quantity is related to singular values. Let $s_1, s_2, \ldots, s_n$ be the singular values of A, and let us assume that s_1 is the largest singular value and s_n is the smallest. We know that the (operator) norm of an operator equals its largest singular value, so

$$\| A \| = s_1, \| A^{-1} \| = \frac{1}{s_n},$$

so

$$\| A \| . \| A^{-1} \| = \frac{s_1}{s_n}.$$

In other words, the condition number of a matrix equals to the ratio of the largest and the smallest singular values.

We deduced above that $\frac{\| \Delta x \|}{\| x \|} \leq \| A^{-1} \| . \| A \| . \frac{\| \Delta b \|}{\| b \|}$. It is not hard to see that this estimate is sharp, i.e. that it is possible to pick the right side b and the error Δb such that we have equality

$$\frac{\| \Delta x \|}{\| x \|} = \| A^{-1} \| . \| A \| . \frac{\| \Delta b \|}{\| b \|}.$$

We just put $b = v_1$ and $\Delta b = a w_n$, where v_1 is the first column of the matrix V, and w_n is the nth column of the matrix W in the singular value decomposition $A = W \Sigma V^*$. Here α can be any scalar.

A matrix is called *well conditioned* if its condition number is not too big. If the condition number is big, the matrix is called ill conditioned. What is "big" here depends on the problem: with what precision you can find your right side, what precision is required for the solution, etc. Effective rank of a matrix. Theoretically, the rank of a matrix is easy to compute: one just needs to row reduce matrix and count pivots.However, in practical applications not everything is so easy. The main reason is that very often we do not know the exact matrix, we only know its approximation up to some precision.

Moreover, even if we know the exact matrix, most computer programs introduce round–o errors in the computations, so effectively we cannot distinguish between a zero pivot and a very small pivot.

A simple naive idea of working with round–off errors is as follows. When computing the rank (and other objects related to it, like column space, kernel, etc) one simply sets up a tolerance (some small number) and if the pivot is smaller than the tolerance, count it as zero. The advantage of this approach is its simplicity, since it is very easy to programme. However, the main disadvantage is that is is impossible to see what the tolerance is responsible for. For example, what do we lose is we set the tolerance equal to 10^{-6}? How much better will 10^{-8} be?

While the above approach works well for well conditioned matrices, it is not very reliable in the general case.

A better approach is to use singular values. It requires more computations, but gives much better results, which are easier to interpret. In this approach we also set up some small number as a tolerance, and then perform singular value decomposition. Then we simply treat singular values smaller than the tolerance as zero. The advantage of this approach is that we can see what we are doing. The singular values are the half–axes of the ellipsoid $A(B)$ (B is the closed unit ball), so by setting up the tolerance we just deciding how "thin" the ellipsoid should be to be considered "flat".

Structure of Orthogonal Matrices

An orthogonal matrix U with det $U = 1$ is often called a *rotation*. The theorem below explains this name.

Theorem. Let U be an orthogonal operator in $\mathbb{R}^n$. *Suppose that* $\det U = 1$. *Then there exists an orthonormal basis* $v_1, v_2, \ldots, v_n$ *such that the matrix of U in this basis has the block diagonal form*

$$\begin{pmatrix} R_{\varphi_1} & & & & 0 \\ & R_{\varphi 2} & & & \\ & & \ddots & & \\ & & & R_{\varphi k} & \\ 0 & & & & I_{n-2k} \end{pmatrix},$$

where R_{jk} *are 2–dimensional rotations,*

$$R_{\varphi_k} = \begin{pmatrix} \cos\varphi_k & -\sin\varphi_k \\ \sin\varphi_k & \cos\varphi_k \end{pmatrix}$$

and I_{n-2k} *stands for the identity matrix of size* $(n-2k) \times (n-2k)$.

Proof. We know that if p is a polynomial with real coefficient and λ is its complex root, $p(\lambda) = 0$, then $\bar{\lambda}$ is a root of p as well, $p(\bar{\lambda}) = 0$ (this can easily be checked by plugging $\bar{\lambda}$ into $p(z) = \sum_{k=0}^{n} a_k z^k$). Therefore, all complex eigenvalues of a real matrix A can be split into pairs λ_k, $\bar{\lambda}_k$.

We know, that eigenvalues of a unitary matrix have absolute value 1, so all complex eigenvalues of A can be written as $\lambda_k = \cos \alpha k + i \sin \alpha k$, $\bar{\lambda}_k = \cos \alpha_k + i \sin \alpha_k$.

Fix a pair of complex eigenvalues λ and $\bar{\lambda}$, and let $u \in \mathbb{C}^n$ be the eigenvector of U, $Uu = \lambda_u$. Then $U\bar{u} = \bar{\lambda}\,\bar{u}$. Now, split u into real and maginary parts, i.e. define

$$x_k := \text{Re } u = (u + \bar{u})/2, \ y = \text{Im } u = (u - \bar{u})/(2i),$$

so $u = x + iy$ (note, that x, y are real vectors, i.e. vectors with real entries).

Then

$$Ux = U\frac{1}{2}(u+\bar{u}) = \frac{1}{2}(\lambda u + \overline{\lambda u}) = \text{Re}(\lambda u)$$

Similarly,

$$Uy = \frac{1}{2i}U(u-\bar{u}) = \frac{1}{2i}(\lambda u - \overline{\lambda u}) = \text{Im}\,(\lambda u).$$

Since $= \cos \alpha + i \sin \alpha$, we have

$\lambda u = (\cos \alpha + i \sin \alpha)\,(x + iy) = ((\cos \alpha)x - (\sin \alpha)y) + i((\cos \alpha)y + (\sin \alpha)x)$.

so

$$Ux = \text{Re}(\lambda u) = (\cos \alpha)\,x - (\sin \alpha)y, \ Uy = \text{Im}(u) = (\cos \alpha)y + (\sin \alpha)x.$$

In other word, U leaves the 2–dimensional subspace E_λ spanned by the vectors x, y invariant and the matrix of the restriction of U onto this subspace is the rotation matrix

$$R_{-\alpha} = \begin{pmatrix} \cos\alpha & \sin\alpha \\ -\sin\alpha & \cos\alpha \end{pmatrix}.$$

Note, that the vectors u and u (eigenvectors of a unitary matrix, corresponding to dierent eigenvalues) are orthogonal, so by the Pythagorean Theorem

$$\| x \| = \| y \| = \frac{\sqrt{2}}{2}\| u \|.$$

It is easy to check that $x \perp y$, so x, y is an orthogonal basis in E_λ. If we multiply each vector in the basis x, y by the same non–zero number, we do not change matrices of linear transformations, so without loss of generality we can assume that $\| x \| = \| y \| = 1$ i.e. that x, y is an orthogonal basis in E_λ.

Let us complete the orthonormal system $v_1 = x$, $v_2 = y$ to an orthonormal basis in $\mathbb{R}^n$. Since $UE_\lambda \subset E_\lambda$ E_λ, i.e. E is an invariant subspace of U, the matrix of U in this basis has the block triangular form

$$\left(\begin{array}{c|c} R_{-\alpha} & * \\ \hline 0 & U_1 \end{array}\right)$$

where 0 stands for the $(n-2) \times 2$ block of zeroes.

Since the rotation matrix $R_{-\alpha}$ is invertible, we have $UE_\lambda = E_\lambda$. Therefore

$$U^*E_l = U^{-1}E_l = E_l$$

so the matrix of U in the basis we constructed is in fact block diagonal,

$$\left(\begin{array}{c|c} R_{-\alpha} & 0 \\ \hline 0 & U_1 \end{array}\right).$$

Since U is unitary

$$\left(\begin{array}{c|c} I & 0 \\ \hline 0 & U_1^*U_1 \end{array}\right),$$

so, since U_1 is square, it is also unitary.

If U_1 has complex eigenvalues we can apply the same procedure to decrease its size by 2 until we are left with a block that has only real eigenvalues. Real eigenvalues can be only +1 or –1, so in some orthonormal basis the matrix of U has the form

$$\begin{pmatrix} R_{-\alpha_1} & & & & & 0 \\ & R_{-\alpha_2} & & & & \\ & & \ddots & & & \\ & & & R_{-\alpha_d} & & \\ & & & & -I_r & \\ 0 & & & & & I_l \end{pmatrix};$$

here I_r and I_l are identity matrices of size $r \times r$ and $l \times l$ respectively. Since $\det U = 1$, the multiplicity of the eigenvalue –1 (i.e. r) must be even.

Note, that the 2×2 matrix $-I_2$ can be interpreted as the rotation through the angle π'. Therefore, the above matrix has the form given in the conclusion of the theorem with '$\varphi_k = -\alpha_k$ or '$\varphi_k = \pi$

Let us give a dierent interpretation of Theorem. Define T_j to be a rotation thorough φ_j in the plane spanned by the vectors vj , $vj + 1$. Then Theorem simply says that U is the composition of the rotations T_j , $j = 1, 2, ..., k$. Note, that because the rotations T_j act in mutually orthogonal planes, they commute, i.e. it does not matter in what order we take the composition. So, the theorem can be interpreted as follows:

Any rotation in $\mathbb{R}^n$ can be represented as a composition of at most $n/2$ commuting planar rotations.

If an orthogonal matrix has determinant –1, its structure is described by the following theorem.

Theorem. Let U be an orthogonal operator in $\mathbb{R}^n$, *and let* $\det U = -1$. *Then there exists an orthonormal basis* $v_1, v_2, \ldots, v_n$ *such that the matrix of U in this basis has block diagonal form*

$$\begin{pmatrix} R_{\varphi_1} & & & & 0 \\ & R_{\varphi 2} & & & \\ & & \ddots & & \\ & & & R_{\varphi k} & \\ 0 & & & & I_{n-2k} \end{pmatrix},$$

where $r = n - 2k - 1$ *and* $R_{\varphi k}$ *are 2–dimensional rotations,*

$$R_{\varphi k} = \begin{pmatrix} \cos\varphi_k & -\sin\varphi_k \\ \sin\varphi_k & \cos\varphi_k \end{pmatrix}$$

and I_{n-2k} *stands for the identity matrix of size* $(n-2k) \times (n-2k)$.

The modification that one should make to the proof of Theorem are pretty obvious.

Note, that it follows from the above theorem that an orthogonal 2×2 matrix U with determinant -1 is always a reflection.

Let us now fix an orthonormal basis, say the standard basis in $\mathbb{R}^n$. We call an *elementary rotation*[2] a rotation in the $x_j - x_k$ plane, i.e. a linear transformation which changes only the coordinates x_j and x_k, and it acts on these two coordinates as a plane rotation.

Theorem. Any rotation U (i.e. an orthogonal transformation U with $\det U = 1$*) can be represented as a product at most* $n(n-1)/2$ *elementary rotations.*

To prove the theorem we will need the following simple lemmas.

Lemma. Let $x = (x_1, x_2)^T \in \mathbb{R}^2$. *There exists a rotation R of* R_2 *which moves the vector x to the vector* $(a, 0)^T$, *where* $a = \sqrt{x_1^1 + x_2^2}$.

One can just draw a picture or/and write a formula for R_α.

Lemma. Let $x = (x_1, x_2, \ldots, x_n)^T \in \mathbb{R}^n$. *There exist* n^{-1} *elementary rotations* $R_1, R_2, \ldots, R_{n-1}$ *such that* $R_{n-1} \ldots, R_2 R_1 x = (a, 0, 0, \ldots, 0)^T$, *where* $a = \sqrt{x_1^2 + x_2^2 + \ldots + x_n^2}$.

Proof. The idea of the proof of the lemma is very simple. We use an elementary rotation R_1 in the $x_n - 1 - x_n$ plane to "kill" the last coordinate of x. Then use an elementary rotation R_2 in $x_{n-2} - x_{n-1}$ plane to "kill" the coordinate number $n-1$ of $R_1 x$ (the rotation R_2 does not change the last coordinate, so the last coordinate of $R_2 R_1 x$ remains zero), and so on.

For a formal proof we will use induction in n. The case $n = 1$ is trivial, since any vector in $\mathbb{R}^1$ has the desired form. The case $n = 2$ is treated.

Assuming now that Lemma is true for $n-1$, let us prove it for n. There exists a 2×2 rotation matrix R_α such that

$$R_\alpha \begin{pmatrix} x_{n-1} \\ x_n \end{pmatrix} = \begin{pmatrix} a_{n-1} \\ 0 \end{pmatrix},$$

where $a_{n-1} = \sqrt{x_{n-1}^2 + x_n^2}$. So if we define the $n \times n$ elementary rotation R_1 by

$$R_1 = \begin{pmatrix} I_{n-2} & 0 \\ 0 & R_\alpha \end{pmatrix}$$

(I_{n-2} is $(n-2) \times (n-2)$ identity matrix), then

$$R_{1x} = (x_1, x_2, \ldots, x_{n-2}, a_{n-1}, 0)^T.$$

We assumed that the conclusion of the lemma holds for $n - 1$, so there exist $n - 2$ elementary rotations (let us call them $R_2, R_3, \ldots, R_{n-1}$) in $\mathbb{R}^{n-1}$ which transform the vector $(x_1, x_2, \ldots, x_{n-1}, a_{n-1})^T \in \mathbb{R}^{n-1}$ to the vector $(a, 0, \ldots, 0)^T \in \mathbb{R}^{n-1}$. In other words

$$R_{n-1}, \ldots, R_3R_2(x_1, x_2, \ldots, x_{n-1}, a_{n-1})^T = (a, 0, \ldots, 0)^T.$$

We can always assume that the elementary rotations $R_2, R_3, \ldots, R_{n-1}$ act in $\mathbb{R}^n$, simply by assuming that they do not change the last coordinate.

Then

$$R_{n-1}, \ldots, R_3R_2R_x = (a, 0, \ldots, 0)^T \in \mathbb{R}^n.$$

Let us now show that $a = \sqrt{x_1^2 + x_2^2 + \ldots + x_n^2}$. It can be easily checked directly, but we apply the following indirect reasoning. We know that orthogonal transformations preserve the norm, and we know that $a \geq 0$.

But, then we do not have any choice, the only possibility for a is $a = \sqrt{x_1^2 + x_2^2 + \ldots + x_n^2}$.

Lemma. Let A be an $n \times n$ matrix with real entries. There exist elementary rotations $R_1, R_2, \ldots, R_{n'}$, $n \leq n(n-1)/2$ such that the matrix $B = R_{N'}, \ldots, R_2R_1A$ is upper triangular, and, moreover, all its diagonal entries except the last one $B_{n,n}$ are non–negative.

Proof. We will use induction in n. The case $n = 1$ is trivial, since we can say that any 1×1 matrix is of desired form.

Let us consider the case $n = 2$. Let a1 be the first column of A. There exists a rotation R which "kills" the second coordinate of a_1, making the first coordinate non–negative. Then the matrix $B = RA$ is of desired form.

Let us now assume that lemma holds for $(n-1) \times (n-1)$ matrices, and we want to prove it for $n \times n$ matrices. For the $n \times n$ matrix A let a1 be its first column. We can find $n - 1$ elementary rotations (say $R_1, R_2, \ldots, R_{n-1}$ which transform a_1 into $(a, 0, \ldots, 0)^T$. So, the matrix $R_{n-1}, \ldots, R_2R_1A$ has the following block triangular form

$$R_{n-1} \ldots R_2R_1A = \begin{pmatrix} a & * \\ 0 & A_1 \end{pmatrix},$$

where A_1 is an $(n-1) \times (n-1)$ block.

We assumed that lemma holds for $n-1$, so A_1 can be transformed by at most $(n-1)(n-2)/2$ rotations into the desired upper triangular form. Note, that these rotations act in $\mathbb{R}^{n-1}$ (only on the coordinates $x_2, x_3, ..., x_n$), but we can always assume that they act on the whole $\mathbb{R}^n$ simply by assuming that they do not change the first coordinate. Then, these rotations do not change the vector $(a, 0, ..., 0)^T$ (the first column of $R_{n-1}, ..., R_2R_1A$), so the matrix A can be transformed into the desired upper triangular form by at most

$$n-1+(n-1)(n-2)/2 = n(n-1)/2$$

elementary rotations.

Proof. There exist elementary rotations $R_1, R_2, ..., R_N$ such that the matrix

$$U_1 = R_N, ..., R_2R_2U$$

is upper triangular, and all diagonal entries, except may be the last one, are non–negative.

Note, that the matrix U_1 is orthogonal. Any orthogonal matrix is normal, and we know that an upper triangular matrix can be normal only if it is diagonal. Therefore, U_1 is a diagonal matrix.

We know that an eigenvalue of an orthogonal matrix can either be 1 or –1, so we can have only 1 or –1 on the diagonal of U_1. But, we know that all diagonal entries of U_1, except may be the last one, are non–negative, so all the diagonal entries of U_1, except may be the last one, are 1. The last diagonal entry can be ±1.

Since elementary rotations have determinant 1, we can conclude that

$$\det U_1 = \det U = 1,$$

so the last diagonal entry also must be 1. So $U_1 = I$, and therefore U can be represented as a product of elementary rotations

$$U = R_1^{-1}R_2^{-1}\ldots R_N^{-1}.$$

Here we use the fact that the inverse of an elementary rotation is an elementary rotation as well.

Orientation

Motivation. In Figures 3 orthonormal bases in $\mathbb{R}^2$ and $\mathbb{R}^3$ respectively. In each figure, the basis *b*) can be obtained from the standard basis a) by a rotation, while it is impossible to rotate the standard basis to get the basis *c*) (so that e_k goes to $v_k \forall k$).

You have probably heard the word "orientation" before, and you probably know that bases (*a*) and (*b*) have positive orientation, and orientation of the bases (*c*) is negative. You also probably know some rules to determine the orientation, like the right hand rule from physics. So, if you can see a basis, say in $\mathbb{R}^3$, you probably can say what orientation it has. But what if you only given coordinates of the vectors v_1, v_2, v_3? Of course, you can try to draw a picture to visualize the vectors, and then to see what the orientation is.

But this is not always easy. Moreover, how do you "explain" this to a computer?

It turns out that there is an easier way. Let us explain it. We need to check whether it is possible to get a basis v_1, v_2, v_3 in $\mathbb{R}^3$ by rotating the standard basis e_1, e_2, e_n.

$U_{ek} = vk, k = 1, 2, 3;$

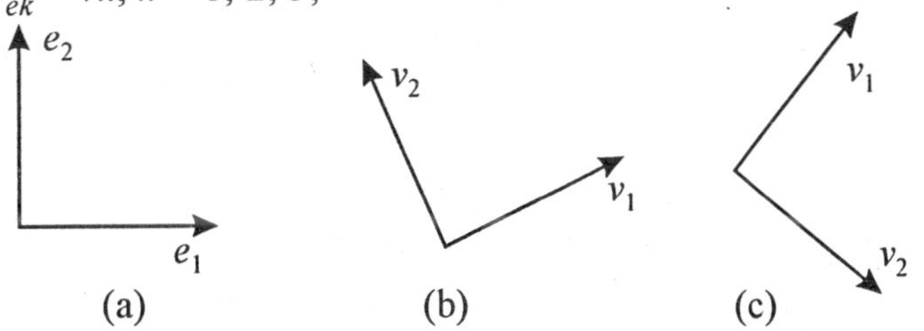

Fig. Standard Bases $\mathbb{R}^2$

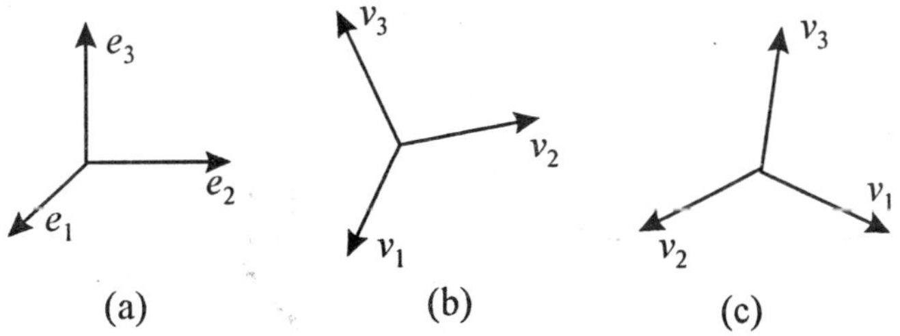

Fig. Orientation in $\mathbb{R}^3$

There is unique linear transformation U such that its matrix (in the standard basis) is the matrix with columns v_1, v_2, v_3. It is an orthogonal matrix (because it transforms an orthonormal basis to an orthonormal basis), so we need to see when it is rotation. Theorems give us the answer: the matrix U is a rotation if and only if det $U = 1$. Note, that (for 3×3 matrices) if det $U = -1$, then U is the composition of a rotation about some axis and a reflection in the plane of rotation, *i.e.* in the plane orthogonal to this axis. This gives us a motivation for the formal definition below.

Definition. Let a and b be two bases in a real vector space X. We say that the bases a and b have the same orientation, if the change of coordinates matrix $[I]_{B,A}$ has positive determinant, and say that they have dierent orientations if the determinant of $[I]_{B,A}$ is negative. Note, that since

$$[I]_{A,B} = [I]_{B,A}^{-1}$$

one can use the matrix $[I]_{A,B}$ in the definition.

We usually assume that the standard basis $e_1, e_2, ..., e_n$ in $\mathbb{R}^n$ has positive orientation. In an abstract space one just needs to fix a basis and declare that its orientation is positive.

If an orthonormal basis $v_1, v_2, ..., v_n$ in $\mathbb{R}^n$ has positive orientation (i.e. the same orientation as the standard basis) This equation show that the basis $v_1, v_2, ..., v_n$ is obtained from the standard basis by a rotation.

Continuous Transformations of Bases and Orientation

Definition. We say that a basis a = $\{a_1, a_2, ..., a_n\}$ can be continuously transformed to A basis

$$b = \{b_1, b_2, ..., b_n\}$$

if there exists a continuous family of bases

$$V(t) = \{v_1(t), v_2(t), ..., v_n(t)\}, t \in [a, b]$$

such that

$$v_k(a) = a_k, v_k(b) = b_k, k = 1, 2 ..., n.$$

"Continuous family of bases" mean that the vector–functions $v_k(t)$ are continuous (their coordinates in some bases are continuous functions) and, which is essential, the system $v_1(t), v_2(t), ..., v_n(t)$ is a basis for all $t \in [a, b]$. Note, that performing a change of variables, we can always assume, if necessary that

$$[a, b] = [0, 1].$$

Theorem. Two bases $A = \{a_1, a_2, ..., a_n\}$ *and* $B = \{b_1, b_2, ..., b_n\}$ *have the same orientation, if and only if one of the bases can be continuously transformed to the other.*

Proof. Suppose the basis A can be continuously transformed to the basis B, and let $V(t)$, $t \in [a, b]$ be a continuous family of bases, performing this transformation. Consider a matrix–function $V(t)$ whose columns are the coordinate vectors $[v_k(t)]_A$ of $v_k(t)$ in the basis A.

Clearly, the entries of $V(t)$ are continuous functions and $V(a) = I$, $V(b) = [I]_{A,B}$. Note, that because $V(t)$ is always a basis, det $V(t)$ is never zero. Then, the Intermediate Value Theorem asserts that det $V(a)$ and det $V(b)$ has the same sign. Since det $V(a)$ = det $I = 1$, we can conclude that

$$\det[I]A,B = \det V(b) > 0,$$

so the bases A and B have the same orientation.

To prove the oppqsite implication, i.e. the "only if" part of the theorem, one needs to show that the identity matrix I can be continuously transformed through invertible matrices to any matrix B satisfying det $B > 0$. In other words, that there exists a continuous matrix–function $V(t)$ on an interval $[a, b]$ such that for all $t \in [a, b]$ the matrix $V(t)$ is invertible and such that

$$V(a) = I, V(b) = B.$$

Chapter 8

Bilinear and Quadratic Forms

Main Definition

Bilinear forms on R^n. A bilinear form on R^n is a function $L = L(x, y)$ of two arguments $x, y \in \mathbb{R}^n$ which is linear in each argument, i.e. such that

1. $L(\alpha x_1 + \beta x_2, y) = \alpha L(x_1, y) + \beta L(x_2, y)$;
2. $L(x, \alpha y_1 + \beta y_2) = \alpha L(x, y_1) + \beta L(x, y_2)$.

One can consider bilinear form whose values belong to an arbitrary vector space, but in this book we only consider forms that take real values.

If $x = (x_1, x_2, ..., x_n)^T$ and $y = (y_1, y_2, ..., y_n)^T$, a bilinear form can be written as

$$L(x, y) = \sum_{j,k=1}^{n} a_{j,k} x_k y_j,$$

or in matrix form

$$L(x, y) = (Ax, y)$$

where

$$A = \begin{pmatrix} a_{1,1} & a_{1,2} & \cdots & a_{1,n} \\ a_{2,1} & a_{2,2} & \cdots & a_{2,n} \\ \vdots & \vdots & \ddots & \vdots \\ a_{n,1} & a_{n,2} & \cdots & a_{n,n} \end{pmatrix}.$$

The matrix A is determined uniquely by the bilinear form L.

Quadratic forms on $\mathbb{R}^n$. There are several equivalent definition of a quadratic form. One can say that a quadratic form on R^n is the "diagonal" of a bilinear form L, i.e. that any quadratic form Q is defined by $Q[x] = L(x, x) = (Ax, x)$.

Another, more algebraic way, is to say that a quadratic form is a *homogeneous polynomial of degree 2*, i.e. that $Q[x]$ is a polynomial of n variables $x_1, x_2, ..., x_n$ having only terms of degree 2. That means that only terms ax_k^2 and cx_jx_k are allowed.

There many ways (in fact, infinitely many) to write a quadratic form $Q[x]$ as $Q[x] = (Ax, x)$. For example, the quadratic form $Q[x] = x_1^2 + x_2^2 - 4x_1x_2$ on $\mathbb{R}^2$ can be represented as (Ax, x) where A can be any of the matrices

$$\begin{pmatrix} 1 & -4 \\ 0 & 1 \end{pmatrix}, \begin{pmatrix} 1 & 0 \\ -4 & 1 \end{pmatrix}, \begin{pmatrix} 1 & -1 \\ -2 & 1 \end{pmatrix}.$$

In fact, any matrix A of form

$$\begin{pmatrix} 1 & a-4 \\ -a & 1 \end{pmatrix}$$

will work.

But if we require the matrix a to be symmetric, then such a matrix is unique:

Any quadratic form $Q[x]$ on $\mathbb{R}^n$ admits unique representation $Q[x] = (Ax, x)$ where a is A (real) symmetric matrix.

For example, for the quadratic form

$$Q[x] = x_1^2 + 3x_2^2 + 5x_3^2 + 4x_1x_2 - 16x_2x_3 + 7x_1x_3$$

on $\mathbb{R}^3$, the corresponding symmetric matrix A is

$$\begin{pmatrix} 1 & 2 & -8 \\ 2 & 3 & 3.5 \\ -8 & 3.5 & 5 \end{pmatrix}.$$

Quadratic forms on $\mathbb{C}^n$. One can also define a quadratic form on $\mathbb{C}^n$ (or any complex inner product space) by taking a self–adjoint transformation $A = A^*$ and defining Q by $Q[x] = (Ax, x)$. While our main examples will be in R^n, all the theorems are true in the setting of Cn as well. Bearing this in mind, we will always use a instead of A^T

Diagonalization of Quadratic Forms

You have probably met quadratic forms before, when you studied second order curves in the plane. May be you even studied the second order surfaces in $\mathbb{R}^3$.

We want to present a unified approach to classification of such objects. Suppose we are given a set in $\mathbb{R}^3$ defined by the equation $Q[x] = 1$, where Q is some quadratic form. If Q has some simple form, for example if the corresponding matrix is diagonal, i.e. if $Q[x] = a_1x_1^2 + a_2x_2^2 + \ldots + a_nx_n^2$, we can easily visualize this set, especially if $n = 2, 3$. In higher dimensions, it is also possible, if not to visualize, then to understand the structure of the set very well.

So, if we are given a general, complicated quadratic form, we want to simplify it as much as possible, for example to make it diagonal. The standard way of doing that is the change of variables. Orthogonal diagonalization. Let us have a quadratic form $Q[x] = (Ax, x)$ in $\mathbb{R}^n$. Introduce new variables $y = (y_1, y_2, ..., y_n)^T \in \mathbb{R}^n$, with $y = S^{-1} x$, where S is some invertible $n \times n$ matrix, so $x = Sy$.

Then,

$$Q[x] = Q[Sy] = (ASy, Sy) = (S^*ASy, y),$$

so in the new variables y the quadratic form has matrix S^*AS.

So, we want to find an invertible matrix S such that the matrix S^*AS is diagonal. Note, that it is dierent from the diagonalization of matrices we had before: we tried to represent a matrix a as $A = SDS^{-1}$, so the matrix $D = S^{-1}AS$ is diagonal. However, for orthogonal matrices U, we have $U = U^{-1}$, and we can orthogonally diagonalize symmetric matrices. So we can apply orthogonal diagonalization we studied before to the quadratic forms.

Namely, we can represent the matrix a as

$$A = UDU^* = UDU^{-1}.$$

Recall, that D is a diagonal matrix with eigenvalues of A on the diagonal, and U is the matrix of eigenvectors (we need to pick an orthogonal basis of eigenvectors). Then

$$D = U^*AU,$$

so in the variables

$$y = U^{-1} x$$

the quadratic form has diagonal matrix.

Let us analyse the geometric meaning of the orthogonal diagonalization. The columns $u_1, u_2, ..., u_n$ of the orthogonal matrix U form an orthonormal basis in $\mathbb{R}^n$, let us call this basis S. The change of coordinate matrix $[I]_{S,B}$ from this basis to the standard one is exactly U. We know that

$$y = (y_1, y_2, ..., y_n)^T = Ax,$$

so the coordinates $y_1, y_2, ..., y_n$ can be interpreted as coordinates of the vector x in the new basis $u_1, u_2, ..., u_n$. So, orthogonal diagonalization allows us to visualize very well the set $Q[x] = 1$, or a similar one, as long as we can visualize it for diagonal matrices.

Example. Consider the quadratic form of two variables (i.e. quadratic form on $\mathbb{R}^2$),

$$Q(x, y) = 2x^2 + 2y^2 + 2xy.$$

Let us describe the set of points $(x, y)^T \in \mathbb{R}^2$ satisfying $Q(x, y) = 1$.

The matrix A of Q is

$$A = \begin{pmatrix} 2 & 1 \\ 1 & 2 \end{pmatrix}.$$

Orthogonally diagonalizing this matrix we can represent it as

$$A = \begin{pmatrix} 3 & 0 \\ 0 & 1 \end{pmatrix} U^*, \text{ where } U = \frac{1}{\sqrt{2}} \begin{pmatrix} 1 & -1 \\ 1 & 1 \end{pmatrix}$$

or, equivalently

$$U^* A U = \begin{pmatrix} 3 & 0 \\ 0 & 1 \end{pmatrix} =: D.$$

The set $\{y : (Dy, y) = 1\}$ is the ellipse with half–axes $1/\sqrt{3}$ and 1. Therefore the set $\{x \in \mathbb{R}^2 : (Ax, x) = 1\}$, is the same ellipse only in the basis $\left(\frac{1}{\sqrt{2}}, \frac{1}{\sqrt{2}}\right)^T, \left(-\frac{1}{\sqrt{2}}, \frac{1}{\sqrt{2}}\right)^T$, or, equivalently, the same ellipse, rotated $\pi/4$.

Non–orthogonal diagonalization. Orthogonal diagonalization involves computing eigenvalues and eigenvectors, so it may be dicult to do without computers for $n > 2$. There is another way of diagonalization based on completion of squares, which is easier to do without computers.

Let us again consider the quadratic form of two variables, $Q[x] = 2x_1^2 + 2x_1x_2 + 2x_2^2$ (it is the same quadratic form as in the above example, only here we call variables not x, y but x_1, x_2). Since

$$2\left(x_1 + \frac{1}{2}x_2\right)^2 = 2\left(x_1^2 + 2x_1\frac{1}{2}x_2 + \frac{1}{4}x_2^2\right) = 2x_1^2 + 2x_1x_2 + \frac{1}{2}x_2^2$$

(note, that the first two terms coincide with the first two terms of Q), we get

$$2x_1^2 + 2x_1x_2 + 2x_1^2 = 2\left(x_1 + \frac{1}{2}x_2\right)^2 + \frac{3}{2}x_2^2 = 2y_1^2 + \frac{3}{2}y_2^2,$$

where $y_1 = x_1 + x_1 + \frac{1}{2}x_2$ and $y_2 = x_2$.

The same method can be applied to quadratic form of more than 2 variables. Let us consider, for example, a form $Q[x]$ in $\mathbb{R}^3$,

$$Q[x] = x_1^2 - 6x_1x_2 + 4x_1x_3 - 6x_2x_3 + 8x_2^2 - 3x_3^2.$$

Considering all terms involving the first variable x_1 (the first 3 terms in this case), let us pick a full square or a multiple of a full square which has exactly these terms (plus some other terms).

Since

$$(x_1 - 3x_2 + 2x_3)^2 = x_1^2 - 6x_1x_2 + 4x_1x_3 - 12x_2x_3 + 9x_2^2 + 4x_3^2$$

we can rewrite the quadratic form as

$$(x_1 - 3x_2 + 2x_3)^2 = x_2^2 - 6x_2x_3 + 7x_3^2.$$

Note, that the expression $-x_2^2 + 6x_2x_3 - 7x_3^2$ involves only variables x_2 and x_3. Since

$$-(x_2 - 3x_3)^2 = -(x_2^2 - 6x_2x_3 + 9x_3^2) = -x_2^2 + 6x_2x_3 - 9x_3^2$$

we have

$$-x_2^2 + 6x_2x_3 - 7x_3^2 = -(x_2 - 3x_3)^2 + 2x_3^2.$$

Thus we can write the quadratic form Q as

$$Q[x](x_1 - 3x_2 + 2x_3)^2 - (x2 - 3x_3)^2 + 2x_3^2 = y_1^2 +_2^2 +2_3^2$$

where

$$y_1 = x_1 - \ 3x_2 + \ 2x_3, y_2 = x_2 - \ 3x_3, y_3 = x_3.$$

There is another way of performing non–orthogonal diagonalization of a quadratic form. The idea is to perform row operations on the matrix a of the quadratic form. The difference with the row reduction (Gauss–Jordan elimination) is that after each row operation we need to perform the same column operation, the reason for that being that we want to make the matrix S^*AS diagonal.

Let us explain how everything works on an example. Suppose we want to diagonalize a quadratic form with matrix

$$A = \begin{pmatrix} 1 & -1 & 3 \\ -1 & 2 & 1 \\ 3 & 1 & 1 \end{pmatrix}.$$

We augment the matrix A by the identity matrix, and perform on the augmented matrix $(A|I)$ row/column operations. After each row operation we have to perform on the matrix a the same column operation. We get

$$\left(\begin{array}{ccc|ccc} 1 & -1 & 3 & 1 & 0 & 0 \\ -1 & 2 & 1 & 0 & 1 & 0 \\ 3 & 1 & 1 & 0 & 0 & 1 \end{array}\right) \begin{matrix} \\ +R_1 \\ \\ \end{matrix} \sim \left(\begin{array}{ccc|ccc} 1 & -1 & 3 & 1 & 0 & 0 \\ 0 & 1 & 4 & 1 & 1 & 0 \\ 3 & 1 & 1 & 0 & 0 & 1 \end{array}\right) \sim$$

$$\left(\begin{array}{ccc|ccc} 1 & 0 & 3 & 1 & 0 & 0 \\ 0 & 1 & 4 & 1 & 1 & 0 \\ 3 & 4 & 1 & 0 & 0 & 1 \end{array}\right)_{-3R_1} \sim \left(\begin{array}{ccc|ccc} 1 & 0 & 3 & 1 & 0 & 0 \\ 0 & 1 & 4 & 1 & 1 & 0 \\ 0 & 4 & -8 & -3 & 0 & 1 \end{array}\right) \sim$$

$$\left(\begin{array}{ccc|ccc} 1 & 0 & 0 & 1 & 0 & 0 \\ 0 & 1 & 4 & 1 & 1 & 0 \\ 0 & 4 & -8 & -3 & 0 & 1 \end{array}\right)_{-4R_2} \sim \left(\begin{array}{ccc|ccc} 1 & 0 & 0 & 1 & 0 & 0 \\ 0 & 1 & 4 & 1 & 1 & 0 \\ 0 & 0 & -24 & -7 & -4 & 1 \end{array}\right) \sim$$

$$\left(\begin{array}{ccc|ccc} 1 & 0 & 0 & 1 & 0 & 0 \\ 0 & 1 & 0 & 1 & 1 & 0 \\ 0 & 0 & -24 & -7 & -4 & 1 \end{array}\right).$$

Note, that we perform column operations only on the left side of the augmented matrix. We get the diagonal D matrix on the left, and the matrix S^* on the right, so $D = S^*AS$,

$$\begin{pmatrix} 1 & 0 & 0 \\ 0 & 1 & 0 \\ 0 & 0 & -24 \end{pmatrix} = \begin{pmatrix} 1 & 0 & 0 \\ 1 & 1 & 0 \\ -7 & -4 & 1 \end{pmatrix} \begin{pmatrix} 1 & -1 & 3 \\ -1 & 2 & 1 \\ 3 & 1 & 1 \end{pmatrix} \begin{pmatrix} 1 & 1 & -7 \\ 0 & 1 & -4 \\ 0 & 0 & 1 \end{pmatrix}.$$

Let us explain why the method works. A row operation is a left multiplication by an elementary matrix. The corresponding column operation is the right multiplication by the transposed elementary matrix. Therefore, performing row operations $E_1, E_2, \ldots, E_N$ and the same column operations we transform the matrix A to

$$E_N \ldots E_2 E_1 A E_1^* E_2^* \ldots E_N^* = EAE^*.$$

As for the identity matrix in the right side, we performed only row operations on it, so the identity matrix is transformed to

$$E_N \ldots E_2 E_1 I = EI = E.$$

If we now denote $E^* = S$ we get that S^*AS is a diagonal matrix, and the matrix $E = S$ is the right half of the transformed augmented matrix.

In the above example we got lucky, because we did not need to interchange two rows. This operation is a bit tricker to perform. It is quite simple if you know what to do, but it may be hard to guess the correct row operations. Let us consider, for example, a quadratic form with the matrix

$$A = \begin{pmatrix} 0 & 1 \\ 1 & 0 \end{pmatrix}$$

If we want to diagonalize it by row and column operations, the simplest idea would be to interchange rows 1 and 2. But we also must to perform the same column operation, i.e. interchange columns 1 and 2, so we will end up with the same matrix.

So, we need something more non–trivial. The identity

$$2x_1x_2 = \frac{1}{2}[(x_1 + x_2)^2 - (x_1 - x_2)^2]$$

gives us the idea for the following series of row operations:

$$\left(\begin{array}{cc|cc} 0 & 1 & 1 & 0 \\ 1 & 0 & 0 & 1 \end{array}\right)_{-\frac{1}{2}R_1} \sim \left(\begin{array}{cc|cc} 0 & 1 & 1 & 0 \\ 1 & -1/2 & -1/2 & 1 \end{array}\right) \sim$$

$$\left(\begin{array}{cc|cc} 0 & 1 & 1 & 0 \\ 1 & -1 & -1/2 & 1 \end{array}\right)^{+R_1} \sim \left(\begin{array}{cc|cc} 1 & 0 & 1/2 & 1 \\ 1 & -1 & 1/2 & 1 \end{array}\right) \sim$$

$$\left(\begin{array}{cc|cc} 1 & 0 & 1/2 & 1 \\ 0 & -1 & -1/2 & 1 \end{array}\right).$$

Non–orthogonal diagonalization gives us a simple description of a set $Q[x] = 1$ in a non–orthogonal basis. It is harder to visualize, then the representation given by the orthogonal diagonalization. However, if we are not interested in the details, for example if it is sucient for us to know that the set is ellipsoid (or hyperboloid, etc), then the non–orthogonal diagonalization is an easier way to get the answer.

Silvester's Law of Inertia

As we discussed above, there many ways to diagonalize a quadratic form. Note, that a resulting diagonal matrix is not unique. For example, if we got a diagonal matrix

$$D = \text{diag}\{\lambda_1, \lambda_2, ..., \lambda_n\},$$

we can take a diagonal matrix

$$S = \text{diag}\{s_1, s_2, ..., s_n\}, s_k \in \mathbb{R}, s_k \neq 0$$

and transform D to

$$S^* DS = diag\{s_1^2\lambda_1, s_2^2\lambda_2, ... s_n^2\lambda_n\}$$

This transformation changes the diagonal entries of D. However, it does not change the *signs* of the diagonal entries. And this is always the case

Namely, the famous Silvester's Law of Inertia states that:

For a Hermitian matrix A (i.e. for a quadratic form $Q[x] = (Ax, x)$) and any its diagonalization $D = S^*AS$, the number of positive (negative, zero) diagonal entries of D depends only on A, but not on a particular choice of diagonalization. Here we of course assume that S is an invertible matrix, and D is a diagonal one. The idea of the proof of the Silvester's Law of Inertia is to express the number of positive (negative, zero) diagonal entries of a diagonalization $D = S^*AS$ in terms of A, not involving S or D.

We will need the following definition.

Definition. Given an $n \times n$ symmetric matrix $A = A^*$ (a quadratic form $Q[x] = (Ax, x)$ on $\mathbb{R}^n$) we call a subspace $E \subset \mathbb{R}^n$ positive (resp. *negative*, resp. *neutral*) if

$$(Ax, x) > 0 \text{ (resp. } (Ax, x) < 0\text{, resp. } (Ax, x) = 0) \text{ for all } x \in E, x \neq 0.$$

Sometimes, to emphasize the role of a we will say A–positive (A negative, A–neutral).

Theorem. Let A be an $n \times n$ symmetric matrix, and let $D = S^*AS$ be its diagonalization by an invertible matrix S. Then the number of positive (resp. negative, resp. zero) diagonal entries of D coincides with the maximal dimension of an A–positive (resp. A – negative, resp. A – neutral) subspace.

The above theorem says that if r_+ is the number of positive diagonal entries of D, then there exists an A – positive subspace E of dimension r_+, but it is impossible to find a positive subspace E with $\dim E > r_+$.

We will need the following lemma, which can be considered a particular case of the above theorem.

Lemma. Let D be a diagonal matrix $D = diag\{\lambda_1, \lambda_2, ..., \lambda_n\}$. Then the number of positive (resp. negative, resp. zero) diagonal entries of D coincides with the maximal dimension of an D – positive (resp. D – negative, resp. D – neutral) subspace.

Proof. By rearranging the standard basis in $\mathbb{R}^n$ (changing the numeration) we can always assume without loss of generality that the positive diagonal entries of D are the first r + diagonal entries.

Consider the subspace E_+ spanned by the first r_+ coordinate vectors $e_1, e_2, ..., e_r+$. Clearly E_+ is a D–positive subspace, and $\dim E_+ = r_+$.

Let us now show that for any other D–positive subspace E we have $\dim E \le r_+$. Consider the orthogonal projection $P = P_{E+}$,

$$Px = (x_1, x_2, ..., x_r+, 0 ..., 0)^T, x = (x_1, x_2, ..., x_n)^T.$$

For a D – positive subspace E define an operator $T : E \to E_+$ by

$$Tx = Px, \ \forall x \in E.$$

In other words, T is the restriction of the projection P : P is defined on the whole space, but we restricted its domain to E and target space to E_+. We got an operator acting from E to E_+, and we use a dierent letter to distinguish it from P.

Note, that Ker $T = \{0\}$. Indeed, let for $x = (x_1, x_2, ..., xn)^T \in E$ we have $Tx = Px = 0$. Then, by the definition of P

$$x_1 = x_2 = ... = x_r + = 0,$$

and therefore

$$(Dx,x) = \sum_{k=r_+ +1}^{n} \lambda_k x_k^2 \le 0 \ (\lambda_k \le 0 \ for \ k > r_+).$$

But x belongs to a D – positive subspace E, so the inequality $(Dx, x) \le 0$ holds only for $x = 0$.

Let us now apply the Rank Theorem. First of all, rank $T = \dim \text{Ran } T \le \dim E_+ = r_+$ because Ran $T \subset E_+$. By the Rank Theorem, dimKer T + rank T = dim E. But we just proved that Ker $T = \{0\}$, i.e. that dim Ker $T = 0$, so

$$\dim E = \text{rank } T \le \dim E_+ = r_+.$$

To prove the statement about negative entries, we just apply the above reasoning to the matrix –D. The case of zero entries is treated similarly, but even simpler.

Proof. Let $D = S^*AS$ be a diagonalization of A. Since

$$(Dx, x) = (S^* ASx, x) = (ASx, Sx)$$

it follows that for any D – positive subspace E, the subspace SE is an Apositive subspace. The same identity implies that for any A – positive subspace F the subspace $S^{-1}F$ is D – positive.

Since S and S^{-1} are invertible transformations, $\dim E = \dim SE$ and $\dim F = \dim S^{-1}F$. Therefore, for any D positive subspace E we can find an A – positive subspace (namely SE) of the same dimension, and vice versa: for any A – positive subspace F we can find a D – positive subspace (namely $S^{-1}F$) of the same dimension. Therefore the maximal possible dimensions of a A – positive and a D – positive subspace coincide, and the theorem is proved. The case of negative and zero diagonal entries treated similarly.

Positive Definite Forms

Minimax characterization of eigenvalues and the Silvester's criterion of positivity

Definition. A quadratic form Q is called

- Positive definite if $Q[x] > 0$ for all $x \neq 0$.
- Positive semidefinite if $Q[x] \geq 0$ for all x.
- Negative definite if $Q[x] < 0$ for all $x \neq 0$.
- Negative semidefinite if $Q[x] \leq 0$ for all x.
- Indefinite if it take both positive and negative values, i.e. if there exist vectors x_1 and x_2 such that $Q[x_1] > 0$ and $Q[x_2] < 0$.

Definition. A symmetric matrix $A = A^*$ is called positive definite (negative definite, etc.) if the corresponding quadratic form $Q[x] = (Ax, x)$ is positive definite (negative definite, etc.).

Theorem. Let $A = A^*$. Then

1. *A is positive definite i all eigenvalues of a are positive.*
2. *A is positive semidefinite i all eigenvalues of a are non–negative.*
3. *A is negative definite i all eigenvalues of a are negative.*
4. *A is negative semidefinite i all eigenvalues of a are non–positive.*
5. *A is indefinite i it has both positive and negative eigenvalues.*

Proof. The proof follows trivially from the orthogonal diagonalization. Indeed, there is an orthonormal basis in which matrix of a is diagonal, and for diagonal matrices the theorem is trivial.

Remark. Note, that to find whether a matrix (a quadratic form) is positive definite (negative definite, etc) one does not have to compute eigenvalues. By Silvester's Law of Inertia it is sucient to perform an arbitrary, not necessarily orthogonal diagonalization $D = SAS$ and look at the diagonal entries of D. Silvester's criterion of positivity. It is an easy exercise to see that a 2×2 matrix

$$A = \begin{pmatrix} a & b \\ b & c \end{pmatrix}$$

is positive definite if and only if

$$a > 0 \text{ and } \det A = ac - b^2 > 0$$

Indeed, if a > 0 and $\det A = ac - b^2 > 0$, then $c > 0$, so trace $A = a + c > 0$. So we know that if λ_1, λ_2 are eigenvalues of a then $\lambda_1\lambda_2 > 0$ ($\det A > 0$) and $1 + 2 = \text{trace}A > 0$. But that only possible if both eigenvalues are positive. So we have proved that conditions imply that A is positive definite. The opposite implication is quite simple. This result can be generalized to the case of $n \times n$ matrices. Namely, for a matrix A

$$A = \begin{pmatrix} a_{1,1} & a_{1,2} & \cdots & a_{1,n} \\ a_{2,1} & a_{2,2} & \cdots & a_{2,n} \\ \vdots & \vdots & \ddots & \vdots \\ a_{n,1} & a_{n,2} & \cdots & a_{n,n} \end{pmatrix}$$

A let us consider its all *upper left submatrices*

$$A_1 = (a_{1,1}), A_2 = \begin{pmatrix} a_{1,1} & a_{1,2} \\ a_{2,1} & a_{2,2} \end{pmatrix}, A_3 = \begin{pmatrix} a_{1,1} & a_{1,2} & a_{1,3} \\ a_{2,1} & a_{2,2} & a_{2,3} \\ a_{3,1} & a_{3,2} & a_{3,3} \end{pmatrix}, \ldots, A_n = A$$

Theorem. (Silvester's Criterion of Positivity). *A matrix $A = A^*$ is positive definite if and only if*

$$\det A_k > 0 \text{ for all } k = 1, 2, \ldots, n.$$

First of all let us notice that if $A > 0$ then $A_k > 0$ also (can you explain why?). Therefore, since all eigenvalues of a positive definite matrix are positive, $\det A_k > 0$ for all k.

One can show that if $\det A_k > 0$ $\forall k$ then all eigenvalues of A are positive by analyzing diagonalization of a quadratic form using row and column operations. The key here is the observation that if we perform row/column operations in natural order (i.e. first subtracting the first row/column from all other rows/columns, then subtracting the second row/columns from the rows/columns 3, 4, ..., n, and so on...), and if we are not doing any row interchanges, then we automatically diagonalize quadratic forms A_k as well. Namely, after we subtract first and second rows and columns, we get diagonalization of A_2; after we subtract the third row/column we get the diagonalization of A_2, and so on.

Since we are performing only row replacement we do not change the determinant. Moreover, since we are not performing row exchanges and performing the operations in the correct order, we preserve determinants of A_k. Therefore, the condition $\det A_k > 0$ guarantees that each new entry in the diagonal is positive.

Of course, one has to be sure that we can use only row replacements, and perform the operations in the correct order, i.e. that we do not encounter any pathological situation. If one analyzes the algorithm, one can see that the only bad situation that can happen is the situation where at some step we have zero in the pivot place. In other words, if after we subtracted the first k rows and columns and obtained a diagonalization of A_k, the entry in the $k + 1$st row and $k + 1$st column is 0.

We will need the following characterization of eigenvalues of a hermitian matrix. Minimax characterization of eigenvalues. Let us recall that the codimension of a subspace $E \subset X$ is by the definition the dimension of its orthogonal complement, $\operatorname{codim} E = \dim(E^\perp)$. Since for a subspace $E \subset X$, $\dim X = n$ we have $\dim E + \dim E^\perp = n$, we can see that $\operatorname{codim} E = \dim X - \dim E$. Recall that the trivial subspace $\{0\}$ has dimension zero, so the whole space X has codimension 0.

Theorem. (Minimax characterization of eigenvalues). *Let $A = A^*$ be an $n \times n$ matrix, and let $\lambda_1, \lambda_2 \geq \ldots \geq \lambda_n$ be its eigenvalues taken in the decreasing order. Then*

$$\lambda_k = \max_{\substack{E:\\ \dim E=k}} \min_{\substack{x\in E\\ \|x\|=1}} (Ax,x) = \min_{\substack{F:\\ \operatorname{codim} F=k-1}} \max_{\substack{x\in E\\ \|x\|=1}} (Ax,x).$$

Let us explain in more details what the expressions like max min and minmax mean. To compute the first one, we need to consider all subspaces E of dimension k. For each such subspace E we consider the set of all $x \in E$ of norm 1, and find the minimum of (Ax, x) over all such x. Thus for each subspace we obtain a number, and we need to pick a subspace E such that the number is maximal. That is the maxmin. The min max is defined similarly.

Remark. A sophisticated reader may notice a problem here: why do the maxima and minima exist? It is well known, that maximum and minimum have a nasty habit of not existing: for example, the function $f(x) = x$ has neither maximum nor minimum on the open interval (0, 1).

However, in this case maximum and minimum do exist. There are two possible explanations of the fact that (Ax, x) attains maximum and minimum. The first one requires some familiarity with basic notions of analysis: one should just say that the unit sphere in E, i.e. the set $\{x \in E : \| x \| = 1\}$ is compact, and that a continuous function ($Q[x] = (Ax, x)$ in our case) on a compact set attains its maximum and minimum.

Another explanation will be to notice that the function $Q[x] = (Ax, x)$, $x \in E$ is a quadratic form on E. It is not dicult to compute the matrix of this form in some orthonormal basis in E, but let us only note that this matrix is not A: it has to be a $k \times k$ matrix, where $k = \dim E$.

It is easy to see that for a quadratic form the maximum and minimum over a unit sphere is the maximal and minimal eigenvalues of its matrix. As for optimizing over all subspaces, we will prove below that the maximum and minimum do exist.

Proof. First of all, by picking an appropriate orthonormal basis, we can assume without loss of generality that the matrix A is diagonal, $A = \operatorname{diag}\{\lambda_1, \lambda_2, \ldots, \lambda_n\}$.

Pick subspaces E and F, $\dim E = k$, codim $F = k - 1$, i.e. $\dim E = n - k + 1$. Since $\dim E + \dim F > n$, there exists a non–zero vector $x_0 \in E \cap F$. By normalizing it we can assume without loss of generality that $\| x_0 \| = 1$. We can always arrange the eigenvalues in decreasing order, so let us assume that $\lambda_1 \geq \lambda_2 \geq \ldots \geq \lambda_n$. Since x belongs to the both subspaces E and F

$$\min_{\substack{x\in E\\ \|x\|=1}} (Ax,x) \leq (Ax_0, x_0) \leq \max_{\substack{x\in F\\ \|x\|=1}} (Ax,x).$$

We did not assume anything except dimensions about the subspaces E and F, so the above inequality

$$\min_{\substack{x\in E\\ \|x\|=1}} (Ax,x) \leq \max_{\substack{x\in F\\ \|x\|=1}} (Ax,x).$$

holds for all pairs of E and F of appropriate dimensions. Define

$$E_0 := \text{span}\{e_1, e_2, ..., e_k\},\ F_0 := \text{span}\{e_k, e_{k+1}, e_{k+2}, ..., e_n\}.$$

Since for a self–adjoint matrix B, the maximum and minimum of (Bx, x) over the unit sphere $\{x : \| x \| = 1\}$ are the maximal and the minimal eigenvalue respectively (easy to check on diagonal matrices), we get that

$$\min_{\substack{x\in E_0\\ \|x\|=1}}(Ax,x) = \max_{\substack{x\in F_0\\ \|x\|=1}}(Ax,x) = \lambda_k.$$

It follows from equation that for any subspace E, $\dim E = k$

$$\min_{\substack{x\in E\\ \|x\|=1}}(Ax,x) \le \max_{\substack{x\in F_0\\ \|x\|=1}}(Ax,x) = \lambda_k.$$

and similarly, for any subspace F of codimension $k - 1$,

$$\max_{\substack{x\in F\\ \|x\|=1}}(Ax,x) \ge \min_{\substack{x\in E_0\\ \|x\|=1}}(Ax,x) = \lambda_k.$$

But on subspaces E_0 and F_0 both maximum and minimum are λ_k, so

$$\min\max = \max\min = \lambda_k.$$

Corollary (Intertwining of Eigenvalues). Let $A = A^* = \{a_{j,k}\}_{j,k=1}^{n}$ *be a self–adjoint matrix, and let* $\tilde{A} = \{a_{j,k}\}_{j,k=1}^{n-1}$ *be its submatrix of size* $(n - 1) \times (n - 1)$. *Let* $\lambda_1, \lambda_2, ..., \lambda_n$ *and* $\mu_1, \mu_2, ..., \mu_{n-1}$ *be the eigenvalues of A and A respectively, taken in decreasing order. Then*

$$\lambda_1 \ge \mu_1 \ge \lambda_2 \ge \mu_2 \ge \ldots \ge \lambda_{n-1} \ge \mu_{n-1} \ge \lambda_n.$$

i.e.

$$\lambda_k \ge \mu_k \ge \lambda_{k+1}, k = 1,2,\ldots,n-1$$

Proof. Let $\tilde{X} \subset F^n$ be the subspace spanned by the first $n - 1$ basis vectors, $eX =$ $\text{span}\{e_1, e_2, ..., e_{n-1}\}$. Since $(\tilde{A}x, x) = (Ax, x)$ for all $x \in \tilde{X}$,

$$\mu_k = \max_{\substack{x\subset \tilde{X}\\ \dim E=k}} \min_{\substack{x\in E\\ \|x\|=1}}(Ax,x).$$

To get λ_k we need to get maximum over the set of all subspaces E of F^n, $\dim E = k$, i.e. take maximum over a bigger set (any subspace of $\tilde{X}$ is a subspace of F^n). Therefore

$$\mu_k \le l_k.$$

(the maximum can only increase, if we increase the set).

On the other hand, any subspace $E \subset \tilde{X}$ of codimension $k - 1$ (here we mean codimension in $\tilde{X}$) has dimension $n - 1 - (k - 1) = n - k$, so its codimension in F_n is k. Therefore

$$\mu_k = \min_{\substack{E\subset \tilde{X}\\ \dim E=n-k}} \max_{\substack{x\in E\\ \|x\|=1}}(Ax,x) \le \min_{\substack{E\subset F^n\\ \dim E=n-k}} \max_{\substack{x\in E\\ \|x\|=1}}(Ax,x) = \lambda_{k+1}$$

(minimum over a bigger set can only be smaller).

Proof. If $A > 0$, then $A_k > 0$ for $k = 1, 2, \ldots, n$ as well (can you explain why?). Since all eigenvalues of a positive definite matrix are positive, det $A_k > 0$ for all $k = 1, 2, \ldots, n$.

Let us now prove the other implication. Let det $A_k > 0$ for all k. We will show, using induction in k, that all A_k (and so $A = A_n$) are positive definite.

Clearly A_1 is positive definite (it is 1×1 matrix, so $A_1 = \det A_1$). Assuming that $A_{k-1} > 0$ (and det $A_k > 0$) let us show that A_k is positive definite. Let $\lambda_1, \lambda_2, \ldots, \lambda_k$ and $\mu_1, \mu_2, \ldots, \mu_{k-1}$ be eigenvalues of A_k and A_{k-1} respectively. By Corollary

$$\lambda_j \geq \mu_j > 0 \quad \text{for } j = 1, 2, \ldots, k-1$$

Since det $A_k = \lambda_1\lambda_2 \ldots\lambda_{k-1}\lambda_k > 0$, the last eigenvalue λ_k must also be positive. Therefore, since all its eigenvalues are positive, the matrix A_k is positive definite.

Chapter 9

Advanced Spectral Theory

Cayley–Hamilton Theorem

Theorem (Cayley–Hamilton). Let a be a square matrix, and let $p(\lambda) = det(A - I)$ be its characteristic polynomial. Then $p(A) = 0$.

The proof looks ridiculously simple: plugging A instead of in the definition of the characteristic polynomial, we get

$$p(A) = \det(A - AI) = \det 0 = 0.$$

But this is a wrong proofl To see why, let us analyse what the theorem states. It states, that if we compute the characteristic polynomial

$$\det(A - \lambda I) = p(\lambda) = \sum_{k=0}^{n} c_k \lambda^k$$

and *then* plug matrix A instead of λ to get

$$p(A) = \sum_{k=0}^{n} c_k A^k = c_0 I + c_1 A + \ldots c_n A^n$$

then the result will be zero matrix.

It is not clear why we get the same result if we just plug A instead of λ in the determinant $\det(A - \lambda I)$. Moreover, it is easy to see that with the exception of trivial case of 1×1 matrices we will get a dierent object.

Namely, $A - AI$ is zero *matrix*, and its determinant is just the *number* 0.

But $p(A)$ is a matrix, and the theorem claims that this matrix is the zero *matrix*. Thus we are comparing apples and oranges. Even though in both cases we got zero, these are dierent zeroes: he number zero and the zero matrix!

Let us present another proof, which is based on some ideas from analysis. A "continuous" proof. The proof is based on several observations. First of all, the theorem is trivial for diagonal matrices, and so for matrices similar to diagonal (i.e. for diagonalizable matrices). The second observation is that any matrix can be approximated (as close as we want) by diagonalizable matrices. Since any operator has an upper triangular matrix in some orthonormal basis, we can assume without loss of generality that a is an upper triangular matrix.

We can perturb diagonal entries of A (as little as we want), to make them all dierent, so the perturbed matrix $\tilde{A}$ is diagonalizable (eigenvalues of a a triangular matrix are its diagonal entries, and by Corollary an $n \times n$ matrix with n distinct eigenvalues is diagonalizable). We can perturb the diagonal entries of A as little as we want, so Frobenius norm $\| A - \tilde{A} \|^2$ is as small as we want. Therefore one can find a sequence of diagonalizable matrices A_k such that $A_k \to A$ as $k \to \infty$ for example such that $A_k - A_k \to A$ as $k \to \infty$). It can be shown that the characteristic polynomials $p_k(\lambda) = \det(A_k - \lambda I)$ converge to the characteristic polynomial $p(\lambda) = \det(A - \lambda I)$ of A. Therefore

$$p(A) = \lim_{k \to \infty} p_k(A_k).$$

But as we just discussed above the Cayley–Hamilton Theorem is trivial for diagonalizable matrices, so $p_k(A_k) = 0$. Therefore $p(A) = lim_{k \to \infty}\ 0 = 0$.

This proof is intended for a reader who is comfortable with such ideas from analysis as continuity and convergence. Such a reader should be able to fill in all the details, and for him/her this proof should look extremely easy and natural.

However, for others, who are not comfortable yet with these ideas, the proof definitely may look strange. It may even look like some kind of cheating, although, let me repeat that it is an absolutely correct and rigorous proof (modulo some standard facts in analysis). So, let us resent another, proof of the theorem which is one of the "standard" proofs from linear algebra textbooks.

A "standard" proof. We know, see Theorem, that any square matrix is unitary equivalent to an upper triangular one. Since for any polynomial p we have $p(UAU^{-1}) = Up(A)U^{-1}$, and the characteristic polynomials of unitarily equivalent matrices coincide, it is sucient to prove the theorem only for upper triangular matrices. So, let A be an upper triangular matrix. We know that diagonal entries of a triangular matrix coincide with it eigenvalues, so let $\lambda_1, \lambda_2, \ldots, \lambda_n$ be eigenvalues of A ordered as they appear on the diagonal, so

$$A = \begin{pmatrix} \lambda_1 & & & * \\ & \lambda_2 & & \\ & & \ddots & \\ 0 & & & \lambda_n \end{pmatrix}.$$

The characteristic polynomial $p(z) = \det(A - zI)$ of A can be represented as

$$p(z) = (\lambda_1 - z)(\lambda_2 - z) \ldots (\lambda_n - z) = (-1)^n (z - \lambda_1)(z - \lambda_2) \ldots (z - \lambda_n),$$

so

$$p(A) = (-1)^n (A - \lambda_1 I)(A - \lambda_2 I) \ldots (A - \lambda_n I).$$

Define subspaces

$$E_k := \text{span}\{e_1, e_2, ..., e_k\},$$

where $e_1, e_2, ..., \lambda_n$ is the standard basis in $\mathbb{C}^n$. Since the matrix of A is upper triangular, the subspaces E_k are so–called *invariant* subspaces of the operator A, i.e. $AE_k \subset E_k$ (meaning that $A_v \in E_k$ for all $v \in E_k$). Moreover, since for any $v \in E_k$ and any λ

$$(A - \lambda I)v = Av - \lambda v \in E_k,$$

because both Av and λv are in E_k. Thus $(A - \lambda I)E_k \subset E_k$, i.e. E_k is an invariant subspace of $A - \lambda I$. We can say even more about the the subspace $(A - \lambda_k I)E_k$. Namely,

$$(A - \lambda_k I)_{ek} \in \text{span}\{e_1, e_2, ..., e_{k-1}\},$$

because only the first $k - 1$ entries of the kth column of the matrix of $A - \lambda_k I$ can be non–zero. On the other hand, for $j < k$ we have

$$(A - \lambda_k)e_j \in E_j \subset E_k$$

(because E_j is an invariant subspace of $A - \lambda_k I$).

Take any vector $v \in E_k$. By the definition of E_k it can be represented as a linear combination of the vectors $e_1, e_2, ..., e_k$. Since all vectors $e_1, e_2, ..., e_k$ are transformed by $A - \lambda_k I$ to some vectors in E_{k-1}, we can conclude that

$$(A - \lambda_k I)v \in E_{k-1} \; \forall v \in E_k.$$

Take an arbitrary vector $x \in \mathbb{C}^n = E_n$. Applying inductively with $k = n, n - 1, ..., 1$ we get

$$x_1 := (A - \lambda_n I)x \in E_{n-1},$$
$$x_2 := (A - l_{n-1}I)x_1 = (A - l_{n-1}I)(A - l_n I)\, x \in E_{n-2},$$
$$...$$
$$x_n := (A - l_2 I)x_{n-1} = (A - l_2 I) \;...\; (A - l_{n-1}I)(A - l_n I)x \in E_1.$$

The last inclusion mean that $x_n = \alpha e_1$. But $(A - \lambda_1 I)e_1 = 0$, so

$$0 = (A - \lambda_1 I)x_n = (A - \lambda_1 I)(A - \lambda_2 I) \;...(A - \lambda_n I)x.$$

Therefore $p(A)x = 0$ for all $x \in \mathbb{C}^n$, which means exactly that $p(A) = 0$.

Spectral Mapping Theorem

Polynomials of operators. Let us also recall that for a square matrix (an operator) A and for a polynomial $p(z) =$ the operator $p(A)$ is defined by substituting a instead of the independent variable,

$$p(A) := \sum_{k=1}^{N} a_k A^k = a_0 I + a_1 A + a_2 A^2 + \ldots + a_N A^N;$$

here we agree that $A^0 = I$.

We know that generally matrix multiplication is not commutative, *i.e.* generally $AB \neq BA$ so the order is essential. However

$$A_k A_j = A^j A^k = A^{k+j},$$

and from here it is easy to show that for arbitrary polynomials p and q

$$p(A)q(A) = q(A)p(A) = R(A)$$

where $R(z) = p(z)q(z)$.

That means that when dealing only with polynomials of an operator A, one does not need to worry about non–commutativity, and act like a is simply an independent (scalar) variable. In particular, if a polynomial $p(z)$ can be represented as a product of monomials

$$p(z) = a(z - z_1)(z - z_2) \ldots (z - z_N),$$

where $z_1, z_2, \ldots, z_N$ are the roots of p, then $p(A)$ can be represented as

$$p(A) = a(A \; - n_1 I)(A - z_N I)$$

Spectral Mapping Theorem. Let us recall that the spectrum $s(A)$ of asquare matrix (an operator) A is the set of all eigenvalues of A (not counting multiplicities.

Theorem (Spectral Mapping Theorem). For a square matrix a and an arbitrary polynomial p

$$\sigma(p(A)) = p(\sigma(A)).$$

In other words, μ is an eigenvalue of p(A) if and only if $\mu = p(\lambda)$ for some eigenvalue of A. Note, that as stated, this theorem does not say anything about multiplicities of the eigenvalues.

Remark. Note, that one inclusion is trivial. Namely, if is an eigenvalue of A, $Ax - x$ for some $x \neq 0$, then $A^k x = \lambda^k x$, and $p(A)x = p(\lambda)x$, so $p(\lambda)$ is an eigenvalue of $p(A)$. That means that the inclusion $p(\sigma(A)) \subset \sigma(p(A))$ is trivial. If we consider a particular case μ = 0 of the above theorem, we get the following corollary.

Corollary. Let a be a square matrix with eigenvalues 1, 2, ..., *n and let p be a polynomial. Then p(A) is invertible if and only if*

$$p(\lambda_k) \neq \; \forall k = 1, 2, \ldots n.$$

Proof. As it was discussed above, the inclusion $p(\sigma(A)) \subset (p(A))$ is trivial.

To prove the opposite inclusion $\sigma(p(A)) \subset p(\sigma(A))$ take a point $\mu \in \sigma(p(A))$. Denote

$$q(z) = p(z) - \mu, \text{ so } q(A) = p(A) - \mu I.$$

Since $\mu \in (p(A))$ the operator $q(A) = p(A) - \mu I$ is not invertible. Let us represent the polynomial $q(z)$ as a product of monomials,

$$q(z) = a(z - z_1)(z - z_2) \ldots (z - z_N).$$

Then, as it was discussed above in Section 2.1, we can represent

$$q(A) = a(A - z_1 I)(A - z_2 I) \ldots (A - z_N I).$$

The operator $q(A)$ is not invertible, so one of the terms $A - z_k I$ must be not invertible (because a product of invertible transformations is always invertible). That means $zk \in$ (A). On the other hand z_k is a root of q, so

$$0 = q(zk) = p(zk) - \mu$$

and therefore $\mu = p(zk)$. So we have proved the inclusion $s(p(A)) \subset p((A))$.

Invariant Subspaces

Definition. Let $A : V \rightarrow V$ be an operator (linear transformation) in a vector space V. a subspace E of the vector space V is called an invariant subspace of the operator A (or, shortly, A – invariant) if $AE \subset E$, *i.e.* if

$$Av \in E \text{ for all } v \in E.$$

If E is A–invariant, then

$$A^2E = A(AE) \subset AE \subset E,$$

i.e. E is A^2– invariant.

Similarly one can show (using induction, for example), that if $AE \subset E$. then

$$A_kE \subset E \ \forall k \geq 1.$$

This implies that $P(A)E \subset E$ for any polynomial p, *i.e.* that any A–invariant subspace E is an invariant subspace of $p(A)$. If E is an A–invariant subspace, then for all $v \in E$ the result Av also belongs to E. Therefore we can treat A as an operator acting on E, not on the whole space V.

Formally, for an A–invariant subspace E we define the so–called restriction $A|_E : E \to E$ of A onto E by

$$(A|_E)v = Av \ \forall v \in E.$$

Here we changed domain and target space of the operator, but the rule assigning value to the argument remains the same.

We will need the following simple lemma

Lemma. Let p be a polynomial, and let E be an A–invariant subspace. Then

$$p(A|_E) = p(A)|_E.$$

Proof. The proof is trivial. If $E_1, E_2, \ldots, E_r$ a basis of A–invariant subspaces, and $AE_k := A|_{Ek}$ are the corresponding restrictions, then, since $AE_k = A_kE_k \subset E_k$, the operators A_k act independently of each other (do not interact), and to analyse action of A we can analyse operators A_k separately.

In particular, if we pick a basis in each subspace E_k and join them to get a basis in V then the operator a will have in this basis the following block–diagonal form

$$A = \begin{pmatrix} A_1 & & & 0 \\ & A_2 & & \\ & & \ddots & \\ 0 & & & A_r \end{pmatrix}$$

(of course, here we have the correct ordering of the basis in V, first we take a basis in E_1,then in E_2 and so on). Our goal now is to pick a basis of invariant subspaces $E_1, E_2, \ldots, E_r$ such that the restrictions A_k have a simple structure. In this case we will get basis in which the matrix of A has a simple structure.

The eigenspaces Ker$(A - \lambda_k I)$ would be good candidates, because the restriction of a to the eigenspace Ker$(A - \lambda_k I)$ is simply $\lambda_k I$. Unfortunately, as we know eigenspaces do not always form a basis (they form a basis if and only if A can be diagonalized. However, the so–called *generalized eigenspaces* will work.

Generalized Eigenspaces

Definition. A vector v is called a generalized eigenvector (corresponding to an eigenvalue) if $(A - \lambda I)^k v = 0$ for some $k \geq 1$.

The collection E_λ of all generalized eigenvectors, together with 0 is called the generalized eigenspace (corresponding to the eigenvalue λ.

In other words one can represent the generalized eigenspace E_λ as

$$E_\lambda = \bigcup_{k\geq 1} \operatorname{Ker}(A-\lambda I)^k.$$

The sequence $\operatorname{Ker}(A - \lambda I)^k$, $k = 1, 2, 3, \ldots$ is an increasing sequence of subspaces, *i.e.*

$$\operatorname{Ker}(A - \lambda I)^k \subset \operatorname{Ker}(A - \lambda I)^{k+1}\ \forall_k \geq k_\lambda.$$

The representation does not look very simple, for it involves an infinite union. However, the sequence of the subspaces $\operatorname{Ker}(A - \lambda I)^k$ stabilizes, *i.e.*

$$\operatorname{Ker}(A - \lambda I)^k \subset \operatorname{Ker}(A - \lambda I)^{k+1}\ \forall k \geq 1.,$$

so, in fact one can take the finite union.

To show that the sequence of kernels stabilizes, let us notice that if for finite–dimensional subspaces E and F we have $E \subsetneq F$ (symbol $E \subsetneq F$ means that $E \subset F$ but $E \neq F$), then $\dim E < \dim F$. Since $\dim\operatorname{Ker}(A - I)^k \leq \dim V < \infty$, it cannot grow to infinity, so at some point

$$\operatorname{Ker}(A - I)^k = \operatorname{Ker}(A - lI)^{k+1}.$$

The rest follows from the lemma below.

Lemma. Let for some k

$$\operatorname{Ker}(A - I)^k = \operatorname{Ker}(A - \lambda I)^{k+1}.$$

Then

$$\operatorname{Ker}(A - I)^{k+r} = \operatorname{Ker}(A - \lambda I)^{k+r+1}\ \forall r \geq 0.$$

Proof. Let $v \in \operatorname{Ker}(A - I)^{k+r+1}$, *i.e.* $(A - \lambda I)^{k+r+1}v = 0$. Then

$$w := (A - lI)^r \in \operatorname{Ker}(A - lI)^{k+1}.$$

But we know that $\operatorname{Ker}(A - \lambda I)^k = \operatorname{Ker}(A - \lambda I)^{k+1}$ so $w \in \operatorname{Ker}(A - \lambda I)^k$, which means $(A - I)^k w = 0$. Recalling the definition of w we get that

$$(A - \lambda I)^{k+r}\, v = (A - \lambda I)^k\, w = 0$$

so $v \in \operatorname{Ker}(A - \lambda I)^{k+r}$. We proved that $\operatorname{Ker}(A - I)^{k+r+1} \subset \operatorname{Ker}(A - \lambda I)^{k+r}$.

The opposite inclusion is trivial.

Definition. The number $d = d(\lambda)$ on which the sequence $\operatorname{Ker}(A - \lambda I)^k$ stabilizes, *i.e.* the number d such that

$$\operatorname{Ker}(A - \lambda I)^{d-1} \subsetneq \operatorname{Ker}(A - \lambda I)^d = \operatorname{Ker}(A - \lambda I)^{d+1}$$

is called the depth of the eigenvalue λ.

It follows from the definition of the depth, that for the generalized eigenspace E_λ

$$(A - \lambda I)^{d(\lambda)} v = \forall v \in E_\lambda.$$

Now let us summarize, what we know about generalized eigenspaces.

1. E is an invariant subspace of A, AE, E.
2. If $d(\lambda)$ is the depth of the eigenvalue , then
$$((A - \lambda I)|_{E\lambda})^{d(\lambda)} = (A|E_\lambda - \lambda I_{E\lambda})^{d(\lambda)} = 0.$$
3. $\sigma(A|_{E\lambda}) = \{\lambda\}$, because the operator $A|E_\lambda - \lambda I_{E\lambda}$, is nilpotent, see 2, and the spectrum of nilpotent operator consists of one point 0,

Now we are ready to state the main result of this section. Let $A : V \to V$.

Theorem. Let (A) consists of r points $\lambda_1, \lambda_2, ..., \lambda_r$, and let $E_k := E_k$ be the corresponding generalized eigenspaces. Then the system of subspace $E_1, E_2, ..., E_r$ is a basis of subspaces in V.

Remark. If we join the bases in all generalized eigenspaces E_k, then by Theorem, we will get a basis in the whole space. In this basis the matrix of the operator a has the block diagonal form $A = \text{diag}\{A_1, A_2, ..., A_r\}$, where $A_k := A|_{Ek}$, $E_k = E_{\lambda e}$. It is also easy to see, that the operators $N_k := A_k - \lambda_k I_{Ek}$ are nilpotent, $N_k^{d_k} = 0$.

Proof. Let m_k be the multiplicity of the eigenvalue λ_k, so $p(z) = \prod_{k=1}^{r} (z - \lambda_k)^{m_k}$ is the characteristic polynomial of A. Define

$$pk(z) = p(z)/(z - \lambda_k)^{m_k} = \prod_{j \neq k} (z - \lambda_j)^{m_j}.$$

Lemma.

$$(A - \lambda_k I)^{m_k} \mid E_k = 0,$$

Proof. There are 2 possible simple proofs. The first one is to notice that $m_k \geq d_k$, where d_k is the depth of the eigenvalue λ_k and use the fact that

$$(A - \lambda_k I)^{d_k} \mid E_k = (A_k - \lambda_k I_{E_k})^{m_k} = 0,$$

where $A_k := A \mid E_k$ (property 2 of the generalized eigenspaces).

The second possibility is to notice that according to the Spectral Mapping Theorem, the operator $P_k(A) \mid E_k = p_k(A_k)$ is invertible. By the Cayley–Hamilton Theorem.

$$0 = p(A) = (A - \lambda_k I)^{m_k} (A),$$

and restriction all operators to E_k we get

$$0 = p(A_k) = (A_k - \lambda_k I_{E_k})^{m_k}\, p_k(A_k),$$

so

$$(A_k - \lambda_k I_{E_k})^{m_k} = p(A_k) p_k(A_k)^{-1} = 0 p_k(A_k)^{-1} = 0.$$

To prove the theorem define

$$q(z) = \sum_{k=1}^{r} p_k(z)$$

Since $p_k(j) = 0$ for $j \neq k$ and $p_k(k) \neq 0$, we can conclude that $q(k) \neq 0$ for all k. Therefore, by the Spectral Mapping Theorem, the operator

$$B = q(A)$$

is invertible.

Note that $BE_k \subset E_k$ (any A – invariant subspace is also $p(A)$ – invariant).

Since B is an invertible operator, $\dim(BE_k) = \dim E_k$, which together with $BE_k \subset E_k$ implies $BE_k = E_k$. Multiplying the last identity by B^{-1} we get that $B^{-1}E_k = E_k$, *i.e.* that E_k is an invariant subspace of B^{-1}.

Note also, that it follows from that

$$p_k(A)\,|_{Ej} = 0 \;\; \forall j \neq k,$$

because $p_k(A) \mid E_j = p_k(A_j)$ and $p_k(A_j)$ contains the factor $(A_j - \lambda_j I_{Ej})m_j = 0$.

Define the operators P_k by

$$P_k = B^{-1} p_k(A).$$

Lemma. For the operators P_k defined above

1. $P_1 + P_2 + \ldots + P_r = I$;
2. $P_k|E_j = 0$ for $j \neq k$;
3. $\text{Ran}P_k \subset E_k$;
4. *Moreover*, $P_k v = v \forall v \in E_k$, so, *in fact Ran* $P_k = E_k$.

Proof. Property 1 is trivial:

$$\sum_{k=1}^{r} p_k = B^{-1} \sum_{k=1}^{r} p_k p P_k(A) = B^{-1}B = 1.$$

Indeed, $p_k(A)$ contains the factor $(A - \lambda_j)m_j$, restriction of which to E_j is zero. Therefore $p_k(A) \mid E_j = 0$ and thus

$$P_k \mid E_j = B^{-1}\, p_k(A) \mid E_j = 0.$$

To prove property 3, recall that according to Cayley–Hamilton Theorem $p(A) = 0$. Since

$$p(z) = (z - \lambda_k)^{m_k}\, p_k(z),$$

we have for $w = p_k(A)v$

$$(A - \lambda_k I)^{m_k}\, w = (A - \lambda_k I)^{m_k}\, pk(A)v = p(A)v = 0.$$

That means, any vector w in Ran $p_k(A)$ is annihilated by some power of $(A - kI)$, which by definition means that Ran $p_k(A) \subset E_k$.

To prove the last property, let us notice that it follows from that for $v \in E_k$

$$p_k(A)v = \sum_{j=1}^{r} p_j(A)v = Bv,$$

which implies $P_k v = B^{-1}Bv = v$. Now we are ready to complete the proof of the theorem. Take $v \in V$ and define

$$v_k = P_k v.$$

Then according to Statement 3 of the above lemma, $v_k \in E_k$, and by statement,

$$v = \sum_{k=1}^{r} v_k,$$

so v admits a representation as a linear combination.

To show that this representation is unique, we can just note, that if v is represented as $v = \sum_{k=1}^{r} v_k, \in E_k$, then it follows from the Statements 2 and 4 of the lemma that

$$P_k v = P_k(v_1 + v_2 + ... + v_r) = P_k v_k = v_k.$$

Geometric Meaning of Algebraic Multiplicity

Proposition. Algebraic multiplicity of an eigenvalue equals to the dimension of the corresponding generalized eigenspace.

Proof. If we joint bases in generalized eigenspaces $E_k = E_{\lambda k}$ to get a basis in the whole space, the matrix of a in any such basis has a block–diagonal form diag$\{A_1, A_2, ..., A_r\}$, where

$$A_k := A|E_k .$$

Operators

$$N_k = A_k - \lambda_k - I_{Ek}$$

are nilpotent, so

$$\sigma(N_k) = \{0\}.$$

Therefore, the spectrum of the operator A_k (recall that

$$A_k = N_k - \lambda_k I)$$

consists of one eigenvalue k of (algebraic) multiplicity

$$n_k = \dim E_k.$$

The multiplicity equals n_k because an operator in a finite–dimensional space V has exactly $\dim V$ eigenvalues counting multiplicities, and A_k has only one eigenvalue.

Note that we are free to pick bases in E_k, so let us pick them in such a way that the corresponding blocks A_k are upper triangular. Then

$$\det(A - \lambda I) = \prod_{k=1}^{r} \det(A_k - \lambda I_{Ek} = \prod_{k=1}^{r} (\lambda_k - \lambda)^{n_k}.$$

But this means that the algebraic multiplicity of the eigenvalue λ_k is

$$n_k = \dim E_{\lambda k}.$$

An important application. The following corollary is very important for dierential equations.

Corollary. Any operator A in V can be represented as A = D + N, where D is diagonalizable (i.e. diagonal in some basis) and N is nilpotent ($N^m = 0$ for some m), and DN = ND.

Proof. As we discussed above, if we join the bases in E_k to get a basis in V, then in this basis A has the block diagonal form $A = \text{diag}\{A_1, A_2, \ldots, A_r\}$, where

$$A_k := A \mid E_k,\ E_k = E_k.$$

The operators $N_k := A_k - \lambda_k I_{E_k}$ are nilpotent, and the operator in the equation.

$$D = \text{diag}\{\lambda_1 I_{E_1},\ \lambda_2 I_{E_2}\ \ldots,\ \lambda_r I_{E_r}\}$$

Notice also that

$$\lambda_k I_{E_k} N_k = N_k \lambda_k I_{E_k}$$

(identity operator commutes with any operator), so the block diagonal operator

$$N = \text{diag}\{N_1, N_2, \ldots, N_r\}$$

commutes with

$$D,\ DN = ND.$$

Therefore, defining N as the block diagonal operator

$$N = \text{diag}\{N_1, N_2, \ldots, N_r\}$$

we get the desired decomposition.

This corollary allows us to compute functions of operators. Let us recall that if p is a polynomial of degree d, then $p(a + x)$ can be computed with the help of Taylor's formula

$$p(a+x) = \sum_{k=0}^{d} \frac{p^{(k)}(a)}{k!} x^k$$

This formula is an algebraic identity, meaning that for each polynomial p we can check that the formula is true using formal algebraic manipulations with a and x and not caring about their nature. Since operators D and N commute,

$$DN = ND,$$

the same rules as for usual (scalar) variables apply to them, and we can write (by plugging D instead of a and N instead of x

$$p(A) = p(D + N) = \sum_{k=0}^{d} \frac{p^{(k)}(D)}{k!} N^k$$

Here, to compute the derivative $p^{(k)}(D)$ we first compute the kth derivative of the polynomial $p(x)$ (using the usual rules from calculus), and then plug D instead of x. But since N is nilpotent,

$$N^m = 0$$

for some m, only first m terms can be non–zero, so

$$p(A) = p(D + N) = \sum_{k=0}^{m-1} \frac{f^{(k)}(D)}{k!} N^k.$$

In m is much smaller than d, this formula makes computation of $p(A)$ much easier.

The same approach works if p is not a polynomial, but an infinite power series. For general power series we have to be careful about convergence of all the series involved, so we cannot say that the formula is true for an arbitrary power series $p(x)$. However, if the radius of convergence of the power series is 1, then everything works fine. In particular, if $p(x) = e^x$, then, using the fact that $(e^x)^t = e^x$ we get.

$$e^A = \sum_{k=0}^{m-1} \frac{e^D}{k!} N^k = e^D = \sum_{k=0}^{m-1} \frac{1}{k!} N^k$$

This formula has important applications in dierential equation. Note, that the fact that

$$ND = DN$$

is essential here!

Structure of Nilpotent Operators

Recall, that an operator a in a vector space V is called *nilpotent* if $A^k = 0$ for some exponent k.

In the previous section we have proved, that if we join the bases in all generalized eigenspaces

$$E_k = E_k$$

to get a basis in the whole space, then the operator a has in this basis A block diagonal form diag$\{A_1, A_2, ..., A_r\}$ and operators A_k ca be represented as

$$A_k = \lambda_k I + N_k,$$

where N_k are nilpotent operators.

In each generalized eigenspace E_k we want to pick up a basis such that the matrix of A_k in this basis has the simplest possible form. Since matrix (in any basis) of the identity operator is the identity matrix, we need to find a basis in which the nilpotent operator N_k has a simple form. Since we can deal with each N_k separately, we will need to consider the following problem:

For a nilpotent operator A find a basis such that the matrix of a in this basis is simple.

Let see, what does it mean for a matrix to have a simple form. It is easy to see that the matrix

$$\begin{pmatrix} 0 & 1 & & & 0 \\ & 0 & 1 & & \\ & & 0 & \ddots & \\ & & & \ddots & 1 \\ 0 & & & & 0 \end{pmatrix}$$

is nilpotent.

These matrices (together with 1×1 zero matrices) will be our "building blocks". Namely, we will show that for any nilpotent operator one can find a basis such that the matrix of the operator in this basis has the block diagonal form diag$\{A_1, A_2, ..., A_r\}$, where each A_k is either a block of form or a 1×1 zero block.

Let us see what we should be looking for. Suppose the matrix of an operator A has in a basis $v_1, v_2, ..., v_p$ the form (4.1). Then

$$Av_1 = 0$$

and

$$Av_{k+1} = v_k,\ k = 1, 2, ..., p - 1.$$

Thus we have to be looking for the chains of vectors $v_1, v_2, ..., v_p$ satisfying the above relations.

Cycles of Generalized Eigenvectors

Definition. Let a be a nilpotent operator. a chain of non–zero vectors $v_1, v_2, ..., v_p$ satisfying relations is called a *cycle of generalized eigenvectors* of A. The vector v_1 is called the *initial vector* of the cycle, the vector v_p is called the *end vector* of the cycle, and the number p is called the length of the cycle.

Remark. A similar definition can be made for an arbitrary operator. Then all vectors vk must belong to the same generalized eigenspace E, and they must satisfy the identities

$$(A - \lambda I)v_1 = 0,\ (A - \lambda I)v_{k+1} = v_k,\ k = 1, 2, ..., p - 1,$$

Theorem. Let a be a nilpotent operator, and let $C_1, C_2, ..., C_r$ be cycles of its generalized eigenvectors, $C_k = v_1^k, v_{p_k}^k, v_{p_k}^k, p_k$ being the length of the cycle C_k. Assume that the initial vectors $v_1^1, v_1^2, \ldots, v_1^r$ are linearly independent. Then no vector belongs to two cycles, and the union of all the vectors from all the cycles is a linearly independent.

Proof. Let

$$n = p_1 + p_2 + ... + p_r$$

be the total number of vectors in all the cycles. We will use induction in n. If $n = 1$ the theorem is trivial.

Let us now assume, that the theorem is true for all operators and for all collection of cycles, as long as the total number of vectors in all the cycles is strictly less than n. Without loss of generality we can assume that the vectors v_j^k span the whole space V, because, otherwise we can consider instead of the operator A its restriction onto the invariant subspace

$$\text{span}\{ v_j^k : k = 1, 2, ..., r,\ 1 \le j \le p_k\}.$$

Consider the subspace RanA. It follows from the relations that vectors

$$v_j^k : k = 1, 2, ..., r,\ 1 \le j \le p_k - 1$$

span Ran A. Note that if $p_k > 1$ then the system $v_1^k, v_1^k, \ldots, v_{pk-1}^k$ is a cycle, and that A annihilates any cycle of length 1. Therefore, we have finitely many cycles, and initial vectors of these cycles are linearly independent, so the induction hypothesis applies, and the vectors

$$v_j^k : k = 1, 2, ..., r, 1 \leq j \leq p_k - 1$$

are linearly independent. Since these vectors also span Ran A, we have *a* basis there. Therefore,

$$\text{rank } A = \dim \text{Ran } A = n - r$$

(we had n vectors, and we removed one vector $v_{p_k}^k$ from each cycle C_k, $k = 1, 2, ..., r$, so we have $n–r$ vectors in the basis $v_j^k : k = 1, 2, ..., r, 1 \leq j \leq p_k - 1$). On the other hand

$$Av_1^k = 0$$

for $k = 1, 2, ..., r$, and since these vectors are linearly independent dim Ker $A \geq r$. By the Rank Theorem.

$$\dim V = \text{rank}A + \dim\text{Ker}A = (n - r) + \dim\text{Ker } A \geq (n - r) + r = n$$

so dim $V \geq n$.

On the other hand V is spanned by n vectors, therefore the vectors

$$v_j^k : k = 1, 2, ..., r, 1 \leq j \leq p_k,$$

form a basis, so they are linearly independent

Jordan Canonical form of a Nilpotent operator.

Theorem. Let $A : V \rightarrow V$ *be a nilpotent operator. Then V has a basis consisting of union of cycles of generalized eigenvectors of the operator A.*

Proof. We will use induction in n where $n = \dim V$. For $n = 1$ the theorem is trivial. Assume that the theorem is true for any operator acting in a space of dimension strictly less than n. Consider the subspace $X =$ Ran A. X is an invariant subspace of the operator A, so we can consider the restriction $A|_X$.

Since A is not invertible, dim Ran $A <$ dim V, so by the induction hypothesis there exist cycles $C_1, C_2, ..., C_r$ of generalized eigenvectors such that their union is a basis in X. Let

$$C_k = v_1^k, v_2^k, \ldots, v_{pk}^k$$

where v_1^k is the initial vector of the cycle.

Since the end vector v_{pk}^k belong to Ran A, one can find a vector v_{pk+1}^k such that

$$Av_{pk+1} = v_{pk}^k.$$

So we can extend each cycle C_k to a bigger cycle $\tilde{C}_k = v_1^k, v_2^k, \ldots v_{pk}^k, v_{p_{k+1}}^k$. Since the initial vectors v_1^k of cycles $\tilde{C}_k, vk = 1, 2, \ldots,$ $v_1^k, v_2^k, \ldots v_{pk}^k, v_{p_{k+1}}^k$ are linearly independent, Theorem implies that the union of these cycles is a linearly independent system. By the definition of the cycle we have $v_1^k \in KerA$, and we assumed that the initial vectors v_1^k, $k =$ 1, 2, ..., r are linearly independent. Let us complete this system to a basis in Ker$A$, *i.e.* let find vectors u_1, u_2, ..., u_q such that the system , u_1, u_2, ..., u_q is a basis in Ker A (it may happen that the system $v_1^k, k = 1,2,\ldots,r$ is already a basis in Ker A, in which case we put $q = 0$ and add nothing).

The vector u_j can be treated as a cycle of length 1, so we have a collection of cycles $\tilde{C}_1, \tilde{C}_2, \ldots, \tilde{C}_r$, u_1, u_2, ..., u_q, whose initial vectors are linearly independent. So, we can apply Theorem to get that the union of all these cycles is a linearly independent system. To show that it is a basis, let us count the dimensions. We know that the cycles C_1, C_2, ..., C_r have

$$\dim \text{Ran } A = \text{rank } A$$

vectors total. Each cycle $\tilde{C}_k$ was obtained from C_k by adding 1 vector to it, so the total number of vectors in all the cycles $\tilde{C}_k$ is rank $A + r$.

We know that

$$\dim \text{Ker } A = r + q$$

(because $v_1^1, v_1^2, \ldots v_1^r, u_1, u_2, \ldots u_q$ is a basis there). We added to the cycles $\tilde{C}_1, \tilde{C}_2, \ldots, \tilde{C}_r$ additional q vectors, so we got

$$\text{rank } A + r + q = \text{rank } A + \dim\text{Ker } A = \dim V$$

linearly independent vectors. But dim$V$ linearly independent vectors is a basis.

Definition. A basis consisting of a union of cycles of generalized eigenvectors of a nilpotent operator a (existence of which is guaranteed by the Theorem) is called a Jordan canonical basis for A. Note, that such basis is not unique.

Corollary. Let A be a nilpotent operator. There exists a basis (a Jordan canonical basis) such that the matrix of A in this basis is a block diagonal diag $\{A_1, A_2, \ldots, A_r\}$, *where all* A_k *(except may be one) are blocks of form, and one of the blocks* A_k *can be zero.*

The matrix of a in a Jordan canonical basis is called the Jordan canonicalform of the operator A. We will see later that the Jordan canonical formis unique, if we agree on how to order the blocks (*i.e.* on how to order the vectors in the basis).

Proof. According to Theorem one can find a basis consisting of a union of cycles of generalized eigenvectors. a cycle of size p gives rise to a $p \times p$ diagonal block, and a cycle of length 1correspond to a 1×1 zero block. We can join these 1×1 zero blocks in one large zero block (because o–diagonal entries are 0).

Dot diagrams. Uniqueness of the Jordan canonical form. There is a good way of visualizing Theorem and Corollary, the socalled dot diagrams. This methods also allows us to answer many natural questions, like "is the block diagonal representation given by Corollary unique?"

Of course, if we treat this question literally, the answer is "no", for we always can change the order of the blocks. But, if we exclude such trivial possibilities, for example by agreeing on some order of blocks (say, if we put all non–zero blocks in decreasing order, and then put the zero block), is the representation unique, or not?

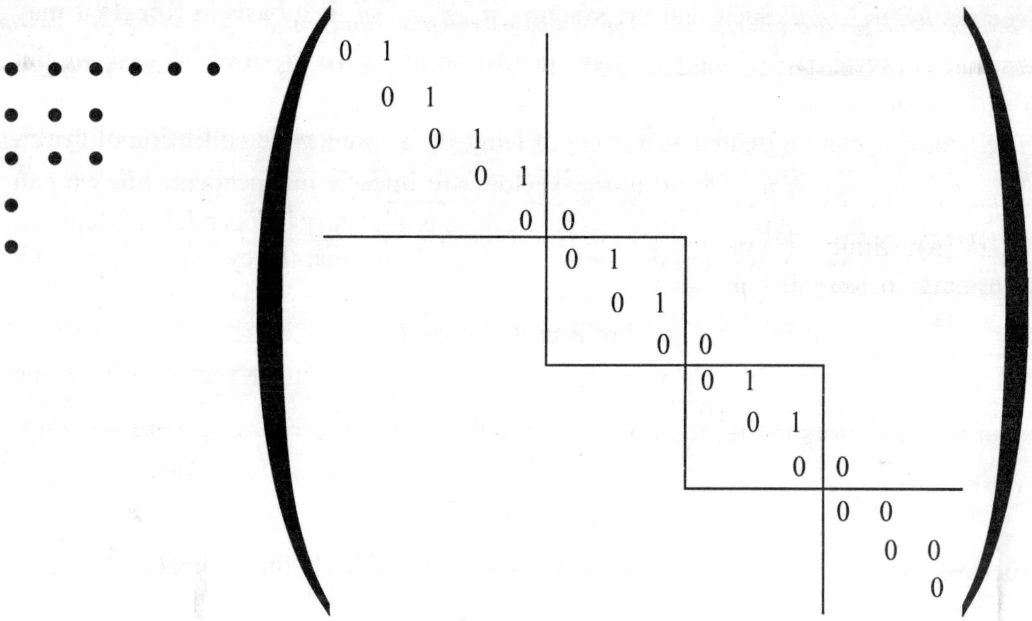

Fig. Dot Diagram and Corresponding Jordan Canonical form of a Nilpotent Operator

To better understand the structure of nilpotent operators, let us draw the so–called dot diagram. Namely, suppose we have a basis, which is a union of cycles of generalized eigenvalues. Let us represent the basis by an array of dots, so that each column represents a cycle. The first row consists of initial vectors of cycles, and we arrange the columns (cycles) by their length, putting the longest one to the left.

On the figure 1 we have the dot diagram of a nilpotent operator, as well as its Jordan canonical form. This dot diagram shows, that the basis has 1 cycle of length 5, two cycles of length 3, and 3 cycles of length 1. The cycle of length 5 corresponds to the 5×5 block of the matrix, the cycles of length 3 correspond to two 3 non–zero blocks. Three cycles of length 1 correspond to three zero entries on the diagonal, which we join in the 3×3 zero block. Here we only giving the main diagonal of the matrix and the diagonal above it; all other entries of the matrix are zero.

If we agree on the ordering of the blocks, there is a one–to–one correspondence between dot diagrams and Jordan canonical forms (for nilpotent operators). So, the question about uniqueness of the Jordan canonical form is equivalent to the question about uniqueness of the dot diagram. To answer this question, let us analyse, how the operator A transforms the dot diagram. Since the operator A annihilates initial vectors of the cycles, and moves vector v_{k+1} of a cycle to the vector v_k, we can see that the operator a acts on its dot diagram by deleting the first (top) row of the diagram.

The new dot diagram corresponds to a Jordan canonical basis in Ran A, and allows us to write down the Jordan canonical form for the restriction $A|_{\text{Ran } A}$.

Similarly, it is not hard to see that the operator A^k removes the first k rows of the dot diagram. Therefore, if for all k we know the dimensions dimKer(A^k), we know the dot diagram of the operator A. Namely, the number of dots in the first row is dimKerA, the number of dots in the second row is

$$\dim\mathrm{Ker}(A^2) - \dim\mathrm{Ker}\, A,$$

and the number of dots in the kth row is

$$\dim\mathrm{Ker}(A^k) - \dim\mathrm{Ker}(A^{k+1}).$$

But this means that the dot diagram, which was initially defined using a Jordan canonical basis, does not depend on a particular choice of such a basis. Therefore, the dot diagram, is unique.

This implies that if we agree on the order of the blocks, then the Jordan canonical form is unique. Computing a Jordan canonical basis. Let us say few words about computing a Jordan canonical basis for a nilpotent operator. Let p_1 be the largest integer such that $A^{p1} \neq 0$ (so $A^{p1+1} = 0$). That p_1 is the length of the longest cycle.

Computing operators

$$A^k,\ k = 1, 2, \ldots, p_1,$$

and counting dimKer(A^k) we can construct the dot diagram of A. Now we want to put vectors instead of dots and find a basis which is a union of cycles.

We start by finding the longest cycles (because we know the dot diagram, we know how many cycles should be there, and what is the length of each cycle). Consider a basis in the column space Ran(A^{p1}). Name the vectors in this basis $v_1^1, v_1^2, \ldots, v_1^{r_1}$, these will be the initial vectors of the cycles.Then we find the end vectors of the cycles $v_{p_1}^1, v_{p_1}^2, \ldots, v_{p_1}^{r_1}$ by solving the equations

$$A^{p_1} v_{p_1}^k = {}_1^k,\ k = 1, 2, \ldots, r_1.$$

Applying consecutively the operator a to the end vector $v_{p_1}^k$, we get all the vectors v_j^k in the cycle. Thus, we have constructed all cycles of maximal length.

Let p_2 be the length of a maximal cycle among those that are left to find. Consider the subspace Ran(A^{p_2}), and let dim Ran(A^{p_2}) = r_2. Since Ran(A^{p_1}) $\subset$ Ran(A^{p_2}), we can complete the basis $v_1^1, v_1^2, \ldots, v_1^{r_1}$ to a basis $v_1^1, v_1^2, \ldots, v_1^{r_1}, v_1^{r_1+1}, \ldots, v_1^{r_2}$ in Ran (A^{p_2}). Then we find end vectors of the cycles $C_{r_1+1}, \ldots, C_{r_2}$ by solving (for $v_{p_2}^k$) the equations

$$A^{p_1} v_{p_2}^k = v_1^k,\ k = r_1 + 1, r_1 + 2, \ldots, r_2,$$

thus constructing the cycles of length p_2.

Let p_3 denote the length of a maximal cycle among ones left. Then, completing the basis $v_1^1, v_1^2, \ldots, v_1^{r_2}$ in Ker(A^{p_2}) to a basis in Ker(A^{p_3} we construct the cycles of length p_3, and so on.

One final remark: as we discussed above, if we know the dot diagram, we know the canonical form, so after we have found a Jordan canonical basis, we do not need to compute the matrix of a in this basis: we already know it.

Jordan Decomposition Theorem

Theorem. Given an operator A there exist a basis (Jordan canonical basis) such that the matrix of a in this basis has a block diagonal form with blocks of form

$$\begin{pmatrix} \lambda & 1 & & & 0 \\ & \lambda & 1 & & \\ & & \lambda & \ddots & \\ & & & \ddots & 1 \\ 0 & & & & \lambda \end{pmatrix}$$

where λ is an eigenvalue of A. Here we assume that the block of size 1 is just λ.

The block diagonal form from Theorem is called the *Jordan canonical form* of the operator A. The corresponding basis is called a *Jordan canonical basis* for an operator A.

Proof. If we join bases in the generalized eigenspaces

$$E_k = E_{\lambda k}$$

to get a basis in the whole space, the matrix of a in this basis has a block diagonal form diag $\{A_1, A_2, ..., A_r\}$, where

$$A_k = A|_{Ek}.$$

The operators

$$N_k = A_k - \lambda_k I_{Ek}$$

are nilpotent, so by Theorem one can find a basis in E_k such that the matrix of N_k in this basis is the Jordan canonical form of N_k. To get the matrix of A_k in this basis one just puts k instead of 0 on the main diagonal.

First of all let us recall that the computing of eigenvalues is the hardest part, but here we do not discuss this part, and assume that eigenvalues are already computed. For each eigenvalue we compute subspaces

$$\text{Ker}(A - \lambda I)^k, k = 1, 2, \ldots$$

until the sequence of the subspaces stabilizes. In fact, since we have an increasing sequence of subspaces ($\text{Ker}(A - \lambda I)^k \subset \text{Ker}(A - \lambda I)^{k+1}$), then it is sucient only to keep track of their dimension (or ranks of the operators $(A - \lambda I)^k$. For an eigenvalue let $m = m$ be the number where the sequence $\text{Ker}(A - \lambda I)^k$ stabilizes, *i.e.* m satisfies

$$\dim\text{Ker}(A - \lambda I)^{m-1} < \dim\text{Ker}(A - \lambda I)^m = \dim \text{Ker}(A - \lambda I)^{m+1}.$$

Then

$$E_\lambda = \text{Ker}(A - \lambda I)^m$$

is the generalized eigenspace corresponding to the eigenvalue.

After we computed all the generalized eigenspaces there are two possible ways of action. The first way is to find a basis in each generalized eigenspace, so the matrix of the operator a in this basis has the block–diagonal form diag$\{A_1, A_2, ..., A_r\}$, where

$$A_k = A|E_{\lambda_k}.$$

Then we can deal with each matrix A_k separately. The operators

$$N_k = A_k - \lambda_k I$$

are nilpotent, so applying the algorithm described in Section 4.4 we get the Jordan canonical representation for N_k, and putting k instead of 0 on the main diagonal, we get the Jordan canonical representation for the block A_k. The advantage of this approach is that we are working with smaller blocks. But we need to find the matrix of the operator in a new basis, which involves inverting a matrix and matrix multiplication.

Another way is to find a Jordan canonical basis in each of the generalized eigenspaces E_{λ_k} by working directly with the operator A, without splitting it first into the blocks. Again, the algorithm works with a slight modification. Namely, when computing a Jordan canonical basis for a generalized eigenspace E_{λ_k}, instead of considering subspaces Ran$(A_k - \lambda_k I)^j$, which we would need to consider when working with the block A_k separately, we consider the subspaces $(A - \lambda_k I)^j E_{\lambda_k}$.

Chapter 10

Linear Transformations

Euclidean Linear Transformations

By a transformation from $\mathbb{R}^n$ into $\mathbb{R}^m$, we mean a function of the type $T: \mathbb{R}^n \to \mathbb{R}^m$, with domain $\mathbb{R}^n$ and codomain $\mathbb{R}^m$. For every vector $x \in \mathbb{R}^n$, the vector $T(x) \in \mathbb{R}^m$ is called the image of x under the transformation T, and the set

$$R(T) = \{T(x) : x \in \mathbb{R}^n\},$$

of all images under T, is called the range of the transformation T.

Remark. For our convenience later, we have chosen to use $R(T)$ instead of the usual $T(\mathbb{R}^n)$ to denote the range of the transformation T.

For every $x = (x_1, \ldots, x_n) \in \mathbb{R}^n$, we can write

$$T(x) = T(x_1, \ldots, x_n) = (y_1, \ldots, y_m):$$

Here, for every $i = 1, \ldots, m$, we have

$$y_i = T_i(x_1, \ldots, x_n), \quad (1)$$

where $T_i : \mathbb{R}^n \to \mathbb{R}$ is a real valued function.

Definition. A transformation $T: \mathbb{R}^n \to \mathbb{R}^m$ is called a linear transformation if there exists a real matrix

$$A = \begin{pmatrix} a_{11} & \ldots & a_{1n} \\ \vdots & & \vdots \\ a_{m1} & \ldots & a_{mn} \end{pmatrix}$$

such that for every

$$x = (x_1, \ldots, x_n) \in \mathbb{R}^n,$$

we have

$$T(x_1, \ldots, x_n) = (y_1, \ldots, y_m),$$

where such that for every

$$i\,x = (x_1, \ldots, x_n) \in \mathbb{R}^n$$

we have

$$T(x_1, \ldots, x_n) = (y_1, \ldots, y_m)$$

where

$$(y_1, \ldots, y_{11}x_1 + \ldots + a_{1n}x_n,$$
$$\vdots$$
$$a_{m_1}x_1 + \ldots + a_{mn}{}^x n,$$

or, in matrix notation,

$$\begin{pmatrix} y_1 \\ \vdots \\ y_m \end{pmatrix} = \begin{pmatrix} a_{11} & \cdots & a_{1n} \\ \vdots & & \vdots \\ a_{m1} & \cdots & a_{mn} \end{pmatrix} \begin{pmatrix} y_1 \\ \vdots \\ y_n \end{pmatrix}$$

The matrix A is called the standard matrix for the linear transformation T.

Remarks. (1) In other words, a transformation

$$T : \mathbb{R}^n \to \mathbb{R}^m$$

is linear if the equation (1) for every $i = 1, \ldots, m$ is linear.

(2) If we write $x \in \mathbb{R}^n$ and $y \in \mathbb{R}^m$ as column matrices, then (2) can be written in the form

$$y = Ax,$$

and so the linear transformation T can be interpreted as multiplication of $x \in \mathbb{R}^n$ by the standard matrix A.

Definition. A linear transformation $T : \mathbb{R}^n \to \mathbb{R}^m$ is said to be a linear operator if $n = m$. In this case, we say that T is a linear operator on $\mathbb{R}^n$.

Example. The linear transformation $T : \mathbb{R}^5 \to \mathbb{R}^3$, defined by the equations

$$y_1 = 2x_1 + 3x_2 + 5x_3 + 7x_4 - 9x_5;$$
$$y_2 = 3x_2 + 4x_3 + 2x_5;$$
$$y_3 = x_1 + 3x_3 - 2x_4\,;$$

can be expressed in matrix form as

$$\begin{pmatrix} y_1 \\ y_2 \\ y_3 \end{pmatrix} = \begin{pmatrix} 2 & 3 & 5 & 7 & -9 \\ 0 & 3 & 4 & 0 & 2 \\ 1 & 0 & 3 & -2 & 0 \end{pmatrix} \begin{pmatrix} x_1 \\ x_2 \\ x_3 \\ x_4 \\ x_5 \end{pmatrix}$$

If $(x_1, x_2, x_3, x_4, x_5) = (1{,}0,\ 1{,}0{,}1)$, then

$$\begin{pmatrix} y_1 \\ y_2 \\ y_3 \end{pmatrix} = \begin{pmatrix} 2 & 3 & 5 & 7 & -9 \\ 0 & 3 & 4 & 0 & 2 \\ 1 & 0 & 3 & -2 & 0 \end{pmatrix} \begin{pmatrix} 1 \\ 0 \\ 1 \\ 0 \\ 1 \end{pmatrix} = \begin{pmatrix} -1 \\ 6 \\ 4 \end{pmatrix}.$$

so that

$$T(1, 0, 1, 0, 1) = (-2, 6, 4).$$

Example. Suppose that A is the zero $m \times n$ matrix. The linear transformation $T : \mathbb{R}^n \to \mathbb{R}^m$, where $T(x) = Ix$ for every $x \in \mathbb{R}^n$, is the zero transformation from Rn into $\mathbb{R}^n$. Clearly $T(x) = 0$ for every $x \in \mathbb{R}^n$.

Example. Suppose that I is the identity $n \times n$ matrix. The linear operator T: $\mathbb{R}^n \to \mathbb{R}^m$, where $T(x) = Ix$ for every $x \in \mathbb{R}^n$, is the identity operator on R^n. Clearly $T(x) = x$ for every $x \in \mathbb{R}^n$.

PROPOSITION. *Suppose that T: $\mathbb{R}^n \to \mathbb{R}^m$ is a linear transformation, and that fe*$_1$, ..., *eng is the standard basis for $\mathbb{R}^n$. Then the standard matrix for T is given by*

$$A = (T(e_1) \ldots T_{(en)});$$

Linear Operators on $\mathbb{R}^2$

In this section, we consider the special case when $n = T(x_1,x_3)$, and study linear operators on $\mathbb{R}^2$. For every $x \in \mathbb{R}^2$, we shall write $x = (x_1, x_2)$.

Example. Consider reection across the x_2–axis, so that $T(x_1, x_2) = (-x_1, x_2)$. Clearly we have

$$T(e_1) = \left(\frac{-1}{0}\right)$$

and so it follows from Proposition that the standard matrix is given by

$$A = \begin{pmatrix} -1 & 0 \\ 0 & 1 \end{pmatrix}$$

It is not dicult to see that the standard matrices for reection across the x_1–axis and across the line $x_1 = x_2$ are given respectively by

$$A = \begin{pmatrix} 1 & 0 \\ 0 & -1 \end{pmatrix} \text{ and } A = \begin{pmatrix} 1 & 0 \\ 0 & 1 \end{pmatrix}$$

Also, the standard matrix for reection across the origin is given by

$$A = \begin{pmatrix} -1 & 0 \\ 0 & -1 \end{pmatrix}.$$

We give a summary in the table below:

Linear operator	Equations	Standard matrix
Reection across x_2–axis ny_1	$\begin{cases} y_1 = -x_1 \\ y_2 = x_2 \end{cases}$	$\begin{pmatrix} -1 & 0 \\ 0 & 1 \end{pmatrix}$
Reection across x_1–axis	$\begin{cases} y_1 = x_1 \\ y_2 = -x_2 \end{cases}$	$\begin{pmatrix} 1 & 0 \\ 0 & -1 \end{pmatrix}$
Reection across $x_1 = x_2$	$\begin{cases} y_1 = x_2 \\ y_2 = x_1 \end{cases}$	$\begin{pmatrix} 0 & 1 \\ 0 & 0 \end{pmatrix}$
Reection across origin	$\begin{cases} y_1 = -x_1 \\ y_2 = -x_2 \end{cases}$	$\begin{pmatrix} -1 & 0 \\ 0 & -1 \end{pmatrix}$

Example. For orthogonal projection onto the x_1–axis, we have $T(x_1, x_2) = (x_1, 0)$, with standard matrix

$$A = \begin{pmatrix} 1 & 0 \\ 0 & 0 \end{pmatrix}.$$

Similarly, the standard matrix for orthogonal projection onto the x_2–axis is given by

$$A = \begin{pmatrix} 0 & 0 \\ 0 & 1 \end{pmatrix}.$$

We give a summary in the table below:

Linear operator	Equations	Standard matrix
Orthogonal projection onto x_1–axis	$\begin{cases} y_1 = x_1 \\ y_2 = 0 \end{cases}$	$\begin{pmatrix} 1 & 0 \\ 0 & 0 \end{pmatrix}$
Orthogonal projection onto x_2–axis	$\begin{cases} y_1 = 0 \\ y_2 = x_2 \end{cases}$	$\begin{pmatrix} 0 & 0 \\ 0 & 1 \end{pmatrix}$

Example. For anticlockwise rotation by an angle θ, we have $T(x_1, x_2) = (y_1, y_2)$, where

$$y_1 + iy_2 = (x_1 + ix_2)(\cos\theta + i\sin\theta);$$

and so

$$\begin{pmatrix} y_1 \\ y_2 \end{pmatrix} = \begin{pmatrix} \cos\theta & -\sin\theta \\ \sin\theta & \cos\theta \end{pmatrix} \begin{pmatrix} x_1 \\ x_2 \end{pmatrix}$$

It follows that the standard matrix is given by

$$A = \begin{pmatrix} \cos\theta & -\sin\theta \\ \sin\theta & \cos\theta \end{pmatrix}$$

We give a summary in the table below:

Linear operator	Equations	Standard matrix
Anticlockwise rotation by angle θ	$\begin{cases} y_1 = x_1\cos\theta - x_2\sin\theta \\ y_2 = x_1\cos\theta - x_2\sin\theta \end{cases}$	$\begin{pmatrix} \cos\theta & -\sin\theta \\ \sin\theta & \cos\theta \end{pmatrix}$

Example. For contraction or dilation by a non–negative scalar k, we have

$$T(x_1, x_2) = (kx_1, kx_2),$$

with standard matrix

$$A = \begin{pmatrix} k & 0 \\ 0 & k \end{pmatrix}.$$

The operator is called a contraction if $0 < k < 1$ and a dilation if $k > 1$, and can be extended to negative values of k by noting that for $k < 0$, we have

$$\begin{pmatrix} k & 0 \\ 0 & k \end{pmatrix} = \begin{pmatrix} -1 & 0 \\ 0 & -1 \end{pmatrix}\begin{pmatrix} -k & 0 \\ 0 & -k \end{pmatrix}.$$

This describes contraction or dilation by non–negative scalar $-k$ followed by reection across the origin. We give a summary in the table below:

Linear operator	Equations	Standard matrix
Contraction or dilation by factor k	$\begin{cases} y_1 = kx_1 \\ y_2 = kx_2 \end{cases}$	$\begin{pmatrix} k & 0 \\ 0 & k \end{pmatrix}$

Example. For expansion or compression in the x_1–direction by positive factor k, we have $T(x_1, x_2)$ (kx_1, x_2), with standard matrix

$$A = \begin{pmatrix} k & 0 \\ 0 & 1 \end{pmatrix}.$$

This can be extended to negative values of by noting that for k < 0, we have

$$\begin{pmatrix} k & 0 \\ 0 & 1 \end{pmatrix} = \begin{pmatrix} -1 & 0 \\ 0 & 1 \end{pmatrix}\begin{pmatrix} -k & 0 \\ 0 & 1 \end{pmatrix}.$$

This describes expansion or compression in the x_1–direction by positive factor $-k$ followed by reection across the x_2–axis. Similarly, for expansion or compression in the x_2–direction by non–zero factor k, we have the standard matrix

$$A = \begin{pmatrix} 1 & 0 \\ 0 & k \end{pmatrix}.$$

We give summary in the table below:

Linear operator	Equations	Standard matrix
Expansion or compression in x_1–direction	$\begin{cases} y_1 = kx_1 \\ y_2 = x_2 \end{cases}$	$\begin{pmatrix} k & 0 \\ 0 & 1 \end{pmatrix}$

Expansion or compression in x_2–direction	$\begin{cases} y_1 = x_1 \\ y_2 = kx_2 \end{cases}$	$\begin{pmatrix} 1 & 0 \\ 0 & k \end{pmatrix}$

Example. For shears in the x_1–direction with factor k, we have

$$T(x_1, x_2) = (x_1 + kx_2, x_2),$$

with standard matrix

$$A = \begin{pmatrix} 1 & k \\ 0 & 1 \end{pmatrix}.$$

For the case $k = 1$, we have the following.

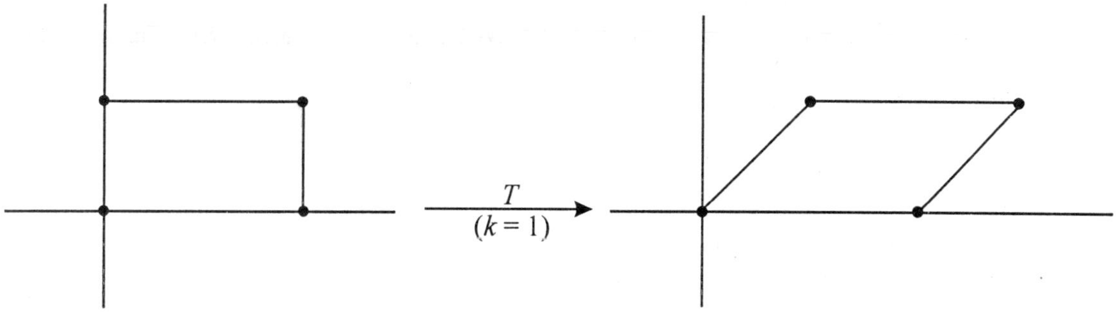

For the case $k = -1$, we have the following.

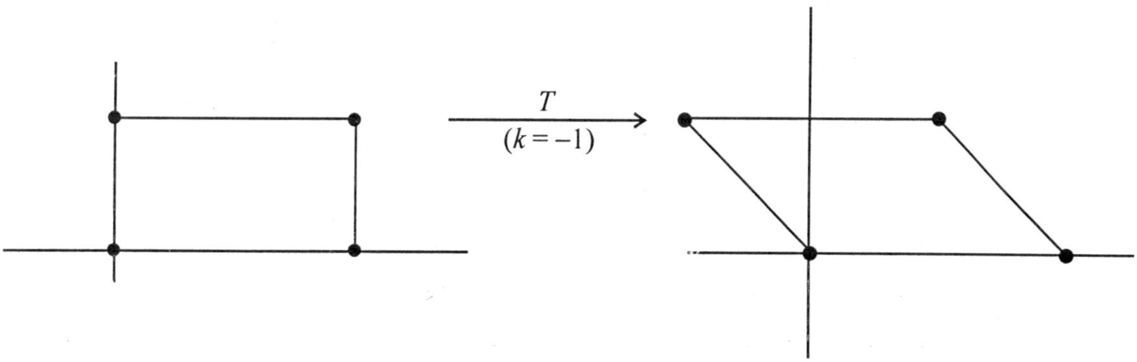

Similarly, for shears in the x_2–direction with factor k, we have standard matrix

$$A = \begin{pmatrix} 1 & 0 \\ k & 1 \end{pmatrix}.$$

We give a summary in the table below:

Linear operator	Equations	Standard matrix
Shear in x_1–direction	$\begin{cases} y_1 = x_1 + kx_2 \\ y_2 = x_2 \end{cases}$	$\begin{pmatrix} 1 & k \\ 0 & 1 \end{pmatrix}$
Shear in x_2–direction	$\begin{cases} y_1 = x_1 + kx_2 \\ y_2 = x_2 \end{cases}$	$\begin{pmatrix} 1 & 0 \\ k & 1 \end{pmatrix}$

Example. Consider a linear operator $T : \mathbb{R}^2 \to \mathbb{R}^2$ which consists of a reection across the x_2–axis, followed by a shear in the x_1–direction with factor 3 and then reection across the x_1–axis. To end the standard matrix, consider the eect of T on a standard basis $\{e_1, e_2\}$ of $\mathbb{R}^2$. Note that

$$e_1 - \begin{pmatrix} 1 \\ 0 \end{pmatrix} \mapsto \begin{pmatrix} -1 \\ 0 \end{pmatrix} \mapsto \begin{pmatrix} -1 \\ 0 \end{pmatrix} \mapsto \begin{pmatrix} -1 \\ 0 \end{pmatrix} = T(e_1),$$

$$e_2 - \begin{pmatrix} 0 \\ 1 \end{pmatrix} \mapsto \begin{pmatrix} 0 \\ 1 \end{pmatrix} \mapsto \begin{pmatrix} 3 \\ 1 \end{pmatrix} \mapsto \begin{pmatrix} 3 \\ -1 \end{pmatrix} = T(e_2).$$

so it follows from Proposition that the standard matrix for T is

$$A = \begin{pmatrix} -1 & 3 \\ 0 & -1 \end{pmatrix}.$$

Let us summarize the above and consider a few special cases. We have the following table of invertible linear operators with $k \neq 0$. Clearly, if A is the standard matrix for an invertible linear operator T, then the inverse matrix A^{-1} is the standard matrix for the inverse linear operator T^{-1}.

Linnear operator T	Standard matrix A	Inverse matrix A^{-1}	Linear operator T^{-1}
Reflection across line $x_1 = x_2$	$\begin{pmatrix} 0 & 1 \\ 1 & 0 \end{pmatrix}$	$\begin{pmatrix} 0 & 1 \\ 1 & 0 \end{pmatrix}$	Reflection across line $x_1 = x_2$
Expansion or compression in x_1-direction	$\begin{pmatrix} k & 0 \\ 0 & 1 \end{pmatrix}$	$\begin{pmatrix} k^{-1} & 0 \\ 0 & 1 \end{pmatrix}$	Expansion or compression
Expansion or compression in x_2-direction	$\begin{pmatrix} 1 & 0 \\ 0 & k \end{pmatrix}$	$\begin{pmatrix} 1 & 0 \\ 0 & k^{-1} \end{pmatrix}$	Expansion or compression in x_2-direction
Shear in x_1-direction	$\begin{pmatrix} 1 & k \\ 0 & 1 \end{pmatrix}$	$\begin{pmatrix} 1 & -k \\ 0 & 1 \end{pmatrix}$	Shear in x_1-direction
Shear in x_2-direction	$\begin{pmatrix} 1 & 0 \\ k & 1 \end{pmatrix}$	$\begin{pmatrix} 1 & 0 \\ -k & 1 \end{pmatrix}$	Shear in x_2-direction

Next, let us consider the question of elementary row operations on 2×2 matrices. It is not dicult to see that an elementary row operation performed on a 2×2 matrix A has the eect of multiplying the matrix A by some elementary matrix E to give the product EA. We have the following table.

Elementary row operation	Elementary matrix E
Interchanging the two rows	$\begin{pmatrix} 0 & 1 \\ 1 & 0 \end{pmatrix}$
Multiplying row 1 by non–zero factor k	$\begin{pmatrix} k & 0 \\ 0 & 1 \end{pmatrix}$
Multiplying row 2 by non–zero factor k	$\begin{pmatrix} 1 & 0 \\ 0 & k \end{pmatrix}$

Adding k times row 2 to row 1 $\begin{pmatrix} 1 & k \\ 0 & 1 \end{pmatrix}$

Adding k times row 1 to row 2 $\begin{pmatrix} 1 & 0 \\ k & 1 \end{pmatrix}$

Now, we know that any invertible matrix A can be reduced to the identity matrix by a finite number of elementary row operations. In other words, there exist a finite number of elementary matrices $E_1, \ldots, E_s$ of the types above with various non–zero values of k such that

$$E_s \ldots E_1 A = I,$$

so that

$$A = E_1^{-1} \ldots E_s^{-1}.$$

We have proved the following result.

Proposition. Suppose that the linear operator $T : \mathbb{R}^2 \to \mathbb{R}^2$ *has standard matrix A, where A is invertible. Then T is the product of a succession of nitely many reections, expansions, compressions and shears.*

In fact, we can prove the following result concerning images of straight lines.

Proposition. Suppose that the linear operator $T : \mathbb{R}^2 \to \mathbb{R}^2$ *has standard matrix A, where A is invertible. Then*

(*a*) *The image under T of a straight line is a straight line;*

(*b*) *The image under T of a straight line through the origin is a straight line through the origin, and*

(*c*) *The images under T of parallel straight lines are parallel straight lines.*

Proof. Suppose that $T(x_1, x_2) = (y_1, y_2)$. Since A is invertible, we have $x = A^{-1}y$, where

$$x = \begin{pmatrix} x_1 \\ x_2 \end{pmatrix} \text{ and } y = \begin{pmatrix} y_1 \\ y_2 \end{pmatrix}$$

The equation of a straight line is given by $\alpha x_1 + \beta x_2 = \gamma$ or, in matrix form, by

$$(\alpha\ \beta)\begin{pmatrix} x_1 \\ x_2 \end{pmatrix} = (\gamma).$$

Hence

$$(\alpha\ \beta)A^{-1}\begin{pmatrix} y_1 \\ y_2 \end{pmatrix} = (\gamma).$$

Let

$$(\alpha'\ \beta') = (\alpha\ \beta)A^{-1}.$$

Then

$$(\alpha' \ \beta') = \begin{pmatrix} y_1 \\ y_2 \end{pmatrix} = (\gamma).$$

In other words, the image under T of the straight line $x_1 + x_2 = \gamma$ is $\gamma' y_1 + \beta' y_2 = \gamma$, clearly another straight line. This proves (*a*). To prove (*b*), note that straight lines through the origin correspond to $\gamma = 0$. To prove (*c*), note that parallel straight lines correspond to dierent values of for the same values of α and β.

Elementary Properties of Euclidean Linear Transformations

In this section, we establish a number of simple properties of euclidean linear transformations.

Proposition. Suppose that $T_1 : \mathbb{R}^n \to \mathbb{R}^m$ *and* $T_2 : \mathbb{R}^m \to \mathbb{R}^k$ *are linear transformations. Then* $T = T_2 \,.\, T_1 : \mathbb{R}^n \to \mathbb{R}^k$ *is also a linear transformation.*

Proof. Since T_1 and T_2 are linear transformations, they have standard matrices A_1 and A_2 respectively. In other words, we have $T_1(x) = A_1 x$ for every $x \in \mathbb{R}^n$ and $T_2(y) = A_2 y$ for every $y \in \mathbb{R}^m$. It follows that $T(x) = T_2(T_1(x)) = A_2 A_1 x$ for every $y \in \mathbb{R}^n$, so that T has standard matrix $A_2 A_1$.

Example. Suppose that $T_1 : \mathbb{R}^2 \to \mathbb{R}^2$ is anticlockwise rotation by $\pi/2$ and $T_2 : \mathbb{R}^2 \to \mathbb{R}^2$ is orthogonal projection onto the x_1–axis. Then the respective standard matrices are

$$A_1 = \begin{pmatrix} 0 & -1 \\ 1 & 0 \end{pmatrix} \text{ and } A_2 = \begin{pmatrix} 1 & 0 \\ 0 & 0 \end{pmatrix}$$

It follows that the standard matrices for $T_2 \,.\, T_1$ and $T_1 \,.\, T_2$ are respectively

$$A_2 A_1 = \begin{pmatrix} 0 & -1 \\ 0 & 0 \end{pmatrix} \text{ and } A_1 A_2 = \begin{pmatrix} 0 & 0 \\ 1 & 0 \end{pmatrix}.$$

Hence $T_2 \,.\, T_1$ and $T_1 \,.\, T_2$ are not equal.

Example. Suppose that $T_1 : \mathbb{R}^2 \to \mathbb{R}^2$ is anticlockwise rotation by and $T_2 : \mathbb{R}^2 \to \mathbb{R}^2$ is anticlockwise rotation by ϕ. Then the respective standard matrices are

$$A_1 = \begin{pmatrix} \cos\theta & -\sin\theta \\ \sin\theta & \cos\theta \end{pmatrix} \text{ and } A_2 = \begin{pmatrix} \cos\phi & -\sin\phi \\ \sin\phi & \cos\phi \end{pmatrix}.$$

It follows that the standard matrix for $T_2 \,.\, T_1$ is

$$A_2 A_1 = \begin{pmatrix} \cos\phi\cos\theta - \sin\phi\sin\theta & -\cos\phi\sin\theta - \sin\phi\cos\theta \\ \sin\phi\cos\theta + \cos\phi\sin\theta & \cos\phi\cos\theta - \sin\phi\sin\theta \end{pmatrix}$$

$$= \begin{pmatrix} \cos(\phi+\theta) & -\sin(\phi+\theta) \\ \sin(\phi+\theta) & \cos(\phi+\theta) \end{pmatrix}.$$

Hence $T_2 \circ T_1$ is anticlockwise rotation by $\phi + \theta$.

Example. The reader should check that in $\mathbb{R}^2$, reaction across the x_1–axis followed by reection across the x_2–axis gives reection across the origin. Linear transformations that map distinct vectors to distinct vectors are of special importance.

Definition. A linear transformation $T : \mathbb{R}^n \to \mathbb{R}^m$ is said to be one–to–one if for every $x', x'' \in \mathbb{R}^n$, we have $x' = x''$ whenever $T(x') = T(x'')$.

Example. If we consider linear operators $T : \mathbb{R}^2 \to \mathbb{R}^2$, then T is one–to–one precisely when the standard matrix A is invertible. To see this, suppose rst of all that A is invertible. If $T(x') = T(x'')$, then $Ax' = Ax''$. Multiplying on the left by A^{-1}, we obtain $x' = x''$. Suppose next that A is not invertible. Then there exists $x \in \mathbb{R}^2$ such that $x \neq 0$ and $Ax = 0$. On the other hand, we clearly have

$$A_0 = 0.$$

It follows that

$$T(x) = T(0),$$

so that T is not one–to–one.

Proposition. Suppose that the linear operator $T : \mathbb{R}^n \to \mathbb{R}^n$ has standard matrix A. Then the following statements are equivalent:

(a) The matrix A is invertible.

(b) The linear operator T is one–to–one.

(c) The range of T is $\mathbb{R}^n$, in other words, $R(T) = \mathbb{R}^n$.

Proof. ((*a*)) ⇒ (*b*)) Suppose that $T(x') = T(x'')$. Then $Ax' = Ax''$. Multiplying on the left by A^{-1} gives $x' = x''$.

((*b*)) ⇒ (*a*)) Suppose that T is one–to–one. Then the system $Ax = 0$ has unique solution $x = 0$ in $\mathbb{R}^n$. It follows that A can be reduced by elementary row operations to the identity matrix I, and is therefore invertible.

((a)) ⇒ (c)) For any $y \in \mathbb{R}^n$, clearly $x = A^{-1}y$ satises $Ax = y$, so that $T(x) = y$.

((c)) ⇒ (a)) Suppose that $\{e_1, \ldots, e_n\}$ is the standard basis for $\mathbb{R}^n$. Let $x_1, \ldots, x_n \in \mathbb{R}^n$ be chosen to satisfy $T(x_j) = e_j$, so that $Ax_j = e_j$, for every $j = 1, \ldots, n$. Write

$$C = (x_1 \ \ldots \ x_n).$$

Then $AC = I$, so that A is invertible.

Definition. Suppose that the linear operator $T : \mathbb{R}^n \to \mathbb{R}^n$ has standard matrix A, where A is invertible. Then the linear operator $T^{-1} : \mathbb{R}^n \to \mathbb{R}^n$, dened by $T^{-1}(x) = A^{-1}x$ for every $x \in \mathbb{R}^n$, is called the inverse of the linear operator T.

Remark. Clearly $T^{-1}(T(x)) = x$ and $T(T^{-1}(x)) = x$ for every $x \in \mathbb{R}^n$.

Example. Consider the linear operator $T:\mathbb{R}^2 \to \mathbb{R}^2$, dened by $T(x) = Ax$ for every $x \in \mathbb{R}^2$, where

$$A = \begin{pmatrix} 1 & 1 \\ 1 & 2 \end{pmatrix}.$$

Clearly A

$$A^{-1} = \begin{pmatrix} 2 & -1 \\ -1 & 1 \end{pmatrix}.$$

Hence the inverse linear operator is $T^{-1} : \mathbb{R}^2 \to \mathbb{R}^2$, dened by $T^{-1}(x) = A^{-1}x$ for every $x \in \mathbb{R}^2$.

Example. Suppose that $T:\mathbb{R}^2 \to \mathbb{R}^2$ is anticlockwise rotation by angle. The reader should check that $T^{-1} : \mathbb{R}^2 \to \mathbb{R}^2$ is anticlockwise rotation by angle $2\pi - \theta$.

Next, we study the linearity properties of euclidean linear transformations which we shall use later to discuss linear transformations in arbitrary real vector spaces.

Proposition. A transformation $T:\mathbb{R}^n \to \mathbb{R}^m$ is linear if and only if the following two conditions are satised:

(a) For every $u,\ v \in \mathbb{R}^n$, we have $T(u + v) = T(u) + T(v)$.

(b) For every $u \in \mathbb{R}^n$ and $c \in \mathbb{R}$, we have $T(cu) = cT(u)$.

Proof. Suppose rst of all that $T:\mathbb{R}^n \to \mathbb{R}^m$ is a linear transformation. Let A be the standard matrix for T. Then for every $u,\ v \in \mathbb{R}^n$ and $c \in \mathbb{R}$, we have

$$T(u + v) = A(u + v) = Au + Av = T(u) + T(v)$$

and

$$T(cu) = A(cu) = c(Au) = cT(u):$$

Suppose now that (a) and (b) hold. To show that T is linear, we need to nd a matrix A such that $T(x) = Ax$ for every $x \in \mathbb{R}^n$. Suppose that $\{e_1, \ldots, e_n\}$ is the standard basis for $\mathbb{R}^n$. As suggested by Proposition 8A, we write

$$A = (\ T(e_1) \ldots T(e_n));$$

where $T(e_j)$ is a column matrix for every $j = 1, \ldots, n$. For any vector

$$x = \begin{pmatrix} x_1 \\ \vdots \\ x_n \end{pmatrix}$$

in $\mathbb{R}^n$, we have

$$Ax = (\ T(e_1)\ \dots\ T(e_n))\begin{pmatrix} x_1 \\ \vdots \\ x_n \end{pmatrix} = x_1T(e_1)+\dots+x_nT(e_n).$$

Using (b) on each summand and then using (a) inductively, we obtain

$$Ax = T(x_1e_1) + \dots + T(x_ne_n) = T(x_1e_1 + \dots + x_ne_n) = T(x)$$

as required.

To conclude our study of euclidean linear transformations, we briey mention the problem of eigenvalues and eigenvectors of euclidean linear operators.

Definition. Suppose that $T:\mathbb{R}^n \to \mathbb{R}^n$ is a linear operator. Then any real number $\lambda \in \mathbb{R}$ is called an eigenvalue of T if there exists a non–zero vector $x \in \mathbb{R}^n$ such that $T(x) = x$. This non–zero vector $x \in \mathbb{R}^n$ is called an eigenvector of T corresponding to the eigenvalue λ.

Remark. Note that the equation $T(x) = x$ is equivalent to the equation $Ax = \lambda x$. It follows that there is no distinction between eigenvalues and eigenvectors of T and those of the standard matrix A. We therefore do not need to discuss this problem any further.

General Linear Transformations

Suppose that V and W are real vector spaces. To define a linear transformation from V into W, we are motivated by Proposition which describes the linearity properties of euclidean linear transformations.

By a transformation from V into W, we mean a function of the type $T : V \to W$, with domain V and codomain W. For every vector $u \in V$, the vector $T(u) \in W$ is called the image of u under the transformation T.

Definition. A transformation $T : V \in W$ from a real vector space V into a real vector space W is called a linear transformation if the following two conditions are satised:

(LT1) For every $u, v \in V$, we have $T(u + v) = T(u) + T(v)$.

(LT2) For every $u \in V$ and $c \in \mathbb{R}$, we have $T(cu) = cT(u)$.

Definition. A linear transformation $T : V \to V$ from a real vector space V into itself is called a linear operator on V.

Example. Suppose that V and W are two real vector spaces. The transformation $T : V \to W$, where $T(u) = 0$ for every $u \in V$, is clearly linear, and is called the zero transformation from V to W.

Example. Suppose that V is a real vector space. The transformation $I : V \to V$, where $I(u) = u$ for every $u \in V$, is clearly linear, and is called the identity operator on V.

Example. Suppose that V is a real vector space, and that $k \in \mathbb{R}$ is fixed. The transformation $T : V \to V$, where $T(u) = ku$ for every $u \in V$, is clearly linear. This operator is called a dilation if $k > 1$ and a contraction if $0 < k < 1$.

Example. Suppose that V is a finite dimensional vector space, with basis $\{w_1, \dots, w_n\}$. Dene a transformation $T : V \to \mathbb{R}^n$ as follows. For every $u \in V$, there exists a unique vector $(\beta_1, \dots, \beta_n) \in \mathbb{R}^n$ such that $u = \beta_1 w_1 + \dots + \beta_n w_n$. We let $T(u) = (\beta_1, \dots, \beta_n)$.

In other words, the transformation T gives the coordinates of any vector $u \in V$ with respect to the given basis $\{w_1, \dots, w_n\}$. Suppose now that

$$v = \lambda_1 w_1 + \dots + \lambda_n w_n$$

is another vector in V. Then

$$u + v = (\beta_1 + \lambda_1)w_1 + \dots + (\beta_n + \lambda_n)w_n,$$

so that

$$T(u + v) = (\beta_1 + \lambda_1, \dots, \beta_n + \beta_n) = (\beta_1, \dots, \beta_n) + (\lambda_1, \dots, \lambda_n) = T(u) + T(v).$$

Also, if $c \in \mathbb{R}$, then $cu = c\beta_1 w_1 + \dots + c\beta_n w_n$, so that

$$T(cu) = (c\beta_1, \dots, c\beta_n) = c(\beta_1, \dots, \beta_n) = cT(u).$$

Hence T is a linear transformation. We shall return to this in greater detail in the next section.

Example. Suppose that P_n denotes the vector space of all polynomials with real coefficients and degree at most n. Dene a transformation $T : P_n \to P_n$ as follows. For every polynomial

$$p = p_0 + p_1 x + \dots + p_n x^n$$

in P_n, we let

$$T(p) = p_n + p_{n-1}x + \dots + p_0 x^n.$$

Suppose now that

$$q = q_0 + q_1 x + \dots + q_n x^n$$

is another polynomial in P_n. Then

$$p + q = (p_0 + q_0) + (p_1 + q_1)x + \dots + (p_n + q_n)x^n;$$

so that

$$\begin{aligned} T(p + q) &= (pn + qn) + (p_{n-1} + q_{n-1})x + \dots + (p_0 + q_0)x^n \\ &= (p_n + p_{n-1}x + \dots + p_0 x^n) + (q_n + q_{n-1}x + \dots + q_0 x^n) = T(p) + T(q). \end{aligned}$$

Also, for any $c \in \mathbb{R}$, we have $cp = cp_0 + cp_1 x + \dots + cp_n x^n$, so that

$$T(cp) = cp_n + cp_{n-1}x + \dots + cp_0 x^n = c(p_n + p_{n-1}x + \dots + p_0 x^n) = cT(p).$$

Hence T is a linear transformation.

Example. Let V denote the vector space of all real valued functions dierentiable everywhere in $\mathbb{R}$, and let W denote the vector space of all real valued functions dened on $\mathbb{R}$. Consider the transformation $T : V \to W$, where $T(f) = f'$ for every $f \in V$. It is easy to check from properties of derivatives that T is a linear transformation.

Example. Let V denote the vector space of all real valued functions that are Riemann integrable over the interval [0, 1]. Consider the transformation $T : V \to \mathbb{R}$, where

$$T(f) = \int_0^1 f(x)dx$$

for every $f \in V$. It is easy to check from properties of the Riemann integral that T is a linear transformation.

Consider a linear transformation $T : V \to W$ from a finite dimensional real vector space V into a real vector space W. Suppose that $\{v_1, ..., v_n\}$ is a basis of V. Then every $u \in V$ can be written uniquely in the form $u = \beta_1 v_1 + ... + \beta_n v_n$, where $\beta_1, ... \beta_n \in \mathbb{R}$. It follows that

$$T(u) = T(\beta_1 v_1 + ... + \beta_n v_n) = T(\beta_1 v_1) + ... + T(\beta_n v_n) = \beta_1 T(v_1) + ... + \beta_n T(v_n).$$

We have therefore proved the following generalization of proposition.

Proposition. Suppose that $T : V \to W$ is a linear transformation from a finite dimensional real vector space V into a real vector space W. Suppose further that $\{v_1, ..., v_n\}$ is a basis of V. Then T is completely determined by $T(v_1), ..., T(v_n)$.

Example. Consider a linear transformation $T : P_2 \to \mathbb{R}$, where $T(1) = 1$, $T(x) = \in$ and $T(x_2) = 3$. Since $\{1, x, x_2\}$ is a basis of P_2, this linear transformation is completely determined. In particular, we have, for example,

$$T(5 - 3x + 2x^2) = 5T(1) - 3T(x) + 2T(x^2) = 5.$$

Example. Consider a linear transformation $T : \mathbb{R}^4 \to \mathbb{R}$, where $T(1, 0, 0, 0) = 1$, $T(1, 1, 0, 0) = 2$, $T(1, 1, 1, 0) = 3$ and $T(1, 1, 1, 1) = 4$. Since $\{(1, 0, 0, 0), (1, 1, 0, 0), (1, 1, 1, 0), (1, 1, 1, 1)\}$ is a basis of $\mathbb{R}^4$, this linear transformation is completely determined. In particular, we have, for example,

$$T(6, 4, 3, 1) = T(2(1\ \ 0, 0, 0) + (1, 1, 0, 0) + 2(1, 1, 1, 0) + (1, 1, 1, 1))$$
$$= 2T(1, 0, 0, 0) + T(1, 1, 0, 0) + 2T(1, 1, 1, 0) + T(1, 1, 1, 1) = 14.$$

We also have the following generalization of PROPOSITION.

Proposition. Suppose that V, W, U are real vector spaces. Suppose further that $T_1 : V \to W$ and $T_2 : W \to U$ are linear transformations. Then $T = T_2\ T_1 : V \to U$ is also a linear transformation.

Proof. Suppose that $u, v \in V$. Then

$$T(u + v) = T_2(T_1(u + v)) = T_2(T_1(u) + T_1(v)) = T_2(T_1(u)) + T_2(T_1(v)) = T(u) + T(v).$$

Also, if $c \in \mathbb{R}$, then

$$T(cu) = T_2(T_1(cu)) = T_2(cT_1(u)) = cT_2(T_1(u)) = cT(u).$$

Hence T is a linear transformation.

Change of Basis

Suppose that V is a real vector space, with basis $B = \{u_1, ..., u_n\}$. Then every vector $u \in V$ can be written uniquely as a linear combination

$$u = \beta 1 u_1 + ... + \beta_n u_n, \text{ where } \beta_1, ..., \beta_n \in \mathbb{R}.$$

It follows that the vector u can be identied with the vector $(\beta_1, ..., \beta_n) \in \mathbb{R}^n$.

Definition. Suppose that $u \in V$ and (3) holds. Then the matrix

$$[u]_B = \begin{pmatrix} \beta_1 \\ \vdots \\ \beta_n \end{pmatrix}$$

is called the coordinate matrix of u relative to the basis $B = \{u_1, ..., u_n\}$.

Example. The vectors

$$u_1 = (1, 2, 1, 0),\ u_2 = (3, 3, 3, 0),\ u_3 = (2, -10, 0, 0),\ u_4 = (-2, 1, -6, 2)$$

are linearly independent in $\mathbb{R}^4$, and so $B = \{u_1, u_2, u_3, u_4\}$ is a basis of $\mathbb{R}^4$. It follows that for any $u = (x, y, z, w) \in \mathbb{R}^4$, we can write

$$u = \beta_1 u_1 + \beta_2 u_2 + \beta_3 u_3 + \beta_4 u_4:$$

In matrix notation, this becomes

$$\begin{pmatrix} x \\ y \\ z \\ w \end{pmatrix} = \begin{pmatrix} 1 & 3 & 2 & -2 \\ 2 & 3 & -10 & 1 \\ 1 & 3 & 0 & -6 \\ 0 & 0 & 0 & 2 \end{pmatrix} \begin{pmatrix} \beta_1 \\ \beta_2 \\ \beta_3 \\ \beta_4 \end{pmatrix},$$

so that

$$[u]B = \begin{pmatrix} \beta_1 \\ \beta_2 \\ \beta_3 \\ \beta_4 \end{pmatrix} = \begin{pmatrix} 1 & 3 & 2 & -2 \\ 2 & 3 & -10 & 1 \\ 1 & 3 & 0 & -6 \\ 0 & 0 & 0 & 2 \end{pmatrix} \begin{pmatrix} x \\ y \\ z \\ w \end{pmatrix}.$$

Remark. Consider a function $\phi : V \to \mathbb{R}^n$, where $(u) = [u]_B$ for every $u \in V$. It is not dicult to see that this function gives rise to a one–to–one correspondence between the elements of V and the elements of $\mathbb{R}^n$. Furthermore, note that

$$[u + v]_B = [u]_B + [v]_B \text{ and } [cu]_B = c[u]_B,$$

so that $\phi(u + v) = \phi(u) + \phi(v)$ and $\phi(cu) = c\phi(u)$ for every u, $v \in V$ and $c \in \mathbb{R}$. Thus is a linear transformation, and preserves much of the structure of V. We also say that V is isomorphic to $\mathbb{R}^n$. In practice, once we have made this identication between vectors and their coordinate matrices, then we can basically forget about the basis B and imagine that we are working in $\mathbb{R}^n$ with the standard basis.

Clearly, if we change from one basis $B = \{u_1, \dots, u_n\}$ to another basis $C = \{v_1, ..., v_n\}$ of V, then we also need to nd a way of calculating $[u]C$ in terms of $[u]B$ for every vector $u \in V$. To do this, note that each of the vectors $v_1, ..., v_n$ can be written uniquely as a linear combination of the vectors $u_1, ..., u_n$. Suppose that for $i = 1, ..., n$, we have

$$v_i = a_1 i u_1 + \dots + a_{ni} u_n, \text{ where } a_{1i}, \dots, a_{ni} \in \mathbb{R},$$

so that

$$[v_i]_B = \begin{pmatrix} a_{1i} \\ \vdots \\ a_{ni} \end{pmatrix}.$$

For every $u \in V$, we can write

$u = \beta_1 u_1 + \ldots + \beta_n u_n = \gamma_1 v_1 + \ldots + \gamma_n v_n$, where $\beta_1, \ldots, \beta_n, \gamma_1, \ldots, \gamma_n \in \mathbb{R}$;

so that

$$[u]_B = \begin{pmatrix} \beta_1 \\ \vdots \\ \beta_n \end{pmatrix} \text{ and } [u]_c = \begin{pmatrix} \gamma_1 \\ \vdots \\ \gamma_n \end{pmatrix}.$$

Clearly

$$\begin{aligned} u &= \gamma_1 v_1 + \ldots + \gamma_n v_n \\ &= \gamma_1(a_{11}u_1 + \ldots + a_{n1}u_n) + \ldots + \gamma_n(a_{1n}u_1 + \ldots + a_{nn}u_n) \\ &= (\gamma_1 a_{11} + \ldots + \gamma n_{a1n})u_1 + \ldots + (\gamma_1 a_{n1} + \ldots + \gamma_n a_{nn})u_n \\ &= \beta_1 u_1 + \ldots + \beta_n u_n: \end{aligned}$$

Hence

$$\begin{aligned} \beta_1 &= \gamma_1 a_{11} + \ldots + \gamma_n a_{1n}, \\ &\vdots \\ \beta_n &= \beta_1 a_{n1} + \ldots + \beta\gamma_n a_{nn}. \end{aligned}$$

Written in matrix notation, we have

$$\begin{pmatrix} \beta_1 \\ \vdots \\ \beta_n \end{pmatrix} = \begin{pmatrix} a_{11} & \ldots & a_{1n} \\ \vdots & & \vdots \\ a_{n1} & (\ldots) & a_{nn} \end{pmatrix} \begin{pmatrix} \gamma_1 \\ \vdots \\ \gamma_n \end{pmatrix}.$$

We have proved the following result.

Proposition. Suppose that $B = \{u_1, \ldots, u_n\}$ and $C = \{v_1, \ldots, v_n\}$ are two bases of a real vector space V. Then for every $u \in V$, we have

$$[u]_B = P[u]_C,$$

where the columns of the matrix

$$P = ([v_1]_B \cdots [v_n]_B)$$

are precisely the coordinate matrices of the elements of C relative to the basis B.

Remark. Strictly speaking, Proposition gives $[u]_B$ in terms of $[u]_C$. However, note that the matrix P is invertible (why?), so that $[u]_C = P^{-1}[u]_B$.

Definition. The matrix P in Proposition is sometimes called the transition matrix from the basis C to the basis B.

Example. We know that with

$$u_1 = (1, 2, 1, 0),\ u_2 = (3, 3, 3, 0),\ u_3 = (2, -10, 0, 0),\ u_4 = (-2, 1, -6, 2),$$

and with

$$v_1 = (1, 2, 1, 0),\ v_2 = (1, -1, 1, 0),\ v_3 = (1, 0, -1, 0),\ v_4 = (0, 0, 0, 2),$$

both $B = \{u_1, u_2, u_3, u_4\}$ and $C = \{v_1, v_2, v_3, v_4\}$ are bases of $\mathbb{R}^4$. It is easy to check that

$$\begin{aligned} v_1 &= u_1, \\ v_2 &= -2u_1 + u_2, \\ v_3 &= 11u_1 - 4u_2 + u_3 \\ v_4 &= -27u_1 + 11u_2 - 2u_3 + u_4, \end{aligned}$$

so that

$$P = ([v_1]_B\ [v_2]_B\ [v_3]_B\ [v_4]_B\ ) = \begin{pmatrix} 1 & -2 & 11 & -27 \\ 0 & 1 & -4 & 11 \\ 0 & 0 & 1 & -2 \\ 0 & 0 & 0 & 1 \end{pmatrix}.$$

Hence $[u]B = P[u]_C$ for every $u \in \mathbb{R}^4$. It is also easy to check that

$$\begin{aligned} u_1 &= v_1, \\ u_2 &= 2v_1 + v_2, \\ u_3 &= -3v_1 + 4v_2 + v_3, \\ u_4 &= -v_1 - 3v_2 + 2v_3 + v_4, \end{aligned}$$

so that

$$Q = ([u_1]_C\ [u_2]_C\ [u_3]_C\ [u_4]_C\ ) = \begin{pmatrix} 1 & 2 & -3 & -1 \\ 0 & 1 & 4 & -3 \\ 0 & 0 & 1 & 2 \\ 0 & 0 & 0 & 1 \end{pmatrix}.$$

Hence $[u]_C = Q[u]_B$ for every $u \in \mathbb{R}^4$. Note that $PQ = I$. Now let $u = (6, -1, 2, 2)$. We can check that

$u = v_1 + 3v_2 + 2v_3 + v_4$, so that

$$[u]_c = \begin{pmatrix} 1 \\ 3 \\ 2 \\ 1 \end{pmatrix}.$$

Then

$$[u]_B = \begin{pmatrix} 1 & -2 & 11 & -27 \\ 0 & 1 & -4 & 11 \\ 0 & 0 & 1 & -2 \\ 0 & 0 & 0 & 1 \end{pmatrix} \begin{pmatrix} 1 \\ 3 \\ 2 \\ 1 \end{pmatrix} = \begin{pmatrix} -10 \\ 6 \\ 0 \\ 1 \end{pmatrix}.$$

Check that $u = -10u_1 + 6u_2 + u_4$.

Example. Consider the vector space P_2. It is not too dicult to check that

$$u_1 = 1 + x,\ u_2 = 1 + x^2,\ u_3 = x + x^2$$

form a basis of P_2. Let

$$u = 1 + 4x - x^2.$$

Then

$$u = \beta_1 u_1 + \beta_2 u_2 + \beta_3 u_3,$$

where

$$1 + 4x - x^2 = \beta_1(1 + x) + \beta_2(1 + x^2) + \beta_3(x + x^2) = (\beta_1 + \beta_2) + (\beta_1 + \beta_3)x + (\beta_2 + \beta_3)x^2,$$

so that

$$\beta_1 + \beta_2 = 1,\ \beta_1 + \beta_3 = 4$$

and

$$\beta_2 + \beta_3 = -1.$$

Hence $(\beta_1, \beta_2, \beta_3) = (3, -2, 1)$. If we write

$$B = \{u_1, u_2, u_3\},$$

then

$$[u]_B = \begin{pmatrix} 3 \\ -2 \\ 1 \end{pmatrix}.$$

On the other hand, it is also not too dicult to check that

$$v_1 = 1,\ v_2 = 1 + x,\ v_3 = 1 + x + x^2$$

form a basis of P_2. Also

$$u = \gamma_1 v_1 + \gamma_2 v_2 + \gamma_3 v_3,$$

where

$$\begin{aligned} 1 + 4x - x^2 &= \gamma_1 + \gamma_2(1 + x) + \gamma_3(1 + x + x^2) \\ &= (\gamma_1 + \gamma_2 + \gamma_3) + (\gamma_2 + \gamma_3)x + \gamma_3 x^2, \end{aligned}$$

so that

$$\gamma_1 + \gamma_2 + \gamma_3 = 1,\ \gamma_2 + \gamma_3 = 4$$

and

$$\gamma_3 = -1.$$

Hence

$$(\gamma_1, \gamma_2, \gamma_3) = (-3, 5, -1).$$

If we write

$$C = \{v_1, v_2, v_3\},$$

then

$$[u]_c = \begin{pmatrix} -3 \\ 4 \\ -1 \end{pmatrix}.$$

Next, note that

$$v_1 = \frac{1}{2}u_1 + \frac{1}{2}u_2 - \frac{1}{2}u_3,$$

$$v_2 = u_1,$$

$$v_3 = \frac{1}{2}u_1 + \frac{1}{2}u_2 + \frac{1}{2}u_3.$$

Hence

$$P = ([v_1]B\ [v_2]B\ [v_3]B) = \begin{pmatrix} 1/2 & 1 & 1/2 \\ 1/2 & 0 & 1/2 \\ -1/2 & 0 & 1/2 \end{pmatrix}.$$

To verify that $[u]_B = P[u]_C$, note that

$$\begin{pmatrix} 3 \\ -2 \\ 1 \end{pmatrix} = \begin{pmatrix} 1/2 & 1 & 1/2 \\ 1/2 & 0 & 1/2 \\ -1/2 & 0 & 1/2 \end{pmatrix}\begin{pmatrix} -3 \\ 5 \\ -1 \end{pmatrix}.$$

Kernel and Range

Consider rst of all a euclidean linear transformation $\mathbb{R}^n \to \mathbb{R}^m$. Suppose that A is the standard matrix for T. Then the range of the transformation T is given by

$$R(T) = \{T(x) : x \in \mathbb{R}^n\} = \{Ax : x \in \mathbb{R}^n\}.$$

It follows that $R(T)$ is the set of all linear combinations of the columns of the matrix A, and is therefore the column space of A. On the other hand, the set

$$\{x \in \mathbb{R}^n : Ax = 0\}$$

is the nullspace of A.

Recall that the sum of the dimension of the nullspace of A and dimension of the column space of A is equal to the number of columns of A. This is known as the Rank–nullity theorem. The purpose of this section is to extend this result to the setting of linear transformations. To do this, we need the following generalization of the idea of the nullspace and the column space.

Definition. Suppose that $T : V \to W$ is a linear transformation from a real vector space V into a real vector space W. Then the set

$$\ker(T) = \{u \in V : T(u) = 0$$

is called the kernel of T, and the set

$$R(T) = \{T(u) : u \in V\}$$

is called the range of T.

Example. For a euclidean linear transformation T with standard matrix A, we have shown that $\ker(T)$ is the nullspace of A, while $R(T)$ is the column space of A.

Example. Suppose that $T : V \to W$ is the zero transformation. Clearly we have

$$\ker(T) = V \text{ and } R(T) = \{0\}.$$

Example. Suppose that $T : V \to V$ is the identity operator on V. Clearly we have

$$\ker(T) = \{0\} \text{ and } R(T) = V.$$

Example. Suppose that $T : \mathbb{R}^2 \to \mathbb{R}^2$ is orthogonal projection onto the x_1–axis. Then $\ker(T)$ is the x_2–axis, while $R(T)$ is the x_1–axis.

Example. Suppose that $T : \mathbb{R}^n \to \mathbb{R}^n$ is one–to–one. Then

$$\ker(T) = \{0\} \text{ and } R(T) = \mathbb{R}^n,$$

Example. Consider the linear transformation $T : V \to W$, where V denotes the vector space of all real valued functions dierentiable everywhere in $\mathbb{R}$, where W denotes the space of all real valued functions dened in $\mathbb{R}$, and where $T(f) = f'$ for every $f \in V$. Then $\ker(T)$ is the set of all dierentiable functions with derivative 0, and so is the set of all constant functions in $\mathbb{R}$.

Example. Consider the linear transformation $T : V \to \mathbb{R}$, where V denotes the vector space of all real valued functions Riemann integrable over the interval [0, 1], and where

$$T(f) = \int_0^1 f(x)dx$$

for every $f \in V$. Then $\ker(T)$ is the set of all Riemann integrable functions in [0, 1] with zero mean, while $R(T) = \mathbb{R}$.

Proposition. Suppose that $T : V \to W$ is a linear transformation from a real vector space V into a real vector space W. Then $\ker(T)$ is a subspace of V, while $R(T)$ is a subspace of W.

Proof. Since $T(0) = 0$, it follows that $0 \in \ker(T)$ V and $0 \in R(T)$ W. For any

$$u, v \in \ker(T),$$

we have

$$T(u + v) = T(u) + T(v) = 0 + 0 = 0,$$

so that $u + v \in \ker(T)$. Suppose further that $c \in \mathbb{R}$. Then

$$T(cu) = cT(u) = c_0 = 0,$$

so that $c_u \in \ker(T)$. Hence $\ker(T)$ is a subspace of V. Suppose next that $w, z \in R(T)$. Then there exist $u, v \in V$ such that $T(u) = w$ and $T(v) = z$. Hence

$$T(u + v) = T(u) + T(v) = w + z,$$

so that $w + z \in R(T)$. Suppose further that $c \in \mathbb{R}$. Then

$$T(cu) = cT(u) = cw,$$

so that $cw \in R(T)$. Hence $R(T)$ is a subspace of W.

To complete this section, we prove the following generalization of the Rank–nullity theorem.

Proposition. Suppose that $T : V \to W$ is a linear transformation from an n–dimensional real vector space V into a real vector space W. Then

$$\dim \ker(T) + \dim R(T) = n.$$

Proof. Suppose first of all that dim ker(T) = n. Then ker(T) = V, and so $R(T) = \{0\}$, and the result follows immediately. Suppose next that dim ker(T) = 0, so that ker(T) = $\{0\}$. If $\{v_1, \ldots, v_n\}$ is a basis of V, then it follows that $T(v_1), \ldots, T(v_n)$ are linearly independent in W, for otherwise there exist $c_1, \ldots, c_n \in \mathbb{R}$, not all zero, such that

$$c_1T(v_1) + \ldots + c_nT(v_n) = 0,$$

so that

$$T(c_1v_1 + \ldots + c_nv_n) = 0,$$

a contradiction since

$$c_1v_1 + \ldots + c_nv_n \neq 0.$$

On the other hand, elements of $R(T)$ are linear combinations of $T(v_1), \ldots, T(v_n)$.

Hence $\dim R(T) = n$,

and the result again follows immediately. We may therefore assume that

$$\dim \ker(T) = r,$$

where $1 \leq r < n$.

Let $\{v_1, \ldots, v_r\}$ be a basis of ker(T). This basis can be extended to a basis $\{v_1, \ldots, v_r, v_r + 1, \ldots, v_n\}$ of V. It suces toshow that

$$\{T(v_{r+1}), \ldots, T(v_n)\}$$

is a basis of $R(T)$. Suppose that $u \in V$. Then there exist $\beta_1, \ldots, \beta_n \in \mathbb{R}$ such that

$$u = \beta_1v_1 + \ldots + \beta_rv_r + \beta_{r+1}\, v_{r+1} + \ldots + \beta_nv_n,$$

so that

$$T(u) = \beta_1T(v_1) + \ldots + \beta_rT(v_r) + \beta_{r+1}T(v_r+1) + \ldots + \beta_nT(v_n)$$
$$= \beta_{r+1}T(v_{r+1}) + \ldots + \beta_nT(v_n),$$

It follows that spans $R(T)$. It remains to prove that its elements are linearly independent. Suppose that $c_{r+1}, \ldots, c_n \in \mathbb{R}$ and

$$c_{r+1}T(v_{r+1}) + \ldots + c_nT(v_n) = 0.$$

We need to show that

$$c_{r+1} = \ldots = c_n = 0.$$

By linearity, it follows from equation that $T(c_{r+1}v_{r+1} + \ldots + c_nv_n) = 0$, so that

$$c_{r+1}v_{r+1} + \ldots + c_nv_n \in \ker(T).$$

Hence there exist $c_1, \ldots, c_r \in \mathbb{R}$ such that

$$c_{r+1}v_{r+1} + \ldots + c_nv_n = c_1v_1 + \ldots + c_rv_r;$$

so that

$$c_1v_1 + \ldots + c_rv_r - c_{r+1}v_{r+1} - \ldots - c_nv_n = 0.$$

Since $\{v_1, \ldots, v_n\}$ is a basis of V, it follows that

$$c_1 = \ldots = c_r = c_{r+1} = \ldots = c_n = 0.$$

Remark. We sometimes say that dim $R(T)$ and dim ker(T) are respectively the rank and the nullity of the linear transformation T.

Inverse Linear Transformations

In this section, we generalize some of the ideas first discussed in Section.

Definition. A linear transformation $T : V \to W$ from a real vector space V into a real vector space W is said to be one–to–one if for every $u', u'' \in V$, we have $u' = u''$ whenever $T(u') = T(u'')$.

Proposition. Suppose that $T : V \to W$ is a linear transformation from a real vector space V into a real vector space WF. Then T is one–to–one if and only if $\ker(T) = \{0\}$.

Proof. ($\Rightarrow$) Clearly $0 \in \ker(T)$. Suppose that $\ker(T) \neq \{0\}$. Then there exists a non–zero $v \in \ker(T)$. It follows that $T(v) = T(0)$, and so T is not one–to–one.

(($\Leftarrow$) Suppose that $\ker(T) = \{0\}$. Given any $u', u'' \in V$, we have

$$T(u') - T(u') = T(u' - u'') = 0$$

if and only if $u' - u'' = 0$, in other words, if and only if $u' = u''$. We have the following generalization of PROPOSITION.

Proposition. Suppose that $T : V \to V$ is a linear operator on a finite–dimensional real vector space V. Then the following statements are equivalent.

(a) The linear operator T is one–to–one.

(b) We have $\ker(T) = \{g\}$.

(c) The range of T is V, in other words, $R(T) = V$.

Proof. The equivalence of (a) and (b) is established by PROPOSITION. The equivalence of (b) and (c) follows from PROPOSITION.

Suppose that

$$T : V \to W$$

is a one–to–one linear transformation from a real vector space V into a real vector space W. Then for every $w \in R(T)$, there exists exactly one $u \in V$ such that

$$T(u) = w.$$

We can therefore dene a transformation

$$T^{-1} : R(T) \to V$$

by writing $T^{-1}(w) = u$, where $u \in V$ is the unique vector satisfying $T(u) = w$.

Proposition. Suppose that $T : V \to W$ is a one–to–one linear transformation from a real vector space V into a real vector space W. Then T^{-1}. $R(T) \to V$ is a linear transformation.

Proof. Suppose that $w, z \in R(T)$. Then there exist $u, v \in V$ such that

$$T^{-1}(w) = u$$

and $$T^{-1}(z) = v.$$

It follows that

$$T(u) = w \text{ and } T(v) = z,$$

so that $$T(u + v) = T(u) + T(v) = w + z,$$

whence $T^{-1}(w + z) = u + v = T^{-1}(w) + T^{-1}(z)$.

Suppose further that $C \in \mathbb{R}$.

Then $T(cu) = cw$,

so that $T^{-1}(cw) = cu = cT^{-1}(w)$.

We also have the following result concerning compositions of linear transformations and which requires no further proof, in view of our knowledge concerning inverse functions.

Proposition. Suppose that V, W, U are real vector spaces. Suppose further that $T_1 : V \to W$ and $T_2 : W \to U$ are one–to–one linear transformations. Then

(a) *The linear transformation $T_2 . T_1 : V \to U$ is one–to–one, and*

(b) $(T_2 . T_1)^{-1} = T_1^{-1}.T_2^{-1}$.

Matrices of General Linear Transformations

Suppose that $T : V \to W$ is a linear transformation from a real vector space V to a real vector space W. Suppose further that the vector spaces V and W are finite dimensional, with dim $V = n$ and dim $W = m$. We shall show that if we make use of a basis B of V and a basis C of W, then it is possible to describe T indirectly in terms of some matrix A. The main idea is to make use of coordinate matrices relative to the bases B and C.

Let us recall some discussion in Section. Suppose that

$$B = \{v_1, \ldots, v_n\}$$

is a basis of V. Then every vector $v \in V$ can be written uniquely as a linear combination

$$v = \beta_1 v_1 + \ldots + \beta_n v_n,$$

where $\beta_1, \ldots, \beta_n \in \mathbb{R}$

The matrix

$$[V]_B = \begin{pmatrix} \beta_1 \\ \vdots \\ \beta_n \end{pmatrix}$$

is the coordinate matrix of v relative to the basis B.

Consider now a transformation $\phi : V \to \mathbb{R}^n$, where $\phi(v) = [v]_B$ for every $v \in V$. The proof of the following result is straightforward.

Proposition. Suppose that the real vector space V has basis $B = \{v_1, \ldots, v_n\}$. Then the transformation $\phi : V \to W$, where $\phi(v) = [v]_B$ satises for every $v \in V$, is a one–to–one linear transformation, with range $R(\phi) = \mathbb{R}^n$. Furthermore, the inverse linear transformation $\phi^{-1} : \mathbb{R}^n \to V$ is also one–to–one, with range $R(\mathbb{R}^{n-1}) = V$.

Suppose next that $C = \{w_1, \ldots ,w_m\}$ is a basis of W. Then we can dene a linear transformation $\psi : W \to \mathbb{R}^m$, where $\psi(w) = [w]_C$ for every $w \in W$, in a similar way. We now have the following diagram of linear transformations.

Suppose next that $\{w_1, \ldots, w_m\}$ is basis of W. Then we can define linear transformation $\psi : W \to \mathbb{R}^m$, where $\psi(w) = [w]_c$ for every $w \in W$, in similar way. We now have the following diagram of linear transformations.

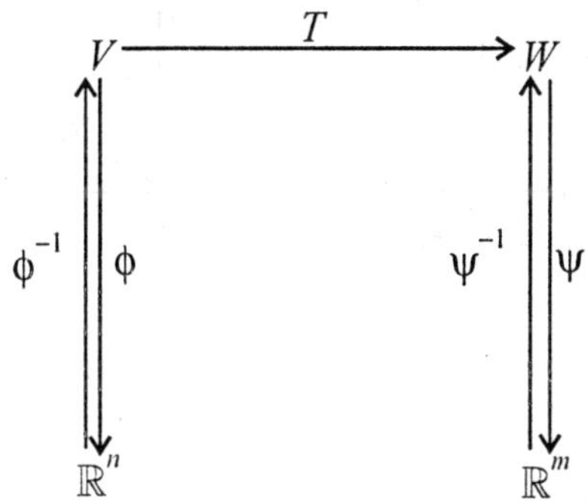

Clearly the composition

$$S = \psi.T \circ T \circ \phi^{-1} : \mathbb{R}^n \to \mathbb{R}^m$$

is a euclidean linear transformation, and can therefore be described in term of a standard matrix A. Our task is to determine this matrix A in terms of T and the bases B and C.

We know from Proposition that

$$A = (S(e_1) \ldots S(e_n)),$$

where $\{e_1, \ldots, e_n\}$ is the standard basis for $\mathbb{R}^n$. For every $j = 1, \ldots, n$, we have

$$S(e_j) = (\psi \circ T \circ \phi^{-1})\ (e_j) = \psi(T(\phi^{-1}(e_j))) = \psi(T(v_j)) = [T(v_j)]c.$$

It follows that

$$A = ([T(v_1)]c \ldots [T(v_n)]_c).$$

Definition. The matrix A given by equatioin is called the matrix for the linear transformation T with respect to the bases B and C. We now have the following diagram of linear transformations.

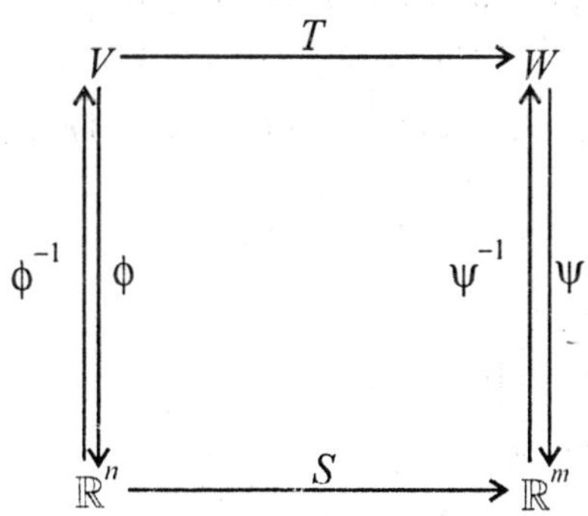

Hence we can write T as the composition

$$T = \psi^{-1} \circ S \circ \phi : V \rightarrow W.$$

For every $v \in V$, we have the following:

$$v \xrightarrow{\phi} [v]_B \xrightarrow{S} A[v]_B \xrightarrow{\psi^{-1}} \psi^{-1}(A[v]_B)$$

More precisely, if $v = \beta_1 v_1 + \ldots + \beta_n v_n$, then

$$[v]_B = \begin{pmatrix} \beta_1 \\ \vdots \\ \beta_n \end{pmatrix} \quad \text{and} \quad [v]_B = A\begin{pmatrix} \beta_1 \\ \vdots \\ \beta_n \end{pmatrix} = \begin{pmatrix} \lambda_1 \\ \vdots \\ \lambda_m \end{pmatrix},$$

say, and so

$$T(v) = \psi^{-1}(A[v]_B) = \gamma_1 w_1 + \ldots + \gamma_m w_m.$$

We have proved the following result.

Proposition. Suppose that $T : V \rightarrow W$ is a linear transformation from a real vector space V into a real vector space W. Suppose further that V and W are finite dimensional, with bases B and C respectively, and that A is the matrix for the linear transformation T with respect to the bases B and C. Then for every $v \in V$, we have $T(v) = w$, where $w \in W$ is the unique vector satisfying $[w]_C = A[v]_B$.

Remark. In the special case when $V = W$, the linear transformation $T : V \rightarrow W$ is a linear operator on T. Of course, we may choose a basis B for the domain V of T and a basis C for the codomain V of T. In the case when T is the identity linear operator, we often choose $B \neq C$ since this represents a change of basis. In the case when T is not the identity operator, we often choose $B = C$ for the sake of convenience, we then say that A is the matrix for the linear operator T with respect to the basis B.

Example. Consider an operator $T : P_3 \rightarrow P_3$ on the real vector space P_3 of all polynomials with real coefficients and degree at most 3, where for every polynomial p(x) in P_3, we have $T(p(x)) = xp'(x)$, the product of x with the formal derivative $p'(x)$ of $p(x)$. The reader is invited to check that T is a linear operator. Now consider the basis $B = \{1, x, x_2, x_3\}$ of P_3. The matrix for T with respect to B is given by

$$A = ([T(1)]_B\, [T(x)]_B\, [T(x^2)]_B\, [T(x^3)]_B\,) = ([0]_B\, [x]_B\, [2x^2]_B\, [3x^3]_B)$$

$$= \begin{pmatrix} 0 & 0 & 0 & 0 \\ 0 & 1 & 0 & 0 \\ 0 & 0 & 2 & 0 \\ 0 & 0 & 0 & 3 \end{pmatrix}.$$

Suppose that $p(x) = 1 + 2x + 4x^2 + 3x^3$. Then

$$[p(x)]_B = \begin{pmatrix} 1 \\ 2 \\ 4 \\ 3 \end{pmatrix} \text{ and } A[p(x)]_B = \begin{pmatrix} 0 & 0 & 0 & 0 \\ 0 & 1 & 0 & 0 \\ 0 & 0 & 2 & 0 \\ 0 & 0 & 0 & 3 \end{pmatrix}\begin{pmatrix} 1 \\ 2 \\ 4 \\ 3 \end{pmatrix} = \begin{pmatrix} 0 \\ 2 \\ 8 \\ 9 \end{pmatrix}.$$

so that $T(p(x)) = 2x + 8x^2 + 9x^3$. This can be easily verified by noting that

$$T(p(x)) = xp'(x) = x(2 + 8x + 9x^2) = 2x + 8x^2 + 9x^3.$$

In general, if $p(x) = p_0 + p_1x + p_2x^2 + p_3x^3$, then

$$[p(x)]_B \begin{pmatrix} p_0 \\ p_1 \\ p_2 \\ p_3 \end{pmatrix} \text{ and } A[p(x)]_B = \begin{pmatrix} 0 & 0 & 0 & 0 \\ 0 & 1 & 0 & 0 \\ 0 & 0 & 2 & 0 \\ 0 & 0 & 0 & 3 \end{pmatrix}\begin{pmatrix} p_0 \\ p_1 \\ p_2 \\ p_3 \end{pmatrix} = \begin{pmatrix} 0 \\ p_1 \\ 2p_2 \\ 3p_3 \end{pmatrix}.$$

so that $T(p(x)) = p_1x + 2p_2x^2 + 3p_3x^3$. Observe that

$$T(p(x)) = xp'(x) = x(p_1x + 2p_2x + 3p_3x) = p_1x + 2p_2x^2 + 3p_3x^3.$$

verifying our result.

Example. Consider the linear operator $T: \mathbb{R}^2 \to \mathbb{R}^2$, given by $T(x_1, x_2)\ 2x_1\ x_2, x_1\ 3x_2)$ for every $(x_1, x_2) \in \mathbb{R}^2$. Consider also the basis $B = (1, 0), 1, 1)$ of $\mathbb{R}^2$. Then the matrix for with respect to is given by

$$A = T(1, 0)]_B[T(1, 1)]_B) = ([(2, 1)]_B\, [(3, 4)]_B = \begin{pmatrix} 1 & -1 \\ 1 & 4. \end{pmatrix}.$$

Suppose that $(x_1, x_2)\ 3, 2)$. Then

$$[(3, 2)]_B = \begin{pmatrix} 1 \\ 2 \end{pmatrix} \text{ and } A[(3,2)]_B = \begin{pmatrix} 1 & -1 \\ 1 & 4 \end{pmatrix}\begin{pmatrix} 1 \\ 2 \end{pmatrix} = \begin{pmatrix} -1 \\ 9 \end{pmatrix},$$

so that $T(3, 2) -(1, 0) +9(1, 1)\ (8, 9)$. This can be easily veried directly. In general, we have

$$[(x_1,x_2)]_B = \begin{pmatrix} x_1 - x_2 \\ x_2 \end{pmatrix} \text{ and } A[x_1,x_2]_B = \begin{pmatrix} 1 & -1 \\ 1 & 4 \end{pmatrix}\begin{pmatrix} x_1 - x_2 \\ x_2 \end{pmatrix} = \begin{pmatrix} x_1 - 2x_2 \\ x_1 + 3x_2 \end{pmatrix},$$

so that $T(x_1, x_2) = (x_1 - 2x_2)(1, 0) + (x_1 + 3x_2)(1, 1) = (2x_1 + x_2, x_1 + 3x_2)$.

Example. Suppose that $T: \mathbb{R}^n \to \mathbb{R}^m$ is a linear transformation. Suppose further that B and C are the standard bases for $\mathbb{R}^n$ and $\mathbb{R}^m$ respectively. Then the matrix for T with respect to B and C is given by

$$A = T(e_1)]_c \ldots T(e_n)]_c = (T(e_1) \ldots T(e_n),$$

so it follows from Proposition that is simply the standard matrix for T.

Suppose now that

$$T_1 \to W$$

and

$$T_2 W \to U$$

are linear transformations, where the real vector spaces V, W, U are finite dimensional, with respective bases

$$B = \{v_1, \ldots, v_n\}, \{w_1, \ldots, w_m\}$$

and

$$D = \{u_1, \ldots, u_k\}.$$

We then have the following diagram of linear transformations.

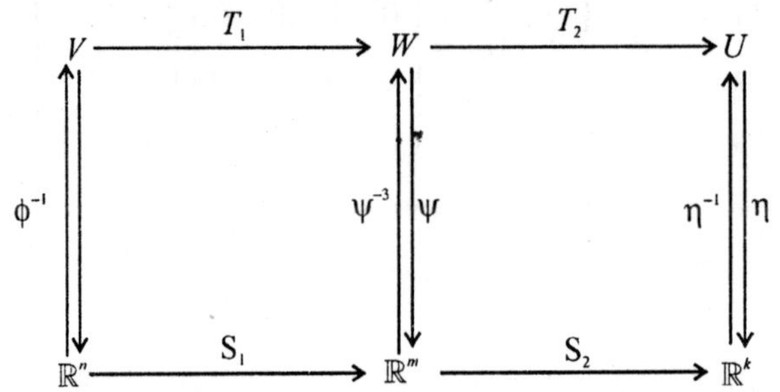

Here $\eta : U \to \mathbb{R}^k$, where $\eta(u) = [u]_D$ for every $u \in U$, is a linear transformation, and

$$S_1 = \psi \circ T_1 \circ \phi^{-1} : \mathbb{R}^n \to \mathbb{R}^m \text{ and } S_2 = \eta \circ T_2 \circ \psi^{-1} : \mathbb{R}^m \to \mathbb{R}^k$$

are euclidean linear transformations. Suppose that A_1 and A_2 for S_1 and S_2, so that they are respectively the matrix for T_1 with respect to B and C and the matrix for T_2 with respect to C and D. Clearly

$$S_2 \circ S_1 = \eta \circ T_2 \circ T_1 \circ \phi^{-1} : \mathbb{R}^n \to \mathbb{R}^k.$$

It follows that A_2A_1 is the standard matrix for $S_2 \circ S_1$, and so is the matrix for $T_2 \circ T_1$ with respect to the bases B and D. To summarize, we have the following result.

Proposition. Suppose that $T_1 : V \to W$ and $T_2 : W \to U$ are linear transformations, where the real vector spaces V, W, U are finite dimensional, with bases B, C, D respectively. Suppose further that A_1 is the matrix for the linear transformation T_1 with respect to the bases B and C, and that A_2 is the matrix for the linear transformation T_2 with respect to the bases C and D. Then A_2A_1 is the matrix for the linear transformation $T_2 \times T_1$ with respect to the bases B and D.

Example. Consider the linear operator $T_1 : P_3 \to P_3$, where for every polynomial $p(x)$ in we have $T_1(p(x)) = xp'(x)$. We have already shown that the matrix for T_1 with respect to the basis $B = \{1, x, x^2, x^3\}$ of P_3 is given by

$$A_1 = \begin{pmatrix} 0 & 0 & 0 & 0 \\ 0 & 1 & 0 & 0 \\ 0 & 0 & 2 & 0 \\ 0 & 0 & 0 & 3 \end{pmatrix}$$

Consider next the linear operator $T_2 : P_3 \to P_3$, where for every polynomial $q(x) = q_0 + q_1x + q_2x^2 + q_3x^3$ in P_3, we have

$$T_2(q(x)) = q(1 + x) = q_0 + q_1(1 + x) + q_2(1 + x)^2 + q_3(1 + x)^3.$$

We have $T_2(1) = 1$, $T_2(x) = 1 + x$, $T_2(x^2) = 1 + 2x + x^2$ and $T_2(x^3) = 1 + 3x + 3x^2 + x^3$, so that the matrix for T_2 with respect to B is given by

$$A_2 = ([T_2(1)]_B\ [T_2(x)]_B\ [T_2(x^2)]_B\ [T_2(x^3)]_B)$$

$$= \begin{pmatrix} 1 & 1 & 1 & 1 \\ 0 & 1 & 2 & 3 \\ 0 & 0 & 1 & 3 \\ 0 & 0 & 0 & 1 \end{pmatrix}$$

Consider now the composition $T = T_2 \circ T_1 : P_3 \to P_3$. Let A denote the matrix for T with respect to B. By Proposition 8*T*, we have

$$A = A_2A_1 = \begin{pmatrix} 1 & 1 & 1 & 1 \\ 0 & 1 & 2 & 3 \\ 0 & 0 & 1 & 3 \\ 0 & 0 & 0 & 1 \end{pmatrix} \begin{pmatrix} 0 & 0 & 0 & 0 \\ 0 & 1 & 0 & 0 \\ 0 & 0 & 2 & 0 \\ 0 & 0 & 0 & 3 \end{pmatrix} = \begin{pmatrix} 0 & 1 & 2 & 3 \\ 0 & 1 & 4 & 9 \\ 0 & 0 & 2 & 9 \\ 0 & 0 & 0 & 3 \end{pmatrix}.$$

Suppose that $p(x) = p_0 + p_1x + p_2x^2 + p_3x^3$. Then

$$[p(x)]_B = \begin{pmatrix} p_0 \\ p_1 \\ p_2 \\ p_3 \end{pmatrix} \quad \text{and} \quad A[p(x)]_B = \begin{pmatrix} 0 & 1 & 2 & 3 \\ 0 & 1 & 4 & 9 \\ 0 & 0 & 2 & 9 \\ 0 & 0 & 0 & 3 \end{pmatrix} \begin{pmatrix} p_0 \\ p_1 \\ p_2 \\ p_3 \end{pmatrix} = \begin{pmatrix} p_1 + 2p_n + 3p_3 \\ p_1 + 4p_2 + 9p_3 \\ 2p_2 + p_3 \\ 3p_3 \end{pmatrix}.$$

so that $T(p(x)) = (p_1 + 2p_2 + 3p_3) + (p_1 + 4p_2 + 9p_3)\,x + (2p_2 + 9p_3)x^2 + 3p_3x^3$. We can check this directly by noting that

$$T(p(x)) = T_2(T_1(p(x))) = T_2(p_1x + 2p_2x^2 + 3p_3x^3)$$
$$= p_1(1 + x) + 2p_2(1 + x)^2 + 3p_3(1 + x)^3$$
$$= (p_1 + 2p_2 + 3p_3) + (p_1 + 4p_2 + 9p_3)x + (2p_2 + 9p_3)x^2 + 3p_3x^3.$$

Example. Consider the linear operator $T : \mathbb{R}^2 \to \mathbb{R}^2$, given by

$$T(x_1, x_2) = (2x_1 + x_2, x_1 + 3x_2)$$

for every $(x_1, x_2) \in \mathbb{R}^2$. We have already shown that the matrix for T with respect to the basis $B = \{(1, 0), (1, 1)\}$ of $\mathbb{R}^2$ is given by

$$A = \begin{pmatrix} 1 & -1 \\ 1 & 4 \end{pmatrix}.$$

Consider the linear operator $T : \mathbb{R}^2 \to \mathbb{R}^2$. By Proposition 8T, the matrix for T_2 with respect to B is given by

$$A^2 = \begin{pmatrix} 1 & -1 \\ 1 & 4 \end{pmatrix}\begin{pmatrix} 1 & -1 \\ 1 & 4 \end{pmatrix} = \begin{pmatrix} 0 & -5 \\ 5 & 15 \end{pmatrix}.$$

Suppose that $(x_1, x_2) \in \mathbb{R}^2$. Then

$$[(x_1, x_2)]_B = \begin{pmatrix} x_1 - x_2 \\ x_2 \end{pmatrix} \text{ and } A_2[x_1, x_2]_B = \begin{pmatrix} 0 & -5 \\ 5 & 15 \end{pmatrix}\begin{pmatrix} x_1 - x_2 \\ x_2 \end{pmatrix} = \begin{pmatrix} -5x_2 \\ 5x_1 + 10x_2 \end{pmatrix},$$

so that $T(x_1, x_2) = -5x_2(1, 0) + (5x_1 + 10x_2)(1, 1) = (5x_1 + 5x_2, 5x_1 + 10x_2)$. The reader is invited to check this directly. A simple consequence of Propositions 8N and 8T is the following result concerning inverse linear transformations.

Proposition. Suppose that $T : V \to V$ is a linear operator on a finite dimensional real vector space V with basis B. Suppose further that A is the matrix for the linear operator T with respect to the basis B. Then T is one–to–one if and only if A is invertible. Furthermore, if T is one–to–one, then A^{-1} is the matrix for the inverse linear operator $T^{-1} : V \to V$ with respect to the basis B.

Proof. Simply note that T is one–to–one if and only if the system

$$Ax = 0$$

has only the trivial solution $x = 0$. The last assertion follows easily from Proposition 8T, since if A' denotes the matrix for the inverse linear operator T^{-1} with respect to B, then we must have

$$A'A = I,$$

the matrix for the identity operator $T^{-1} \circ T$ with respect to B.

Example. Consider the linear operator $T : P_3 \to P_3$, where for every $q(x) = q_0 + q_1x + q_2x^2 + q_3x^3$ in P_3, we have

$$T(q(x)) = q(1 + x) = q_0 + q_1(1 + x) + q_2(1 + x)^2 + q_3(1 + x)^3.$$

We have already shown that the matrix for T with respect to the basis

$$B = \{1 + x, x^2, x^3\}$$

is given by

$$A = \begin{pmatrix} 1 & 1 & 1 & 1 \\ 0 & 1 & 2 & 3 \\ 0 & 0 & 1 & 3 \\ 0 & 0 & 0 & 1 \end{pmatrix}.$$

This matrix is invertible, so it follows that T is one–to–one. Furthermore, it can be checked that

$$A^{-1} = \begin{pmatrix} 1 & -1 & 1 & -1 \\ 0 & 1 & -2 & 3 \\ 0 & 0 & 1 & -3 \\ 0 & 0 & 0 & 1 \end{pmatrix}.$$

Suppose that $p(x) = p_0 + p_1x + p_2x^2 + p_3x^3$. Then

$$[p(x)_B = \begin{pmatrix} p_0 \\ p_1 \\ p_2 \\ p_3 \end{pmatrix} \text{ and } A^{\ 1}[p(x)]_B = \begin{pmatrix} 1 & -1 & 1 & -1 \\ 0 & 1 & -2 & 3 \\ 0 & 0 & 1 & -3 \\ 0 & 0 & 0 & 1 \end{pmatrix},$$

$$\begin{pmatrix} p_0 \\ p_1 \\ p_2 \\ p_3 \end{pmatrix} = \begin{pmatrix} p_0 - p_1 + p_2 - p_3 \\ p_1 - 2p_2 + 3p_2 \\ p_2 - 3p_3 \\ p_3 \end{pmatrix}.$$

so that

$$\begin{aligned} T^{-1}(p(x)) &= (p_0 - p_1 + p_2 - p_3) + (p_1 - 2p_2 + 3p_3)x + (p_2 - 3p_3)x^2 + p_3x^3 \\ &= p_0 + p_1(x - 1) + p_2(x^2 - 2x + 1) + p_3(x^3 - 3x_2 + 3x - 1) \\ &= p_0 + p_1(x - 1) + p_2(x - 1)^2 + p_3(x - 1)^3 = p(x - 1). \end{aligned}$$

Change of Basis

Suppose that V is a finite dimensional real vector space, with one basis $B = \{v_1, \ldots, v_n\}$ and another basis $B' = \{u_1, \ldots, u_n\}$. Suppose that $T : V \to V$ is a linear operator on V. Let A denote the matrix for T with respect to the basis B, and let A' denote the matrix for T with respect to the basis B'. If $v \in V$ and $T(v) = w$, then

$$[w]_B = A[v]_B$$

and

$$[w]_{B'} = A'[v]_{B'}$$

We wish to nd the relationship between A' and A. Recall Proposition J, that if

$$P = ([u_1]B \ldots [u_n]_B)$$

denotes the transition matrix from the basis B' to the basis B, then

$$[v]_B = P[v]_{B'} \text{ and } [w]_B = P[w]_{B'}.$$

Note that the matrix P can also be interpreted as the matrix for the identity operator

$$I : V \to V$$

with respect to the bases B' and B. It is easy to see that the matrix P is invertible, and

$$P^{-1} = ([v_1]B' \ldots [v_n]B')$$

denotes the transition matrix from the basis B to the basis B', and can also be interpreted as the matrix for the identity operator

$$I : V \to V$$

with respect to the bases B and B'. We conclude that,

$$[w]B' = P^{-1} w]_B = P^{-1}A[v]_B = P^{-1}AP[v]B'.$$

Comparing this with (11), we conclude that

$$P^{-1}AP = A'.$$

This implies that

$$A = PA'P^{-1}.$$

Remark. We can use the notation

$$A = [T]_B \text{ and } A' [T]_{B'}$$

to denote that A and A' are the matrices for T with respect to the basis B and with respect to the basis B' respectively. We can also write

$$P = [I]_{B'}, B.$$

to denote that P is the transition matrix from the basis B' to the basis B, so that

$$P^{-1} = [I]_{B'}, B.$$

Then (13) and (14) become respectively

$$[I]_{B', B}[T]_B[I]_{B,B'} = [T]_{B'} \text{ and } [I]_{B, B'} [T]_{B'} [I]_{B',B} = [T]_B.$$

We have proved the following result.

Proposition. Suppose that $T : V \to V$ is a linear operator on a finite dimensional real vector space V, with bases $B = \{v_1, \ldots, v_n\}$ and $B' = \{u_1, \ldots, u_n\}$. Suppose further that A and A' are the matrices for T with respect to the basis B and with respect to the basis B' respectively. Then

$$P^{-1}AP = A' \text{ and } A' = PAP^{-1},$$

where

$$P = ([u_1]_B \cdots [u_n]_B$$

denotes the transition matrix from the basis B' to the basis B.

Remarks. (1) We have the following picture.

(2) The idea can be extended to the case of linear transformations $T : V \to W$ from a finite dimensional real vector space into another, with a change of basis in V and a change of basis in W.

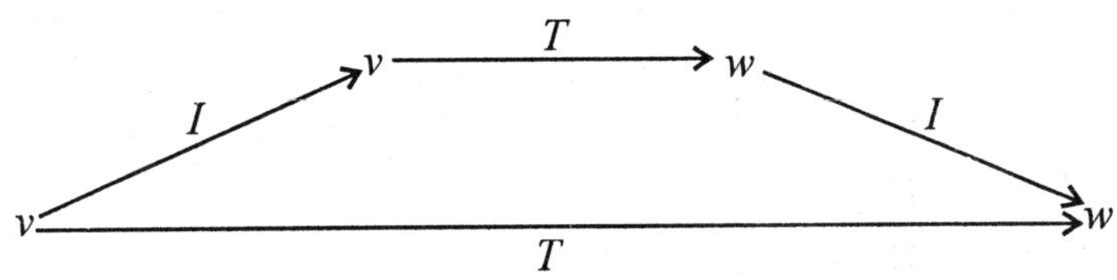

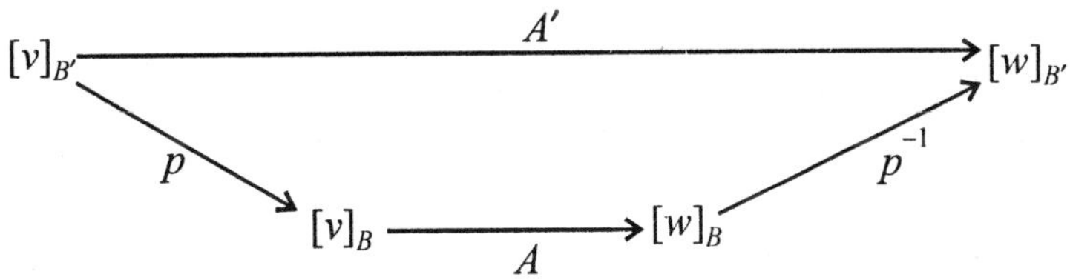

Example. Consider the vector space P_3 of all polynomials with real coefficients and degree at most 3, with bases $B = \{1, x, x^2, x^3\}$ and $B' = \{1, 1 + x, 1 + x + x^2, 1 + x + x^2 + x^3\}$. Consider also the linear operator $T : P_3 \to P_3$, where for every polynomial $p(x) = p_0 + p_1x + p_2x^2 + p_3x^3$, we have $T(p(x)) = (p_0 + p_1) + (p_1 + p_2)x + (p_2 + p_3)x^2 + (p_0 + p_3)x^3$. Let A denote the matrix for T with respect to the basis B. Then

$$T(1) = 1 + x^3,\ T(x) = 1 + x,\ T(x^2) = x + x^2 \text{ and } T(x^3) = x^2 + x^3, \text{ and so}$$

$$A = ([T(1)]_B\ [T(x)]_B\ [T(x^2)]_B\ [T(x^3)]_B) = \begin{pmatrix} 1 & 1 & 0 & 0 \\ 0 & 1 & 1 & 0 \\ 0 & 0 & 1 & 1 \\ 1 & 0 & 0 & 1 \end{pmatrix}.$$

Next, note that the transition matrix from the basis B' to the basis B is given by

$$P = ([1]_B\ [1 + x]_B\ [1 + x + x^2]_B\ [1 + x + x^2 + x^3]_B) = \begin{pmatrix} 1 & 1 & 1 & 1 \\ 0 & 1 & 1 & 1 \\ 0 & 0 & 1 & 1 \\ 0 & 0 & 0 & 1 \end{pmatrix}.$$

It can be checked that

$$P^{-1} = \begin{pmatrix} 1 & -1 & 0 & 0 \\ 0 & 1 & -1 & 0 \\ 0 & 0 & 1 & -1 \\ 0 & 0 & 0 & 1 \end{pmatrix},$$

and so

$$A' = P^{-1}AP = \begin{pmatrix} 1 & -1 & 0 & 0 \\ 0 & 1 & -1 & 0 \\ 0 & 0 & 1 & -1 \\ 0 & 0 & 0 & 1 \end{pmatrix}\begin{pmatrix} 1 & 1 & 0 & 0 \\ 0 & 1 & 1 & 0 \\ 0 & 0 & 1 & 1 \\ 1 & 0 & 0 & 1 \end{pmatrix}\begin{pmatrix} 1 & 1 & 1 & 1 \\ 0 & 1 & 1 & 1 \\ 0 & 0 & 1 & 1 \\ 0 & 0 & 0 & 1 \end{pmatrix} = \begin{pmatrix} 1 & 1 & 0 & 0 \\ 0 & 1 & 1 & 0 \\ -1 & -1 & 0 & 0 \\ 1 & 1 & 1 & 2 \end{pmatrix}$$

is the matrix for T with respect to the basis B_0. It follows that

$$T(1) = 1 - (1 + x + x^2) + (1 + x + x^2 + x^3) = 1 + x^3,$$
$$T(1 + x) = 1 + (1 + x) - (1 + x + x^2) + (1 + x + x^2 + x^3) = 2 + x + x^3,$$
$$T(1 + x + x^2) = (1 + x) + (1 + x + x^2 + x^3) = 2 + 2x + x^2 + x^3,$$
$$T(1 + x + x^2 + x^3) = 2(1 + x + x^2 + x^3) = 2 + 2x + 2x^2 + 2x^3.$$

These can be veried directly.

Eigenvalues and Eigenvectors

Definition. Suppose that $T : V \to V$ is a linear operator on a finite dimensional real vector space V. Then any real number $\lambda \in \mathbb{R}$ is called an eigenvalue of T if there exists a non–zero vector $v \in V$ such that $T(v) = \lambda v$. This non–zero vector $v \in V$ is called an eigenvector of T corresponding to the eigenvalue λ.

The purpose of this section is to show that the problem of eigenvalues and eigenvectors of the linear operator T can be reduced to the problem of eigenvalues and eigenvectors of the matrix for T with respect to any basis B of V. The starting point of our argument is the following theorem, the proof of which is left as an exercise.

Proposition. Suppose that $T : V \to V$ is a linear operator on a finite dimensional real vector space V, with bases B and B'. Suppose further that A and A' are the matrices for T with respect to the basis B and with respect to the basis B' respectively. Then

(*a*) $\det A = \det A'$,

(*b*) A and A' have the same rank,

(*c*) A and A' have the same characteristic polynomial,

(*d*) A and A' have the same eigenvalues, and

(*e*) The dimension of the eigenspace of A corresponding to an eigenvalue λ is equal to the dimension of the eigenspace of A' corresponding to λ.

We also state without proof the following result.

Proposition. Suppose that $T : V \to V$ is a linear operator on a finite dimensional real vector space V. Suppose further that A is the matrix for T with respect to a basis B of V. Then

(*a*) The eigenvalues of T are precisely the eigenvalues of A, and

(*b*) A vector $u \in V$ is an eigenvector of T corresponding to an eigenvalue λ if and only if the coordinate matrix $[u]_B$ is an eigenvector of A corresponding to the eigenvalue λ.

Suppose now that A is the matrix for a linear operator

$$T : V \to V$$

on a finite dimensional real vector space V with respect to a basis

$$B = \{v_1, \ldots, v_n\}.$$

If A can be diagonalized, then there exists an invertible matrix P such that

$$P^{-1}AP = D$$

is a diagonal matrix. Furthermore, the columns of P are eigenvectors of A, and so are the coordinate matrices of eigenvectors of T with respect to the basis B. In other words,

$$P = ([u_1]_B \cdots [u_n]_B),$$

where $B' = \{u_1, \ldots, u_n\}$ is a basis of V consiting of eigenvectors of T. Furthermore, P is the transition matrix from the basis B' to the basis B. It follows that the matrix for T with respect to the basis B' is given by

$$D = \begin{pmatrix} \lambda_1 & & \\ & \ddots & \\ & & \lambda_n \end{pmatrix}.$$

where $\lambda_1, \ldots, \lambda_n$ are the eigenvalues of T.

Example. Consider the vector space P_2 of all polynomials with real coefficients and degree at most 2, with basis $B = \{1, x, x^2\}$. Consider also the linear operator $T : P_2 \to P_2$, where for every polynomial $p(x) = p_0 + p_1x + p_2x^2$, we have

$$T(p(x)) = (5p_0 - 2p_1) + (6p_1 + 2p_2 - 2p_0)x + (2p_1 + 7p_2)x^2.$$

Then

$$T(1) = 5 - 2x,\ T(x) = -2 + 6x + 2x^2$$

and

$$T(x^2) = 2x + 7x^2,$$

so that the matrix for T with respect to the basis B is given by

$$A = ([T(1)]_B\ [T(x)]_B\ [T(x^2)]_B) = \begin{pmatrix} 5 & -2 & 0 \\ -2 & 6 & 2 \\ 0 & 2 & 7 \end{pmatrix}.$$

It is a simple exercise to show that the matrix A has eigenvalues 3, 6, 9, with corresponding eigenvectors

$$x_1 = \begin{pmatrix} 2 \\ 2 \\ -1 \end{pmatrix}, x_2 = \begin{pmatrix} 2 \\ -1 \\ 2 \end{pmatrix}, x_3 = \begin{pmatrix} -1 \\ 2 \\ 2 \end{pmatrix},$$

so that writing ,

$$P = \begin{pmatrix} 2 & 2 & -1 \\ 2 & -1 & 2 \\ -1 & 2 & 2 \end{pmatrix},$$

we have

Now let $B_0 = \{p_1(x), p_2(x), p_3(x)\}$, where

$$[p_1(x)]_B = \begin{pmatrix} 2 \\ 2 \\ -1 \end{pmatrix}, \ [p_2(x)]_B = \begin{pmatrix} 2 \\ -1 \\ 2 \end{pmatrix}, \ [p_3(x)]_B = \begin{pmatrix} -1 \\ 2 \\ 2 \end{pmatrix},$$

Then P is the transition matrix from the basis B' to the basis B, and D is the matrix for T with respect to the basis B'. Clearly

$$p_1(x) = 2 + 2x - x^2, p_2(x) = 2 - x + 2x^2$$

and

$$p_3(x) = -1 + 2x + 2x^2.$$

Note now that

$$T(p_1(x)) = T(2 + 2x - xx^2) = 6 + 6x - 3x^2 = 3p_1(x),$$
$$T(p_2(x)) = T(2 - x + 2x^2) = 12 - 6x + 12x^2 = 6p_2(x),$$
$$T(p_3(x)) = T(-1 + 2x + 2x^2) = -9 + 18x + 18x^2 = 9p_3(x).$$